The Neuropsychiatry of Limbic and Subcortical Disorders

The Neuropsychiatry of Limbic and Subcortical Disorders

Edited by

Stephen Salloway, M.D., M.S.
Paul Malloy, Ph.D.
Jeffrey L. Cummings, M.D.

Washington, DC
London, England

Note: The authors have worked to ensure that all information in this book concerning drug dosages, schedules, and routes of administration is accurate as of the time of publication and consistent with standards set by the U.S. Food and Drug Administration and the general medical community. As medical research and practice advance, however, therapeutic standards may change. For this reason and because human and mechanical errors sometimes occur, we recommend that readers follow the advice of a physician who is directly involved in their care or the care of a member of their family.

Books published by the American Psychiatric Press, Inc., represent the views and opinions of the individual authors and do not necessarily represent the policies and opinions of the Press or the American Psychiatric Association.

Copyright © 1997 American Psychiatric Press, Inc. Exceptions to this copyright are Chapters 1, 9, and 10. These chapters were prepared with government financial support or by government employees and are, therefore, in the public domain.
ALL RIGHTS RESERVED
Manufactured in the United States of America on acid-free paper

Originally published in *The Journal of Neuropsychiatry and Clinical Neurosciences* (Volume 9, Number 3, Summer 1997)

00 99   4 3 2

American Psychiatric Press, Inc.
1400 K Street, N.W., Washington, DC 20005
www.appi.org

Library of Congress Cataloging-in-Publication Data
The neuropsychiatry of limbic and subcortical disorders / edited by Stephen Salloway, Paul Malloy, Jeffrey L. Cummings.
   p. cm.—The journal of neuropsychiatry and clinical neurosciences ISSN 0895-0172 ; v. 9, no. 3)
   "Originally published in: The Journal of Neuropsychiatry and Clinical Neurosciences, volume 9, number 3, summer 1997"—t.p. verso.
   Includes bibliographical references and index.
   ISBN 0-88048-942-1 (alk. paper)
   1. Neuropsychiatry. 2. Limbic system—Diseases. 3. Temporal lobes—Diseases. 4. Neurobehavioral disorders.
I. Salloway, Stephen. II. Malloy, Paul. III. Cummings, Jeffrey L., 1948- IV. Series.
   [DNLM: 1. Limbic System—physiopathology. 2. Temporal Lobe—physiopathology. 3. Thalamus—physiopathology. 4. Organic Mental Disorders—physiopathology. 5. Neurobiology. W1 J0795BD v.9 no.3 1997 / WL 314 N4943 1997]
   RC341.N4355 1997
   616.89—dc21
   DNLM/DLC
   for Library of Congress   97-34481
                                                                                                           CIP

British Library Cataloguing in Publication Data
A CIP record is available from the British Library.

Contents

Contributors vii

Editors ix

Introduction xi

Part 1 Anatomy and Neurochemistry

    1   The Limbic System: An Anatomic, Phylogenetic, and Clinical Perspective   3
Michael S. Mega, M.D., Jeffrey L. Cummings, M.D., Stephen Salloway, M.D., Paul Malloy, Ph.D.

    2   Ventromedial Temporal Lobe Anatomy, With Comments on Alzheimer's Disease and Temporal Injury   19
Gary W. Van Hoesen, Ph.D.

    3   The Thalamus and Neuropsychiatric Illness   31
Arnold B. Scheibel, M.D.

    4   The Accumbens: Beyond the Core–Shell Dichotomy   43
Lennart Heimer, M.D., George F. Alheid, Ph.D., José S. de Olmos, Ph.D., Henk J. Groenewegen, M.D., Ph.D., Suzanne N. Haber, Ph.D., Richard E. Harlan, Ph.D., Daniel S. Zahm, Ph.D.

    5   Neurobiology of Fear Responses: The Role of the Amygdala   71
Michael Davis, Ph.D.

Part 2 Clinical Syndromes

    6   Paroxysmal Limbic Disorders in Neuropsychiatry   95
Stephen Salloway, M.D., M.S., James White, M.D.

    7   Auras and Experiential Responses Arising in the Temporal Lobe   113
Itzhak Fried, M.D., Ph.D.

    8   Neuropsychiatric Symptoms From the Temporolimbic Lobes   123
Michael R. Trimble, M.D., F.R.C.P., F.R.C.Psych., Mario F. Mendez, M.D., Ph.D., Jeffrey L. Cummings, M.D.

    9   The Neurobiology of Emotional Experience   133
Kenneth M. Heilman, M.D.

   10   The Neurobiology of Recovered Memory   143
Stuart M. Zola, Ph.D.

11	The Medial Temporal Lobe in Schizophrenia *Steven E. Arnold, M.D.*	155
12	Limbic-Cortical Dysregulation: A Proposed Model of Depression *Helen S. Mayberg, M.D.*	167
13	The Neurobiology of Drug Addiction *George F. Koob, Ph.D., Eric J. Nestler, M.D., Ph.D.*	179
14	The Neural Substrates of Religious Experience *Jeffrey L. Saver, M.D., John Rabin, M.D.*	195

Index 209

Contributors

George F. Alheid, Ph.D.
Department of Psychiatric Medicine, University of Virginia, Charlottesville, Virginia

Steven E. Arnold, M.D.
Center for Neurobiology and Behavior, Department of Psychiatry, University of Pennsylvania, Philadelphia, Pennsylvania

Jeffrey L. Cummings, M.D.
Departments of Neurology and Psychiatry, UCLA School of Medicine, and the Neurobehavior and Neuropsychiatry Program, Psychiatry Service, West Los Angeles Veterans Affairs Medical Center, Los Angeles, California

Michael Davis, Ph.D.
Ribicoff Research Facilities of the Connecticut Mental Health Center, Department of Psychiatry, Yale University School of Medicine, New Haven, Connecticut

José S. de Olmos, Ph.D.
Instituto de Investigación Médica, Córdoba, Argentina

Itzhak Fried, M.D., Ph.D.
Division of Neurosurgery, UCLA School of Medicine, Los Angeles, California

Henk J. Groenewegen, M.D., Ph.D.
Department of Anatomy and Embryology, Vrije Universiteit, Amsterdam, The Netherlands

Suzanne N. Haber, Ph.D.
Departments of Neurobiology and Anatomy, University of Rochester, Rochester, New York

Richard E. Harlan, Ph.D.
Department of Anatomy, Tulane University, New Orleans, Louisiana

Kenneth M. Heilman, M.D.
Department of Neurology, University of Florida College of Medicine, and Neurology Service, Department of Veterans Affairs Medical Center, Gainesville, Florida

Lennart Heimer, M.D.
Departments of Otolaryngology-Head and Neck Surgery and Neurological Surgery, University of Virginia, Charlottesville, Virginia

George F. Koob, Ph.D.
Department of Neuropharmacology, The Scripps Research Institute, La Jolla, California

Paul Malloy, Ph.D.
Department of Psychology, Butler Hospital, and Department of Psychiatry and Human Behavior, Brown University School of Medicine, Providence, Rhode Island

Helen S. Mayberg, M.D.
Departments of Medicine (Neurology), Psychiatry, and Radiology and the Research Imaging Center, University of Texas Health Science Center at San Antonio, San Antonio, Texas

Michael S. Mega, M.D.
Department of Neurology, UCLA School of Medicine, Los Angeles, California

Mario F. Mendez, M.D., Ph.D.
Department of Neurology, UCLA School of Medicine, and the Neurobehavior and Neuropsychiatry Program, Psychiatry Service, West Los Angeles Veterans Affairs Medical Center, Los Angeles, California

Eric Nestler, M.D., Ph.D.
Laboratory of Molecular Psychiatry, Yale University School of Medicine and Connecticut Mental Health Center, New Haven, Connecticut

John Rabin, M.D.
Department of Neurology, Reed Neurological Research Center, UCLA School of Medicine, Los Angeles, California

Stephen Salloway, M.D., M.S.
Department of Neurology, Butler Hospital, and Departments of Clinical Neurosciences and Psychiatry and Human Behavior, Brown University School of Medicine, Providence, Rhode Island

Jeffrey L. Saver, M.D.
Department of Neurology, Reed Neurological Research Center, UCLA School of Medicine, Los Angeles, California

Arnold B. Scheibel, M.D.
Departments of Neurobiology and Psychiatry and Biobehavioral Sciences and the Brain Research Institute, UCLA Center for the Health Sciences, Los Angeles, California

Michael R. Trimble, M.D., F.R.C.P., F.R.C.Psych.
Institute of Neurology, Queen Square, London, England

Gary W. Van Hoesen, Ph.D.
Departments of Anatomy and Cell Biology and Division of Cognitive Neuroscience, Department of Neurology, University of Iowa, Iowa City, Iowa

James White, M.D.
Epilepsy Division, Department of Neurology, Bowman Gray University School of Medicine, Winston-Salem, North Carolina

Daniel S. Zahm, Ph.D.
Department of Anatomy, St. Louis University School of Medicine, St. Louis, Missouri

Stuart M. Zola, Ph.D.
San Diego Veterans Affairs Medical Center and Departments of Psychiatry and Neurosciences, School of Medicine, University of California, San Diego, La Jolla, California

Editors

Stephen Salloway, M.D., M.S.
Director of Neurology
Butler Hospital
Assistant Professor
Departments of Clinical Neurosciences and
Psychiatry and Human Behavior
Brown University School of Medicine
Providence, Rhode Island

Paul Malloy, Ph.D.
Director of Psychology
Butler Hospital
Associate Professor
Department of Psychiatry and Human Behavior
Brown University School of Medicine
Providence, Rhode Island

Jeffrey L. Cummings, M.D.
Director, Alzheimer's Disease Center
University of California at Los Angeles School of
Medicine
Augustus S. Rose Professor of Neurology and
Professor of Psychiatry
University of California at Los Angeles School of
Medicine, Los Angeles, California

Introduction

Stephen Salloway, M.D., M.S., Paul Malloy, Ph.D., Jeffrey L. Cummings, M.D.

Most neuropsychiatric disorders involve dysfunction of subcortical structures or the limbic or paralimbic cortex.[1,2] Limbic and subcortical brain regions, organized into functional units, mediate fundamental functions such as memory, emotion, motivation, and mood. Limbic and subcortical systems also play a key neurobiological role in other important aspects of human experience, such as substance abuse, reward systems, and religious experience. Dysfunction of temporolimbic systems produces some of the most dramatic and challenging syndromes in clinical medicine.

The contributors to this volume are a diverse group of leading investigators with special expertise in the functional aspects of limbic and subcortical anatomy and its relationship to neuropsychiatric illness. This volume offers an overview of functional limbic anatomy and provides a state-of-the-art report on limbic-related syndromes. This information can serve as a template to be modified as new information becomes available.

The essays in the volume cover a broad range of basic and clinical material at various levels of difficulty, making this collection suitable for medical students, psychiatry and neurology residents, psychology trainees, and upper-level undergraduate and graduate students in the basic and clinical neurosciences as well as experienced clinicians and researchers in these fields. Some of the chapters present complex material requiring careful study and perhaps a second reading. Others include extensive literature reviews and can serve as comprehensive reference sources. Many of the chapters include color illustrations to depict key points.

FUNCTIONAL NEUROANATOMY

Part One of this collection covers the functional neuroanatomy of limbic and subcortical systems. Much has been written recently about frontal-subcortical circuits and human behavior, but the integration of limbic-subcortical pathways into the schema of neurobehavioral networks has not been adequately developed. The chapter by Dr. Mega and colleagues describes the historical development and anatomical boundaries of limbic and paralimbic brain systems, outlining lateral and medial limbic circuits. The lateral limbic circuit, made up of the orbitofrontal-insular-temporopolar cortices, the amygdala, and the dorsomedial thalamic nucleus, processes information concerning the body surface, external world, and personal social interactions. The medial circuit, made up of the hippocampus, limbic diencephalon, and anterior cingulate gyrus, mediates the conscious encoding of experience. The extended amygdala/ventral striatum links limbic and subcortical systems. The lateral and medial orbitofrontal circuits also intersect in the region of the infracallosal cingulate, an area of focus in Dr. Mayberg's model of depression.

Dr. Van Hoesen highlights the afferent and efferent connections of the hippocampus and amygdala to the cortex. He discusses the flow of information in and out of the hippocampus and provides clear documentation of the important role of the early degeneration of the entorhinal cortex in Alzheimer's disease and the often unsuspected mechanical injury to the medial temporal lobe caused by contact with the tentorium cerebelli during closed head injury.

Dr. Heimer helped pioneer the concept of the extended amygdala/ventral striatum. In this volume he has assembled an international team to describe the embryology and functional connections of the nucleus accumbens, a structure that sits at the crossroads between the subcortical striatal system and the limbic system. Dr. Heimer and colleagues describe two parts of the accumbens: a more central striatal core and a limbic shell. The shell is an important component of the limbic striatum/extended amygdala, rich in dopaminergic neurons that participate in substance abuse and reward systems and that mediate psychotic states. The functions of the accumbens are extremely important for those interested in understanding neuropsychiatric illness.

A growing body of evidence implicates the amygdala as central to the modulation of fear responses. Dr. Davis, a recognized expert in this area of research, describes the connections of the amygdala and synthesizes what is known about the specific mechanisms involved in the regulation of fear responses.

The thalamus, long considered a routing and relay station for sensory information from the periphery, has recently become the focus of neuropsychiatric research. For example, the reticular nucleus of the thalamus, through cortical-subcortical feedback, has been shown to exert a gating function on the flow of sensory information to the cortex. In a clear and scholarly style, Dr. Scheibel describes the role of that four thalamic nuclei play as active links in cortical-subcortical pathways regulating behavior.

CLINICAL NEUROPSYCHIATRIC SYNDROMES

The emphasis on neuroanatomy in Part One of this volume provides a foundation for Part Two, which focuses on the limbic system and neuropsychiatric syndromes. The chapters on paroxysmal limbic disorders and temporolimbic syndromes describe the range of often dramatic clinical syndromes—some transient, some enduring—that are seen with temporolimbic dysfunction. These chapters provide the reader with perspective on the broad range of functions subsumed by limbic and subcortical networks and review the differential diagnosis of syndromes that challenge the diagnostic and treatment skills of clinicians from many disciplines.

A number of the chapters discuss the role of limbic-subcortical systems in specific clinical syndromes. Dr. Fried discusses the mechanisms and clinical manifestations of epileptic auras that arise in the limbic system, and he differentiates the persistent auras from the partial complex seizures that often respond to medication or surgery. Dr. Zola provides a framework for understanding the neurobiology of recovered memories—a controversial and compelling topic that has captured the attention of clinicians, forensic experts, and the lay press. Histological abnormalities in the medial temporal lobe in schizophrenia have been reported by a number of workers. Dr. Arnold reviews the evidence of the important role that neuropathological changes in the medial temporal lobe may play in schizophrenia. Drs. Koob and Nestler expand on Dr. Heimer's chapter on the nucleus accumbens by discussing the neurobiology of substance abuse, with a special focus on neurochemical systems in the mesocortical limbic system (especially the ventral striatum/extended amygdala complex) involved in substance abuse and reward systems in the brain.

Finally, three chapters provide a thoughtful framework for understanding neurobiological aspects of complex human experience related in large part to limbic and subcortical systems. Dr. Heilman, known for his work on higher cortical function and hemispheric specialization, reviews previous theories of emotion and offers new views on the anatomy of emotion. He presents a central modular theory involving valence, arousal, and motor activation components. Dr. Mayberg describes her recent work showing that activity in the anterior cingulate gyrus can serve as a marker for antidepressant treatment response. She combines these findings with other functional imaging data from studies on depression to construct a model of the neurobiology of depression, highlighting the pivotal role that the anterior cingulate may play in regulating mood. Drs. Saver and Rabin provide evidence of the role that temporolimbic systems play in the neurobiology of religious experience. They suggest a "limbic marker hypothesis" based on manifestations of religious experience seen in temporolimbic epilepsy, near-death experiences, hallucinogen ingestion, schizophrenia, and affective disorders. Their work describes how the temporolimbic cortex tags internal and external stimuli, integrating and personalizing the processing of sensory information, thoughts, and emotions into religious experience.

We hope that this volume will make a valuable contribution to the literature on limbic-subcortical systems and neuropsychiatric illness.

The editors would like to thank Debbie Javorsky, M.A., for her assistance in organizing the manuscripts and the reviewers for their help in shaping the material. The contents of this volume originally appeared in the summer 1997 special issue of *The Journal of Neuropsychiatry and Clinical Neurosciences*, the official journal of the American Neuropsychiatric Association. Hoechst Marion Roussel Pharmaceuticals provided an unrestricted educational grant that allowed publication of an expanded issue of the Journal.

REFERENCES

1. Salloway S, Cummings J: Subcortical structures and neuropsychiatric illness. The Neuroscientist 1996; 2:66–75
2. Salloway S, Cummings J: Subcortical disease and neuropsychiatric illness. J Neuropsychiatry Clin Neurosci 1994; 6:93–99

PART 1
Anatomy and Neurochemistry

1

The Limbic System

An Anatomic, Phylogenetic, and Clinical Perspective

Michael S. Mega, M.D., Jeffrey L. Cummings, M.D., Stephen Salloway, M.D., Paul Malloy, Ph.D.

A limbus is a margin or border, and, appropriately, the limbic system encompasses the boundary between neurology and psychiatry. The cortical border encircling the brainstem was designated by Thomas Willis in 1664[1] as the *cerebri limbus* 15 years after Descartes proposed the brain as the common ground between the extremes of mind and body. More than 200 years later, Paul Broca, possibly unaware of Willis's term, considered the cingulate gyrus, anterior olfactory region, and hippocampus as a ring of olfactory processing and termed this phylogenetic border zone the *grand lobe limbique*[2] (Figure 1–1A). Broca drew attention to this ring of medial cerebrum by emphasizing two aspects of the limbic lobe: its strong relationship with the olfactory apparatus and its presence as a common denominator among the brains of mammals. Broca conceived the limbic lobe to be an olfactory structure because most of its components received multiple projections from the olfactory system in animals with a well-developed sense of smell.

Information regarding limbic function increased as a result of ablation studies in animals. In 1888, Brown and Schäfer[3] found that a bilateral temporal lobectomy, which included limbic structures, transformed wild, ferocious monkeys into tame and docile animals. Later, Klüver and Bucy[4-6] detailed the specific areas and behaviors involved in this effect. Data originating from neurological clinics provided much evidence for theories concerning the function of the hippocampus and related structures. In 1900, Bechterew[7] found that bilateral lesions of the hippocampus and other limbic structures (the mammillary bodies and anterior thalamus) resulted in derangements of memory function. Both long-term and short-term memory remained relatively intact, but "the process of registration of information" was disturbed.[7] In the early 20th century, Ramón y Cajal[8] systematized the cytoarchitectural features of limbic cortex. He described nuclear structures associated with the limbic lobe, including the amygdala, septal nuclei, hypothalamus, epithalamus, anterior thalamic nuclei, and parts of the basal ganglia, particularly the ventral medial portions. The inclusion of subcortical limbic structures gained support after anatomic studies[9] revealed extensive connections of the anterior thalamus and hypothalamus with the hippocampus and cingulate. In 1927, Cannon[10] proposed that emotion results from the action and reaction of the cerebral cortex and diencephalon. At the same time that the diencephalon discharges downward to the basal ganglia, producing motor impulses of emotional behavior such as crying or laughing, it also discharges upward to the cortex, allowing conscious appreciation of emotional states.

In 1937, Papez[11] combined the results of anatomical studies with the clinical reports of emotional disturbances following lesions to the cingulate and other medial structures to propose a mechanism of emotion based on a medial circuit (Figure 1–1B). The Papez mechanism of emotion is part of a unified concept with three divisions through which "man's volitional energies" flow. First, in Papez's terms, the "stream of movement" is conducted through the dorsal thalamus and the internal capsule to the corpus striatum and then out the central nervous system to the somatic motor neurons. Second, the "stream of thought" arises from the thalamus and ascends the internal capsule to find expression in the lateral cerebral cortex involved in executive function (the ability to organize a solution to a complex problem). Third, there are impulses flowing through medial structures in a "stream of feeling." Papez conceived of a two-way circuit for the flow of emotion, leading to internal or external expression. Information from external sensory receptors arrives at primary sensory cortex and then is relayed to the hippocampal area and onward to the mammillary bodies via the fornix. From the mammillary bodies information spreads throughout the hypothalamus, initiating emotional expression or affect. Impulses are also directed to the anterior nucleus of the thalamus via the mammillothalamic tract and then diffusely projected through the anterior limb of the internal capsule toward the cingulate gyrus. There, Papez conceived, emotion as an internal state or mood is consciously perceived. Process-

FIGURE 1–1. A: *Le grand lobe limbique* common to all mammals as identified by Broca in 1878.[2] B: The medial circuit of Papez, as depicted by MacLean,[14] was proposed in 1937 to support emotional processing.[11] C: On the basis of phylogenetic and cytoarchitectonic evidence, Yakovlev in 1948[13] conceived of three functional zones: the oldest, innermost zone *(black)*, which comprises the visceral system or entopallium; the middle zone *(stippled)*, or mesopallium, including the limbic lobe of Broca (cingulate, retrosplenial, and hippocampal gyri) along with the orbitofrontal and insular cortex; and the most recently developed neocortical outer zone *(white)*, or ectopallium, with its sensory and motor cortices *(lines)*. D: MacLean in 1949[14] included Papez's medial circuit and the basal lateral members of Yakovlev's "middle zone" in a limbic system.

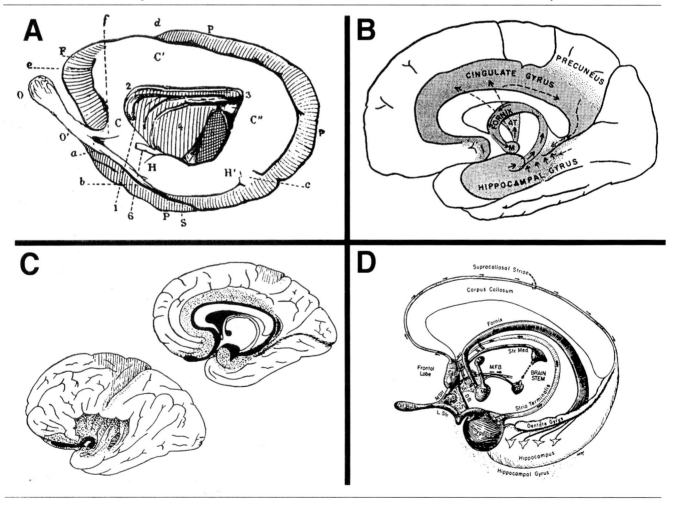

ing within the cingulate provided feedback to the hippocampus via the retrosplenial cortex, completing Papez's circuit. Papez observed that mood may be dissociated from affect, since in decorticate subjects emotional states were displayed but not perceived by the subject. For Papez, the integration of emotional responsiveness with the dorsolateral frontal executive functions occurred in the medial cortex of the cingulate. The circuit of Papez did not find anatomical support for the closing connection of the cingulate to the hippocampus until 1975, when Shipley and Sørensen[12] documented that the presubiculum, which receives a dense cingulate outflow, projects heavily to layer III of the entorhinal cortex—the origin of the perforant pathway into hippocampal pyramidal cells.

In 1948, Yakovlev,[13] without specific reference to Papez, suggested that three lateral cortical regions—the orbital frontal cortex, temporal lobe, and insula—and two subcortical structures—the amygdala and dorsal medial thalamus—played an important role in motivation and emotional expression (Figure 1–1C). In 1952, MacLean linked the medial circuit of Papez with the basal lateral structures of Yakovlev, referring to them for the first time as the *limbic system*[14,15] (Figure 1–1D). With an increase in anatomic studies, the limbic system evolved to include many more areas than were conceived by Papez.

Brodal[16] noted that the term *limbic system* began to include nearly all brain regions and suggested discarding it, but the term has survived, the concept has been refined, and consistent anatomic-clinical correlations have been found. For the purposes of this review, the *limbic system* will be considered as including not only the traditional medial circuit of Papez, but also forebrain structures that *reciprocally* connect with the amygdala, particularly the orbitofrontal region. These structures of the telencephalon are in a unique position to influence neuroendocrine, autonomic, and behavioral mechanisms associated with the selection of relevant external stimuli, the evaluation of internal drives, and stimulus–reward associations. This review provides anatomic, phylogenetic, and clinical perspectives on the limbic system, integrating classical and modern observations relevant to neuropsychiatry. Other chapters in this volume present more detailed data from anatomical, clinical, and imaging studies that continue to refine our conception of limbic disorders and its relevance to neuropsychiatric disorders.

PHYLOGENETIC DEVELOPMENT

Yakovlev[13] proposed three levels of central nervous system function (Table 1–1). He posited a primitive inner core devoted to arousal and autonomic function, which is surrounded by a middle layer that includes the limbic system and basal ganglia, which in turn is encapsulated by the most recent phylogenetic layering of the neocortex and pyramidal system (Figure 1–2A). Each layer subserves different functions. The inner layer contains the reticular core—a feltwork of unmyelinated neurons controlling the basic mechanisms of consciousness, as well as cardiovascular and respiratory functions. The middle layer contains the organized cell groups, partially myelinated, including the basal ganglia and limbic system; its functions concern arousal, communal activities, personality, and emotion. The outer neocortical layer, composed of well-myelinated neurons, subserves fine motor control, detailed sensory processing, praxis, gnosis, and abstract cognition such as language function—the manipulation of symbols according to syntactic rules. Such abstract communication stands in contrast to the more "emotionally charged" information mediated in the middle layer or limbic system. With early mammalian development, rearing of the young was more interactive than in reptilian species. Thus, primal vocalization behavior emerged, such as the sepa-

TABLE 1–1. Yakovlev's three-layered model of brain organization, structure, and function[13]

	Inner Layer	Intermediate Layer	Outer Layer
Neurons	Short, unmyelinated	Long, partially myelinated	Long, well myelinated
Organization	Diffuse	Ganglia, allocortex	Isocortex
Evolution	Invertebrate to reptile	Reptile to early mammal	Mammal to primate
Structure	Reticular core Cranial nerves Periaqueductal gray Hypothalamus	Basal ganglia Limbic thalamus Olfactory paleocortex Hippocampal archicortex	Primary sensory cortices Primary motor cortex Corpus callosum Association cortex
Function	Consciousness Metabolism Respiration Circulation	Motor synergy Arousal, motivation Mood, affect Personality	Motor precision Praxis Language Gnosis

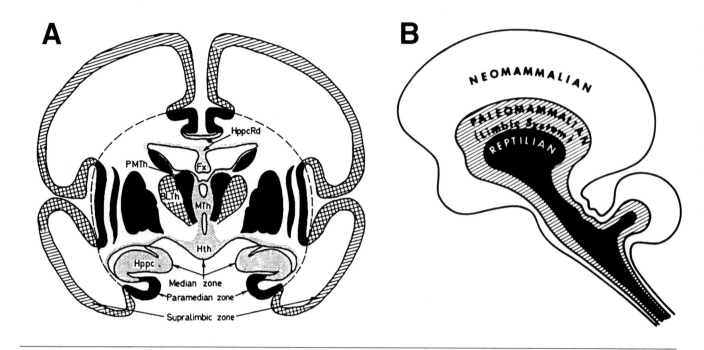

FIGURE 1–2. A: Yakovlev's model of the three phylogenetic zones of brain development, as reflected by myelogenetic stages.[101] B: MacLean's rendition of this evolutionary layering produced a triune brain.[17] BLTh = basolateral thalamus; Fx = fornix; Hppc = hippocampus; HppcRd = hippocampal radiations; Hth = hypothalamus; Mth = medial thalamus; PMTh = paramedial thalamus.

ration cry, while communal bonding and territorial behavior also mirrored the development of the middle limbic layer. During mammalian evolution, with the progressive expansion of the cortical mantle, a developmental progression from three-layered allocortex to six-layered isocortex occurred. The Yakovlev model was elaborated by MacLean into a *triune brain*[17] reflecting the phylogenetic stages of central nervous system development from invertebrate through reptile to early mammals, culminating in the explosive outgrowth of the neocortex in man (Figure 1–2B).

Comparisons of phylogenetic development across mammalian species have revealed two waves of increasing complexity from allocortex to isocortex, first clearly described by Sanides.[18] These two waves originate from two primordial regions within the limbic ring (Figure 1–3). The first developmental wave began in the orbitofrontal region of the olfactory paleocortex in primitive monotremes like the duck-billed platypus, whose cortex is almost entirely olfactory in function. This paleocortical progression spreads ventrolaterally up through the insula, temporal pole, and anterior parahippocampal area. The integration of appetitive drives with aversion or attraction to stimuli dominates paleocortical function. A second nidus of cortical development was centered in the archicortex of the hippocampus and spread posteriorly through the entorhinal and posterior parahippocampal regions, and around the corpus callosum through the cingulate. The archicortex is largely concerned with the integration of information from different sensory modalities—the first step away from thalamic control, as seen in reptiles, toward cortical dominance. By virtue of the connections of these two

FIGURE 1–3. The paralimbic trends of evolutionary cortical development. The phylogenetically older orbitofrontal-centered belt *(red)* extends into the subcallosal cingulate, temporal polar region, and the anterior insula (not shown). The more recent hippocampal-centered trend *(blue)* extends its wave of cortical development dorsally through the posterior and anterior cingulate. Adapted from Mega and Cummings.[29]

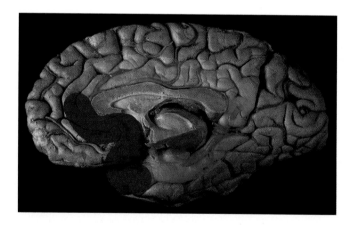

FIGURE 1–4. The phylogenetic progression of the amygdaloid complex. A: *Hemicentetes semispinosus* (an insectivore); B: *Propithecus verreauxi* (a prosimian); C: *Alouatta seniculus* (a simian); and D: *Homo sapiens*. The parvocellular division of the basal nucleus shows a progressive expansion with phylogenetic ascension. BM = basal amygdaloid nucleus pars magnocellularis; BP = basal amygdaloid nucleus pars parvocellularis; C = central amygdaloid nucleus; CO = cortical amygdaloid nucleus; E = entorhinal cortex; HI = hippocampus, I = intercalated cell masses; L = lateral amygdaloid nucleus; M = medial amygdaloid nucleus; PRPI = prepiriform cortex; STR = striatum. Adapted from Stephan and Andy.[21]

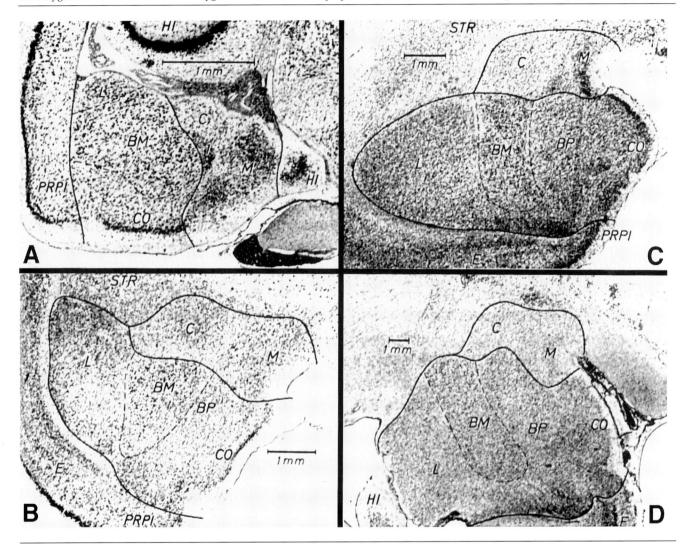

limbic centers and the parallel development of other brain regions linked to them, the behavioral evolution of mammals mirrored the progressive trend toward cytoarchitectural complexity emanating from both the olfactory and hippocampal centers.[19] Mesulam[20] has described these two allocortical trends as *paralimbic belts*.

The olfactory orbitofrontal-centered belt is closely associated with the amygdala. Through mammalian evolution, the paleocortical spread paralleled phylogenetic development of the amygdala. Two basic divisions of the amygdala in higher primates are the anterior cortico-basolateral group, supporting cortical association connections, and the centromedian nuclear group, related to the olfactory bulb, diencephalon, and brainstem. Evolution of the amygdala was heterogeneous; the anterior cortico-basolateral group makes up 75% of the human amygdala and only 54% of the amygdala in insectivores (Figure 1–4).[21] Human ontogeny recapitulates this phylogenetic difference: the centromedian group is the first to develop, followed by the basal and then the lateral nuclei.[22]

The archicortical belt has the hippocampus as its center. Hippocampal evolution was also heterogeneous, with a greater increase in the relative proportion of pyramidal cells in CA1, the cortical relay from the hippocampus, than CA2 or CA3: 14% in basal insectivores compared with 44% in simians.[23] The hippocampal-centered belt emphasizes pyramidal cells and gives rise to the medial supplementary sensory and motor areas.

The dorsal visual system, which processes *location* of objects,[24] is associated with the hippocampal archicortical paralimbic belt. The paleocortical trend emphasizes granular cells as it differentiates into secondary somatosensory, auditory, and visual cortices; it is more directly coupled with the ventral visual system involved in analyzing *features* of objects.[19] Table 1–2 provides an overview of the orbitofrontal and hippocampal paralimbic divisions.

Papez incorrectly conceived his hippocampal circuit as exclusively subserving emotional processes. Bechterew's insight[7] was more accurate: he observed that defects in episodic memory encoding occurred when elements of the hippocampal circuit were damaged. This explicit or conscious encoding of events by the hippocampal circuit stands in contrast to the subconscious processing that may occur in the orbitofrontal circuit when objects are imbued with emotional valence.

In the following sections, we review the anatomy and transmitter systems associated with the two divisions of the limbic system and consider the clinical syndromes relevant to neuropsychiatric disorders associated with limbic damage. Through a combination of anatomical, functional imaging, and clinical observations, a modern conception of the limbic system emerges that continues to aid our interpretation of neuropsychiatric disorders.

ANATOMY AND NEUROTRANSMITTERS

The two paralimbic belts unite cortical and subcortical areas sharing phylogenetic and cytoarchitectural features common to the amygdala/orbitofrontal and hippocampal/cingulate limbic divisions. Because our growing refinement of the brain's connectional anatomy is derived from nonhuman primate tracer studies, all cortical anatomy described in this review is extrapolated from the Walker areas in nonhuman primates to their homologous Brodmann[25] areas on a human brain image. Many areas are homologous, but some are not; the orbitofrontal cortex is such an example.[26,27] The orbitofrontal cortex has two major subdivisions in monkeys: one lateral, containing the lateral portion of areas 11 and 13 and all of 12, and one medial, comprising area 14 and the medial portion of 13 and 11.[28] Inferior portions of areas 25, in the infracallosal prefrontal region, and 24 and 32, in the supracallosal cingulate gyrus, have connectional similarities with the medial orbital division. Brodmann did not find a cytoarchitectural equivalent to his monkey orbitofrontal areas 14, 13, or 12 in humans. In cases where homology is not present, we interpret the animal connectional data with reference to human clinical lesion data to extrapolate the cortical locations (Figure 1–5).[29]

In addition to the hippocampal archicortex, the second limbic arm includes the cingulate gyrus, which is composed of four functional centers[29]: visceral effector, cognitive effector, skeletomotor effector, and sensory processing regions (Figure 1–6). The most anterior of the cingulate centers, the visceral effector in the subcallosal region, overlaps with the medial orbitofrontal limbic division. The two divisions of the limbic system work in concert. Processing in the amygdala/orbitofrontal division concerns the internal relevance that sensory stimuli have for the organism, and it thus facilitates intentional selection, habituation, or episodic encoding of these stimuli by the hippocampal/cingulate division.

Understanding the major reciprocal connections of a cortical region informs us about the functional system containing that region. There are also nonreciprocal connections, or open efferents and afferents, associated with any given region; we limit our attention to the reciprocally connected areas that segregate into general functional systems.

Orbitofrontal Paralimbic Division

The amygdala is the subcortical focus for the orbitofrontal paralimbic division. The primate amygdala[30] contains four deep nuclei from lateral to medial: the lateral nucleus, the basal nucleus (with magnocellular, intermediate, and parvocellular divisions from dorsal to ventral), the accessory basal nucleus (with magnocellular, parvocellular, and ventromedial divisions from lateral to medial), and the paralaminar nucleus on the ventral surface of the amygdala. The remaining part of the amygdala comprises six superficial nuclei (the anterior

TABLE 1–2. The two paralimbic divisions

	Orbitofrontal Division	Hippocampal Division
Evolutionary trend	Paleocortical	Archicortical
Cell type	Granule cell	Pyramidal cell
Structures	Amygdala	Hippocampus
	Anterior parahippocampus	Posterior parahippocampus
	Insula	Retrosplenium
	Temporal pole	Posterior cingulate
	Infracallosal cingulate	Supracallosal cingulate
Function	Implicit processing	Explicit processing
	Visceral integration	Memory encoding
	Visual feature analysis	Visual spatial analysis
	Appetitive drives	Skeletomotor effector
	Social awareness	Attentional systems
	Mood	Motivation

and posterior cortical, medial and central nuclei, the nucleus of the lateral olfactory tract, and the periamygdaloid cortex) and three other nuclei (the anterior amygdaloid area, amygdalohippocampal area, and collections of cell bodies in the fiber bundles within the amygdala forming the intercalated nuclei). The dorsolateral boundaries of the amygdala are indistinct, with collections of cells extending from the superficial central and medial nuclei into the substantia innominata. The central and medial nuclei, along with the bed nucleus of the stria terminalis and the cell columns alongside the stria terminalis, are referred to as the *extended amygdala* (Figure 1–7). A simplification of amygdalar complexity is the functional and anatomic separation of a *cortical-association* division (basal/lateral nuclei) from a *visceral-subcortical* division (central/medial nuclei and extended amygdala). The former has stronger connections with association cortex, whereas the latter focuses projections into the brainstem, hypothalamus, and basal ganglia.[31]

The medial portion of the orbitofrontal paralimbic division (the gyrus rectus and medial orbital gyrus of Brodmann area 11 in humans) has direct reciprocal connections with the medial portion of the basal and the magnocellular division of the accessory basal amygdala.[28,30,32–34] Cortical areas that have reciprocal connections with the medial orbitofrontal cortex influence visceral function when stimulated,[35] probably through their shared amygdalar connections, conveying the visceral state of the organism to the orbitofrontal center. The rostral (agranular) insula,[33,34,36] ventromedial temporal pole area 38,[28,33,37] and infracallosal cingulate areas 25, 24, and 32 are also reciprocally connected with the medial division of the orbitofrontal cortex.[33,38,39] The visceral effector region of the infracallosal cingulate provides motivational input to gustatory, olfactory, and alimentary information from anterior insular processing converging on the medial orbitofrontal cortex. The anterior entorhinal area 36,[28,40] which is also reciprocally connected to the medial portion of the orbitofrontal division, is a paleocortical extension with hippocampal connections. No visual information has direct access to the medial division of the orbitofrontal center, which serves as an integrator of visceral drives while modulating the organism's internal milieu. The major cortical regions reciprocally connected with the medial division of the orbitofrontal paralimbic center are shown in Table 1–3 and Figure 1–8.

The lateral portion of the orbitofrontal paralimbic division (the lateral orbital gyrus of area 11 and the medial inferior frontal gyrus of areas 10 and 47 in humans) is phylogenetically more developed than the medial portion. The lateral portion has direct reciprocal connections with the dorsal and caudal regions of the basal and the magnocellular accessory basal amygdala.[28,30,32–34] The dorsal region of the basal amygdala is the source of

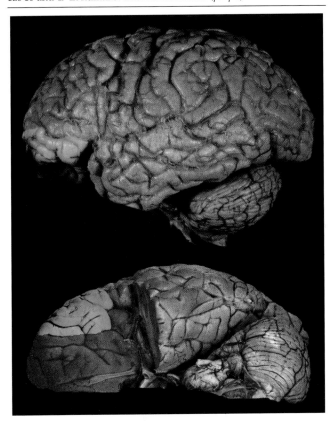

FIGURE 1–5. A: The two portions of the orbitofrontal cortex. The medial portion *(red)* includes the gyrus rectus and medial orbital gyrus of area 11 in humans. The lateral portion *(green)* includes the lateral orbital gyrus of area 11 and the medial inferior frontal gyrus of areas 10 and 47 in humans. The anterior insula *(purple)* is also shown.

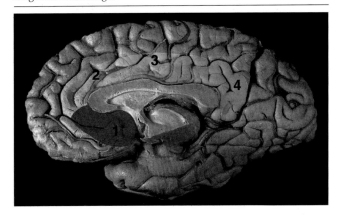

FIGURE 1–6. The four functional divisions of the cingulate. 1: visceral effector region; 2: cognitive effector region; 3: skeletomotor effector region; 4: sensory processing region. Adapted from Mega and Cummings.[29]

FIGURE 1–7. Coronal sections through the right human striatum illustrating the extended amygdala (right image) and the shell portion of the nucleus accumbens (sh Acb; left image). BNST = bed nucleus of stria terminalis; Ce/Me = central and medial amygdaloid nuclei; f = fornix; Slen = sublenticular zone of the extended amygdala; st = stria terminalis.

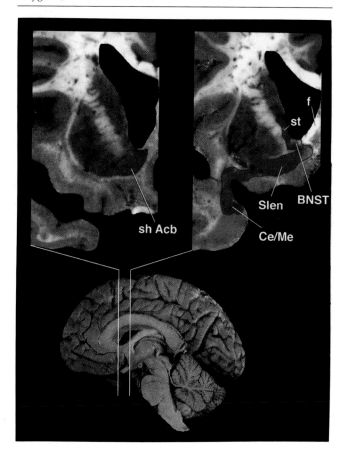

TABLE 1–3. Major reciprocal connections for the medial and lateral portions of the orbitofrontal paralimbic division

Medial Orbitofrontal Portion	Lateral Orbitofrontal Portion
Medial portion of the basal amygdala	Dorsal and caudal portions of the basal amygdala
Accessory basal (mc) amygdala	Accessory basal (mc) amygdala
Infracallosal areas 25, 24, and 32	Supracallosal areas 24 and 32
Ventromedial temporal pole area 38	Dorsolateral temporal pole area 38
Rostral (agranular) insula	Inferior temporal cortex TE area 20
Anterior entorhinal area 36	Supplementary eye field in dorsal portion of area 6

Note: mc = magnocellular.

FIGURE 1–8. Major cortical reciprocal areas connected to the medial portion *(red)* and lateral portion *(green)* of the orbitofrontal limbic division.

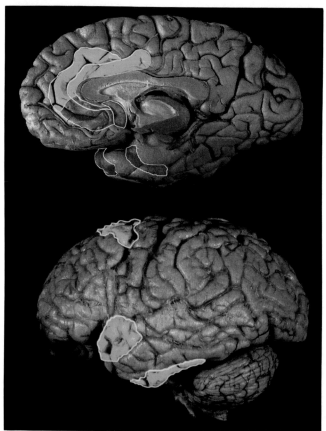

projections to the ventral visual processing system in the inferior temporal cortex. Reciprocal connections occur with the supracallosal cingulate areas 24 and 32,[33,38,39] a region that assists in the dorsolateral attentional system and effects cognitive engagement. The auditory association cortex of dorsolateral temporal pole area 38[28,33,37] is also reciprocally connected with the lateral orbitofrontal cortex. The lateral orbital cortex is a gateway for highly processed sensory information into the orbitofrontal paralimbic division. Reciprocal connections with the inferior temporal cortex TE area 20[28] (the last processing step for the ventral visual system devoted to object feature analysis) and the supplementary eye fields in the dorsal portion of area 6[28] highlight the control over sensory processing occurring in the lateral orbitofrontal cortex. The major cortical regions reciprocally connected with the lateral portion of the orbitofrontal paralimbic division are shown in Table 1–3 and Figure 1–8.

Hippocampal Paralimbic Division

Papez proposed that multimodal sensory information processed in the hippocampus is carried to the mammillary bodies via the fornix and enters the anterior nucleus of the thalamus via the mammillothalamic tract. From there it is projected into the cingulate gyrus, through the retrosplenial cortex, and then back to the hippocampus. Papez's circuit enables the conscious encoding of experience. The hippocampal archicortical developmental wave spreads into the posterior cingulate and then forward into the anterior cingulate. The major reciprocal connections of these two cingulate areas provide two portals through which limbic influence can affect other regions; the diverse affected areas are organized into functional networks.

The posterior cingulate (Brodmann areas 23 and 29/30) is a nexus for sensory and mnemonic networks within the hippocampal paralimbic belt. Functional imaging data during episodic memory encoding tasks implicate the posterior cingulate in the consolidation of declarative memory[41–44] and in associative learning during classical conditioning.[45] The connections of the posterior cingulate focus on the dorsolateral cortex. The posterior cingulate has major reciprocal connections with the posterior parahippocampal and perirhinal areas 36 and 35 as well as the presubiculum.[33,39,46] These connections modulate the multimodal efferents entering the entorhinal layer III cells that give rise to the perforant pathway into the hippocampus. The dorsal visual system of the inferior parietal lobe 7a[33,47,48] (dedicated to spatial processing)[24] and the frontal eye fields in area 8[33,49] also have bidirectional connections with the posterior cingulate. Reciprocal connections with lateral prefrontal area 46[33,49] allow an interaction between executive and sensory/mnemonic processing that may mediate perceptual working memory tasks. The major reciprocal connections of the posterior cingulate are shown in Table 1–4 and Figure 1–9A.

The orbitofrontal and hippocampal paralimbic belts intersect in the infracallosal cingulate region of Brodmann area 24. Cytoarchitectural development progresses from anterior to posterior and from inferomedial to dorsal across area 24.[50] The major pathway for information flow is the cingulum bundle. The cingulum contains the efferents and afferents of the cingulate to the hippocampus, basal forebrain, amygdala, and all cortical areas, as well as fibers of passage between hippocampus and prefrontal cortex and from the median raphe to the dorsal hippocampus.[51] Robust connections with the sensory processing region of the posterior cingulate overlap with the reciprocal amygdalar connections present in the anterior cingulate. Thus, area 24 is a nexus in the distributed networks subserving internal motivating drives and externally directed attentional mechanisms.[47] Papez's initial conception of the cingulate as the "seat of dynamic vigilance by which emotional experiences are endowed with an emotional consciousness"[11] is supported by this anatomic organization. The integration of both divisions of limbic processing in the anterior cingulate influences the dorsolateral executive cortex as Papez predicted: "The sensory excitations which reach the lateral cortex through the internal capsule receive their emotional coloring from the concurrent processes of hypothalamic origin which irradiate them from the gyrus cinguli."[11]

Major reciprocal connections with the cognitive effector region (supracallosal areas 24 and 32) are with the basal amygdala (magnocellular and parvocellular divisions).[30,32–34] The amygdala provides internal affective input to areas 24 and 32. Prefrontal areas 8, 9, 10, and 46 also have reciprocal connections.[33,49] The anterior cingulate is more developed in its cytoarchitecture and has well-developed connections with phylogenetically more recent neocortex of dorsolateral prefrontal areas 8, 9, 10, and 46, devoted to executive function. Areas reciprocally connected to the anterior cingulate share a general similarity with the orbitofrontal center. They include the caudal (dysgranular) orbitofrontal area 12 in monkeys (equivalent to area 47 in humans),[33,38,39] which provides processed sensory information concerning object feature analysis; the anterior inferior temporal pole area 38,[33,37] an auditory association area of the superior temporal gyrus; and the rostral (agranular) insula,[33,34,36]

TABLE 1–4. Major reciprocal connections for the posterior and anterior cingulate cortex

Posterior Cingulate	Anterior Cingulate
Posterior parahippocampal areas 35 and 36	Anterior parahippocampal areas 35 and 36
Presubiculum	Basal amygdala
Prefrontal area 46	Prefrontal areas 8, 9, 10, and 46
Caudal parietal area 7	Caudal orbitofrontal cortex area 47
Frontal eye fields in area 8	Inferior temporal pole area 38
	Rostral insula

FIGURE 1–9. Major cortical reciprocal areas connected to the posterior (A) and anterior (B) cingulate cortex.

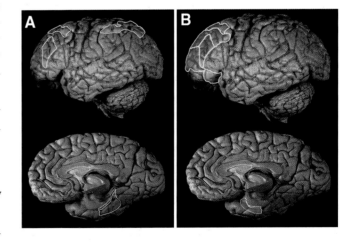

TABLE 1–5. The subcortical limbic structures and the related cortical limbic regions

	Amygdala	Striatum	Pallidum	Thalamus
Orbitofrontal cortex	Basal nucleus (mc) Extended amygdala	Ventromedial Caudate	Dorsomedial GPi Rostromedial SNr	Anterior (mc) Medial portion of mediodorsal (mc)
Hippocampus and posterior cingulate	Accessory basal nucleus (mc) Periamygdaloid cortex	Dorsal caudate	Dorsolateral GPi	Mediodorsal Anterior Medial pulvinar
Anterior cingulate	Basal nucleus (pc and mc)	Ventral striatum	Ventral pallidum	Midline nuclei Dorsal portion of mediodorsal (mc)

Note: mc = magnocellular; pc = parvocellular; SNr = substantia nigra reticulata; GPi = globus pallidus interna.

a transitional paralimbic region integrating visceral alimentary input with olfactory and gustatory afferents.[52] Reciprocal connections with the anterior parahippocampal areas 35 and 36[33,39,46] allow attentional influence over multimodal sensory afferents entering the hippocampus. The major reciprocal connections of the anterior cingulate are shown in Table 1–4 and Figure 1–9B.

Subcortical Limbic Structures

Subcortical limbic structures have major connections with the amygdala, orbitofrontal, and anterior cingulate regions. Frontal-subcortical loops[53] interact with cortical regions and function in effector mechanisms enabling the organism to act on the environment. Five main frontal-subcortical circuits have been initially described.[54] They define frontal-subcortical regions into functional modules. The two frontal-subcortical circuits with major limbic structures are the anterior cingulate subcortical circuit, required for motivated behavior, and the orbitofrontal circuit, allowing the integration of emotional information into contextually appropriate behavioral responses. The amygdala links all subcortical limbic structures in a common system providing visceral input that potentiates emotional associations. In addition to the amygdala, the subcortical limbic structures include the ventral striatum (ventromedial caudate, ventral putamen, nucleus accumbens, and olfactory tubercle); the ventral pallidum; and the ventral anterior, mediodorsal, and midline thalamic nuclei (Table 1–5).

Anterior Cingulate Subcortical Circuit: The ventral striatum receives input from Brodmann area 24[55] and provides input to the rostromedial globus pallidus interna and ventral pallidum (the region of the globus pallidus inferior to the anterior commissure) as well as the rostrodorsal substantia nigra.[56] The ventral pallidum projects to the entire mediolateral range of the substantia nigra pars compacta, the medial subthalamic nucleus and its extension into the lateral hypothalamus,[57] and the midline nuclei of the thalamus and the magnocellular mediodorsal thalamus.[57] The anterior cingulate circuit closes with major projections from the midline thalamic nuclei[58] and relatively fewer from the dorsal section of the magnocellular mediodorsal thalamus[59,60] back to the anterior cingulate cortex. The midline thalamic nuclei provide reticular activating system relays to the medial frontal cortical wall.

Orbitofrontal Subcortical Circuit: The ventromedial caudate receives input from the medial and lateral orbitofrontal cortex.[55] This portion of the caudate projects directly to the most medial portion of the mediodorsal globus pallidus interna and to the rostromedial substantia nigra pars reticulata.[61] Neurons are sent from the globus pallidus and substantia nigra to the medial section of the magnocellular division of the ventral anterior thalamus as well as an inferomedial sector of the magnocellular division of the mediodorsal thalamus.[55,62] These thalamic regions close this frontal subcortical limbic loop with projections back to the medial and lateral orbitofrontal cortex.[62] Figure 1–10 shows the subcortical limbic structures involved in the anterior cingulate and orbitofrontal subcortical limbic circuits contrasted with the dorsolateral frontal circuit structures.

Limbic Chemoarchitecture

Neurotransmitters central to limbic processing arise from the inner core of Yakovlev[13] and are modulatory in action. Serotonin, acetylcholine, dopamine, and norepinephrine are the primordial modulating transmitters, in contrast to the fast signaling transmitters concentrated in the outer layer (including aspartate, glutamate, and γ-aminobutyric acid). The limbic distribution of the neuromodulators is being mapped and their receptor are being subtypes targeted with "designer" pharmaceuticals.

Serotonergic System: Serotonin is supplied by projections from the raphe nuclei. Synapses are well defined in the "basket system" innervating the limbic belt from M fibers of the median raphe (area B8); diffusely projecting D fibers to frontal cortex have indistinct synapses on the "varicose system" arising from the rostrolateral portion of area B7 in the dorsal raphe. Of the many serotonin (5-hydroxytryptamine; 5-HT) receptor subtypes, the $5\text{-}HT_{1A}$ and $5\text{-}HT_3$ subtypes predominate in the limbic belt and the $5\text{-}HT_2$ subtype predominates in neocortex. The $5\text{-}HT_{1A}$ and $5\text{-}HT_2$ receptors are G protein coupled to control adenosine 3',5'-cyclic monophosphate (cAMP) and open K^+ channels ($5\text{-}HT_{1A}$ agonist: buspirone); or stimulate phosphoinositide hydrolysis via phospholipase C, a probable antagonistic focus for lithium,[63] and close K^+ channels ($5\text{-}HT_2$ agonist: lysergic acid diethylamide; antagonist: clozapine). The $5\text{-}HT_3$ receptor (antagonist: ondansetron) is a ligand-gated cation channel that also modulates the release of acetylcholine and dopamine.

Serotonergic systems are involved in mood disorders and obsessional syndromes. Patients dying from violent suicide have increased serotonin receptors in the frontal cortex.[64] Depressed patients at risk for suicide have decreased serotonergic metabolites in their cerebral spinal fluid.[65] Parkinson's patients with depression and frontal cognitive impairment, when successfully treated with serotonin reuptake inhibitors, have an increase in metabolism in the basal medial frontal region on [^{18}F]fluorodeoxyglucose positron emission tomography (FDG-PET).[66] Successful treatment of patients with obsessive-compulsive tendencies, using serotonin reuptake inhibitors or behavioral therapy, results in a decrease in hypermetabolism to normal levels in the orbitofrontal cortex and caudate nucleus.[67] The anatomy of depression implicates the medial orbitofrontal cortex in general and the serotonin system in particular.

Cholinergic System: Acetylcholine has two sources supplying the forebrain: one in the brainstem—the pedunculopontine and laterodorsal tegmental system—and the other in the basal forebrain, which includes the septum (Ch1), the vertical (Ch2) and horizontal (Ch3) diagonal band of Broca, and the basal nucleus of Meynert (Ch4).[68] The brainstem cholinergic system supplies the basal ganglia and most of the thalamus, whereas cholinergic fibers from the basal forebrain innervate portions of the ventroanterior, mediodorsal, and midline thalamus as well as the limbic structures and neocortex. Muscarinic (M_{1-5}) and nicotinic ($N\alpha_{2-8}$, β_{2-4}) receptors occur throughout the brain, with the M_1 receptor predominating in the hippocampus. The mus-

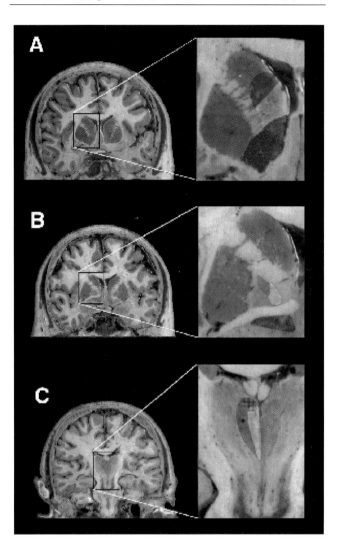

FIGURE 1–10. The subcortical anatomy of the frontal subcortical circuits. Images illustrate the general segregated anatomy of the dorsolateral *(blue)*, orbitofrontal *(green)*, and anterior cingulate *(red)* circuits in the striatum (A), pallidum (D), and mediodorsal thalamus (C).

carinic receptors are G protein coupled; the nicotinic receptors are ligand-gated ion channels. Acetylcholine facilitates thalamic activation of the cortex and assists in the septal hippocampal pathway supporting mnemonic function. The basal nucleus of Meynert has an organized cortical projection pattern; undergoing degeneration early in Alzheimer's disease, its cortical targets, even if spared by plaques or tangles, may become deactivated. Ablation of the posterior basal nucleus of Meynert could result in hypometabolism on FDG-PET in the parietal cortex due to the disconnection of cholinergic activation.[69] Behavioral improvement in patients with Alzheimer's disease, particularly a resolution of apathy[70] associated with anterior cingulate hypoperfusion,[71] occurs with cholinesterase inhibitor therapy.

General cognitive improvement may also occur with drugs promoting cholinergic function.[72]

Dopaminergic System: Dopamine is manufactured in and projected from the substantia nigra pars compacta. The ventral tegmental area and medial portion of the ventral tier of the substantia nigra pars compacta provide dopaminergic innervation to the cortical and subcortical limbic forebrain.[73–75] The severity of dementia in Parkinson's disease is related to decreased cortical dopamine[76,77] and ventral tegmental area cell loss.[78] There are two dopamine receptor families. The D_1 family (D_1 and D_5—antagonist: bromocriptine) are G protein coupled and stimulate adenylate cyclase; the D_2 family (D_{2-4}—antagonist: butyrophenone neuroleptics) are probably all G protein coupled as well and inhibit adenylate cyclase. The D_3 subtype is distributed to limbic regions similar to the D_4 receptor, which may have a greater concentration in the amygdala and frontal cortex. The D_5 receptor is found in the hippocampus and hypothalamus, and the D_2 subtype is found in sensorimotor striatal regions. This regional difference in dopamine receptor subtypes has instigated the development of dopamine receptor antagonists specific to limbic regions, such as the D_4 antagonist clozapine, with fewer motor side effects for the treatment of psychosis. Subcortical lesions of the dopaminergic fibers in the medial forebrain bundle or the ventral pallidum may disrupt the function of the anterior cingulate, resulting in profound apathy, or even akinetic mutism, that responds to agents such as bromocriptine. Motivation and attention rely on a distributed network involving the cingulate, dorsolateral prefrontal, and inferior parietal cortices; a balance between dopamine and norepinephrine may subserve the normal functioning of this distributed attentional network.

Norepinephrine System: Norepinephrine is projected from a dorsal pathway from the locus coeruleus to the entire cortex and hippocampus as well as spinal cord and cerebellum; a ventral pathway arises below the locus coeruleus to innervate the brainstem and hypothalamus. All the norepinephrine receptors are linked to G proteins. The α_{1A-C} receptors activate phospholipase C, the α_{2A-C} inhibit adenylate cyclase, and the β_{1-3} stimulate adenylate cyclase. In the brain, β_2 receptors predominate in the cerebellum, whereas the β_1 receptor is found cortically. A balance between the levels of norepinephrine and dopamine may set the signal-to-noise ratio of the attentional system, enabling the selective filtering of sensory stimuli. Attention deficit disorders, either congenital or acquired by frontal injury, often respond to norepinephrine and dopaminergic pharmacological manipulation with bromocriptine or methylphenidate.

TABLE 1–6. The clinical manifestations and regional localization of hypofunctional, hyperfunctional, and dysfunctional limbic syndromes

Syndrome and Clinical Manifestation	Regional Localization
Hypolimbic syndromes	
Depression	Medial orbitofrontal circuit
Apathy	Anterior cingulate circuit
Amnesia	Archicortical structures
Klüver-Bucy	Amygdala/temporal pole
Hyperlimbic syndromes	
Mania	Medial right diencephalon
Obsessions/compulsions	Orbitofrontal circuit
Limbic epilepsy	Paleocortical structures
Rage	Hypothalamus/amygdala
Dysfunctional limbic syndromes	
Utilization behavior	Lateral orbitofrontal cortex
Social disdecorum	Lateral orbitofrontal circuit
Anxiety/panic	Medial orbitofrontal cortex
Psychosis	The limbic system

CLINICAL SYNDROMES

The distinction between the orbitofrontal/amygdalar division of the limbic system, which supports emotional associations and appetitive drives, and the hippocampal/cingulate limbic division, which supports mnemonic and attentional processes, aids our interpretation of limbic system disorders. This anatomy can inform the field of neuropsychiatry. Psychiatric disorders may be reinterpreted within a brain-based framework of limbic dysfunction and divided into three general groups: decreased, increased, and distorted limbic syndromes (Table 1–6).

Hypolimbic Syndromes

Hypofunctioning of the medial portion of the orbitofrontal cortex is associated with depressive symptoms in patients with Parkinson's disease, Huntington's disease, and primary affective disorders. The medial orbitofrontal cortex, along with the medial infracallosal region, has efferent and afferent connections similar to those of visceromotor centers and the phylogenetically older magnocellular basolateral amygdala.[29] This amygdalar region may subserve the synthesis of internal mood and visceral functions. Future studies should evaluate other pathologic correlates between depression and amygdala/orbitofrontal abnormalities.

Structural or functional lesions of the anterior cingulate above the corpus callosum cause a loss of motivation, producing such profound apathy that patients may become akinetic and mute. A loss of motivation may manifest in the cognitive or motor domains. Anterior cingulate hypofunction produces transcortical motor aphasia[79] or a "loss of the will"[80] to engage in previously enjoyable pastimes.[81] When functional or structural lesions extend posteriorly into the skeletomotor effector region, a loss of spontaneous motor output will also occur.

The hippocampal/cingulate division unites the dorsolateral spatial attentional network with Papez's circuit, to encode objects in the environment. Amnesia—the inability to encode new information—occurs when functional or structural lesions affect the hippocampus or other members of Papez's circuit. Encoding defects are present in patients with lesions of the fornix, mammillary bodies, anterior or dorsomedial thalamus, posterior cingulate, and entorhinal cortex—all members of Papez's classical medial circuit. The hypofunctioning of this more phylogenetically recent archicortical limbic division results in declarative memory abnormalities: episodes of experience can no longer be encoded. This explicit process is contrasted to the implicit stimulus-response associations retained by the phylogenetically older amygdala/orbitofrontal limbic division.

When the amygdala, and probably the adjacent temporal cortex, are damaged, the Klüver-Bucy syndrome[4-6] may ensue. Patients usually have only a portion of the symptoms associated with the classic lesion seen in nonhuman primates—placidity, hyperorality, hypersexuality, hypermetamorphosis (compulsive environmental exploration), and visual agnosia. Loss of premorbid aggressive tendencies is the common postsurgical sequel of bilateral amygdalectomy in humans.[82] The loss of implicit visceral or affective associations with sensory stimuli is the result of human[83] lesions of the amygdala.

Hyperlimbic Syndromes

Mania frequently results from right medial diencephalic lesions[84] that may disrupt hypothalamic circuits or disturb modulating transmitters traversing the adjacent medial forebrain bundle. Mania has also been observed in patients with orbitofrontal lesions, caudate dysfunction in basal ganglia disorders, and lesions of the thalamus.[85-87] Given the strong hypothalamic-amygdala-orbitofrontal connections and the increase in appetitive drives that often accompany it, mania may reflect hyperfunctioning of the paleocortical limbic division.

Obsessive-compulsive symptomatology frequently occurs in basal ganglia disorders including parkinsonism, encephalitis lethargica, manganese intoxication, and Gilles de la Tourette syndrome. Functional imaging studies of patients with idiopathic obsessive-compulsive disorder reveal an abnormal hyperfunctioning of the orbital frontal-subcortical circuit.[67,88,89] This hyperactivity normalizes with either behavioral or pharmacological intervention.[67] Although lateral division activity has also correlated with ritualization urges,[89] the urge might be generated by the medial division and given an environmental target by the lateral division. Thus, surgical resection of the subcallosal cingulate and the medial division of the orbitofrontal cortex provides the best treatment for medically refractory obsessive-compulsive patients.[90]

The interictal syndrome of Gastaut-Geschwind[91,92] comprises personality changes that include hypergraphia, hyposexuality, conversationality, obsessionality, and hyperreligiosity or an increased sense of individual destiny. Patients are viscous or "sticky," being unaware of social cues for discourse termination. Three subgroups have been described:[93] one with primarily altered physiologic drives, another with philosophical preoccupation, and a third with predominantly altered interpersonal skills. These features may represent hyperfunctioning orbitofrontal divisions—the medial portion resulting in altered physiologic drives, the lateral portion in altered interpersonal skills, and the amygdala in a sense of awe.

Rage attacks result from ventromedial hypothalamic lesions in humans.[94] Episodic explosive behavior may be related to hypothalamic or visceral-amygdalar abnormalities, since the best calming effect in such cases may be achieved from lesions involving the lateral part of the corticomedial nuclei.[95] Violent outbursts respond to amygdalectomy[96] or posterior hypothalamotomy,[97] which produce the placidity commonly seen in the Klüver-Bucy syndrome.

Dysfunctional Limbic Syndromes

Utilization and imitation behavior frequently result from lateral orbitofrontal pathology.[98] Patients' automatic imitation of the examiner's gestures and dependence on objects in their environment result from dysfunction in appropriately integrating internal urges with ventral visual object processing. This integration occurs in the lateral portion of the orbitofrontal paralimbic division. The lateral orbital cortex also mediates empathic, civil, and socially appropriate behavior. Thus, patients with dysfunction in this region will also find it difficult to observe the rules of decorum during social interaction. The social disdecorum they demonstrate is often sexual in nature, or other appetitive urges may be difficult to control. All of these behaviors implicate dysfunction at the level of externally directed lateral orbitofrontal integration.

Anxiety and panic disorders result from dysfunction in integrating the visceral-amygdalar functions with the internal state of the organism. This integration occurs in the medial portion of the orbitofrontal paralimbic division. The strong visceral components accompanying these disorders and the dominant role the amygdala plays in building stimulus-response associations implicate the paleocortical division. Functional imaging studies also reveal increased activity in the amygdala[99] and temporal pole[100] in patients with anxiety disorders.

The occurrence of psychotic symptoms in patients may implicate dysfunction in both the implicit integration of affect, drives, and object associations (supported by the paleocortical limbic division) and the functions of explicit sensory processing, encoding, and attentional control (supported by the archicortical limbic division).

SUMMARY

The limbic system is the border zone where psychiatry meets neurology. We have provided a model of limbic function that combines phylogenetic, anatomic, functional, and clinical data to interpret diseases relevant to neuropsychiatry. Understanding the development and organization of the limbic system informs us about how best to approach our patients. The implicit integration of affect, drives, and object associations is the function of the paleocortical limbic division; explicit sensory processing, encoding, and attentional control are the functions of the archicortical limbic division. The two divisions work in concert to integrate thought, feeling, and action.

REFERENCES

1. Willis T: Cerebri anatome. London, Martzer and Alleftry, 1664
2. Broca P: Anatomie comparée des circonvolutions cérébrales: le grand lobe limbique et la scissure limbique dans la série des mammifères [Anatomic comparison of cerebral convolutions: the great limbic lobe and limbic sulci in a series of mammals]. Rev Anthropol Ser 1878; 21:384–498
3. Brown S, Schäfer EA: An investigation into the functions of the occipital and temporal lobes of the monkey's brain. Philos Trans R Soc Lond 1888; 179:303–327
4. Klüver H, Bucy PC: "Psychic blindness" and other symptoms following bilateral temporal lobectomy in rhesus monkeys. Am J Physiol 1937; 119:352–353
5. Klüver H, Bucy PC: An analysis of certain effects of bilateral temporal lobectomy in the rhesus monkey, with special reference to "psychic blindness." J Psychol 1938; 5:33–54
6. Klüver H, Bucy PC: Preliminary analysis of functions of the temporal lobes in monkeys. Arch Neurol Psychiatry 1939; 42:979–1000
7. Bechterew W: Demonstration eines gehirns mit Zestörung der vorderen und inneren Theile der Hirnrinde beider Schläfenlappen [Demonstration of a brain with the frontal portion of the skull damaged]. Neurologisches Centralblatt 1900; 20:990–991
8. Ramón y Cajal S: Studien über die Hirnrinde des Menschen [Studies on the brain of man]. Leipzig, JA Barth, 1900–1906
9. Clarke WEI, Boggon RH: On the connections of the anterior nucleus of the thalamus. J Anat 1933; 67:215–226
10. Cannon WB: The James-Lange theory of emotion: a critical examination and an alternate theory. Am J Psychol 1927; 39:10–124
11. Papez JW: A proposed mechanism of emotion. Arch Neurol Psychiatry 1937; 38:725–733
12. Shipley MT, Sørensen KE: Evidence for an ipsilateral projection from the subiculum to the deep layers of the presubicular and entorhinal cortices. Exp Brain Res 1975; 23(suppl 1):190
13. Yakovlev PI: Motility, behavior, and the brain. J Nerv Ment Dis 1948; 107:313–335
14. MacLean PD: Psychosomatic disease and the "visceral brain": recent developments bearing on the Papez theory of emotion. Psychosom Med 1949; 11:338–353
15. MacLean PD: Some psychiatric implications of physiological studies on the frontotemporal portion of the limbic system (visceral brain). Electroencephalogr Clin Neurophysiol 1952; 4:407–418
16. Brodal A: Neurological Anatomy in Relation to Clinical Medicine. New York, Oxford University Press, 1969
17. MacLean PD: The triune brain, emotion and scientific bias, in The Neurosciences: Second Study Program, edited by Schmitt FO. New York, Rockefeller University Press, 1970, pp 336–349
18. Sanides F: Comparative architectonics of the neocortex of mammals and their evolutionary interpretation. Ann NY Acad Sci 1969; 167:404–423
19. Pandya DN, Yeterian EH: Architecture and connections of cortical association areas, in Cerebral Cortex, vol 4, edited by Peters A, Jones EG. New York, Plenum, 1985, pp 3–55
20. Mesulam M-M: Patterns in behavioral neuroanatomy: association areas, the limbic system, and hemispheric specialization, in Behavioral Neurology, edited by Mesulam M-M. Philadelphia, FA Davis, 1985, pp 1–70
21. Stephan H, Andy OJ: Quantitative comparison of the amygdala in insectivores and primates. Acta Anat (Basel) 1977; 98:130–153
22. Humphrey T: The development of the human amygdaloid complex, in The Neurobiology of the Amygdala, edited by Elftheriou BE. New York, Plenum, 1972, pp 21–80
23. Stephan H: Evolutionary trends in limbic structures. Neurosci Biobehav Rev 1983; 7:367–374
24. Posner MI, Walker JA, Friedrich FA, et al: How do the parietal lobes direct covert attention? Neuropsychologia 1987; 25:135–145
25. Brodmann K: Vergleichende Lokalisationslehre der Grosshirnrinde in ihren Prinzipien dargestellt auf Grund des Zellenbaues [Localization in the cerebral cortex]. Leipzig, Barth, 1909
26. Carmichael ST, Price JL: Architectonic subdivisions of the orbital and medial prefrontal cortex in the macaque monkey. J Comp Neurol 1994; 346:366–402
27. Zald DH, Kim SW: Anatomy and function of the orbital frontal cortex, I: anatomy, neurocircuitry, and obsessive-compulsive disorder. J Neuropsychiatry Clin Neurosci 1996; 8:125–138
28. Carmichael ST, Price JL: Limbic connections of the orbital and medial prefrontal cortex in macaque monkeys. J Comp Neurol 1995; 363:615–641
29. Mega MS, Cummings JL: The cingulate and cingulate syndromes, in Contemporary Behavioral Neurology, edited by Trimble MR, Cummings JL. Boston, Butterworth-Heinemann, 1997, pp 189–214
30. Amaral DG, Price JL, Pitkänen A, et al: Anatomical organization of the primate amygdaloid complex, in The Amygdala, edited by Aggleton JP. New York, Wiley-Liss, 1992, pp 1–66
31. Heimer L, de Olmos J, Alheid GF, et al: "Perestroika" in the basal

forebrain: opening the border between neurology and psychiatry. Prog Brain Res 1991; 87:109–165
32. Amaral DG, Price JL: Amygdalo-cortical projections in the monkey (Macaca fascicularis). J Comp Neurol 1984; 230:465–496
33. Vogt BA, Pandya DN: Cingulate cortex of the rhesus monkey, II: cortical afferents. J Comp Neurol 1987; 262:271–289
34. Müller-Preuss P, Jürgens U: Projections from the "cingular" vocalization area in the squirrel monkey. Brain Res 1976; 103:29–43
35. Kaada BR, Pribram KH, Epstein JA: Respiratory and vascular responses in monkeys from temporal pole, insula, orbital surface and cingulate gyrus. J Neurophysiol 1949; 12:347–356
36. Mufson EJ, Mesulam M-M: Insula of the old world monkey, II: afferent cortical input and comments on the claustrum. J Comp Neurol 1982; 212:23–37
37. Moran MA, Mufson EJ, Mesulam M-M: Neural inputs to the temporopolar cortex of the rhesus monkey. J Comp Neurol 1987; 256:88–103
38. Morecraft RJ, Geula C, Mesulam M-M: Cytoarchitecture and neural afferents of orbitofrontal cortex in the brain of the monkey. J Comp Neurol 1992; 323:341–358
39. Pandya DN, Van Hoesen GW, Mesulam M-M: Efferent connections of the cingulate gyrus in the rhesus monkey. Exp Brain Res 1981; 42:319–330
40. Insausti R, Amaral DG, Cowan WM: The entorhinal cortex of the monkey, II: cortical afferents. J Comp Neurol 1987; 264:356–395
41. Grasby PM, Frith CD, Friston KJ, et al: Functional mapping of brain areas implicated in auditory-verbal memory function. Brain 1993; 116:1–20
42. Grasby PM, Firth CD, Friston K, et al: Activation of the human hippocampal formation during auditory-verbal long-term memory function. Neurosci Lett 1993; 163:185–188
43. Fletcher PC, Firth CD, Grasby PM, et al: Brain systems for encoding and retrieval of auditory-verbal memory: an in vivo study in humans. Brain 1995; 118:401–416
44. Shallice T, Fletcher P, Firth CD, et al: Brain regions associated with acquisition and retrieval of verbal episodic memory. Nature 1994; 368:633–635
45. Molchan SE, Sunderland T, McIntosh AR, et al: A functional anatomical study of associative learning in humans. Proc Natl Acad Sci USA 1994; 91:8122–8126
46. Baleydier C, Mauguière F: The duality of the cingulate gyrus in monkey: neuroanatomical study and functional hypothesis. Brain 1980; 103:525–554
47. Morecraft RJ, Geula C, Mesulam M-M: Architecture of connectivity within a cingulofronto-parietal neurocognitive network. Arch Neurol 1993; 50:279–284
48. Cavada C, Goldman-Rakic PS: Posterior parietal cortex in rhesus monkey, I: parcellation of areas based on distinctive limbic and sensory corticocortical connections. J Comp Neurol 1989; 287:393–421
49. Morecraft RJ, Van Hoesen GW: A comparison of frontal lobe afferents to the primary, supplementary and cingulate cortices in the rhesus monkey (abstract). Society for Neuroscience Abstracts 1991; 17:1019
50. Vogt BA, Nimchinsky EA, Vogt LJ, et al: Human cingulate cortex: surface features, flat maps, and cytoarchitecture. J Comp Neurol 1995; 359:490–506
51. Vogt BA: Structural organization of cingulate cortex: areas, neurons, and somatodendritic transmitter receptors, in Neurobiology of Cingulate Cortex and Limbic Thalamus: A Comprehensive Handbook, edited by Vogt BA, Gabriel M. Boston, Birkhäuser, 1993, pp 19–70
52. Mesulam M-M, Mufson EJ: The insula of Reil in man and monkey: architectonics, connectivity, and function, in Cerebral Cortex, edited by Jones EG, Peters AA. New York, Plenum, 1985, pp 179–226
53. Mega MS, Cummings JL: Frontal-subcortical circuits and neuropsychiatric disorders. J Neuropsychiatry Clin Neurosci 1994; 6:358–370
54. Alexander GE, DeLong MR, Strick PL: Parallel organization of functionally segregated circuits linking basal ganglia and cortex. Annu Rev Neurosci 1986; 9:357–381
55. Selemon LD, Goldman-Rakic PS: Longitudinal topography and interdigitation of corticostriatal projections in the rhesus monkey. J Neurosci 1985; 5:776–794
56. Haber SN, Lynd E, Klein C, et al: Topographic organization of the ventral striatal efferent projections in the rhesus monkey: an anterograde tracing study. J Comp Neurol 1990; 293:282–298
57. Haber SN, Lynd-Balta E, Mitchell SJ: The organization of the descending ventral pallidal projections in the monkey. J Comp Neurol 1993; 329:111–128
58. Vogt BA, Pandya DN: Cingulate cortex of the rhesus monkey, I: cytoarchitecture and thalamic afferents. J Comp Neurol 1987; 262:256–270
59. Giguere M, Goldman-Rakic PS: Mediodorsal nucleus: areal, laminar, and tangential distribution of afferents and efferents in the frontal lobe of rhesus monkey. J Comp Neurol 1988; 277:195–213
60. Goldman-Rakic PS, Porrino LJ: The primate mediodorsal (MD) nucleus and its projection to the frontal lobe. J Comp Neurol 1985; 242:535–560
61. Johnson TN, Rosvold HE: Topographic projections on the globus pallidus and substantia nigra of selectively placed lesions in the precommissural caudate nucleus and putamen in the monkey. Exp Neurol 1971; 33:584–596
62. Ilinsky IA, Jouandet ML, Goldman-Rakic PS: Organization of the nigrothalamocortical system in the rhesus monkey. J Comp Neurol 1985; 236:315–330
63. Snyder SH: Second messengers and affective illness: focus on the phosphoinositide cycle. Pharmacopsychiatry 1992; 25:25–28
64. Mann JJ, Stanley M, McBride PA, et al: Increased serotonin$_2$ and β-adrenergic receptor binding in the frontal cortices of suicide victims. Arch Gen Psychiatry 1986; 43:954–959
65. Brown GL, Linnoila MI: CSF serotonin metabolite (5-HIAA) studies in depression, impulsivity, and violence. J Clin Psychiatry 1990; 51:31–41
66. Mayberg H, Mahurin RK, Brannan SK, et al: Parkinson's depression: discrimination of mood-sensitive and mood-insensitive cognitive deficits using fluoxetine and FDG PET (abstract). Neurology 1995; 45(suppl 4):A166
67. Baxter LR, Schwartz JM, Bergman KS, et al: Caudate glucose metabolic rate changes with both drug and behavior therapy for obsessive-compulsive disorder. Arch Gen Psychiatry 1992; 49:681–689
68. Mesulam M-M, Mufson EJ, Levey AI, et al: Cholinergic innervation of cortex by the basal forebrain: cytochemistry and cortical connections of the septal area, diagonal band nuclei, nucleus basalis (substantia innominata), and hypothalamus in the rhesus monkey. J Comp Neurol 1983; 214:170–197
69. Friedland RP, Brun A, Budinger TF: Pathological and positron emission tomographic correlations in Alzheimer's disease. Lancet 1985; 1:228–230
70. Kaufer DI, Cummings JL, Christine D: Effect of tacrine on behavioral symptoms in Alzheimer's disease: an open-label study. J Geriatr Psychiatry Neurol 1996; 9:1–6
71. Craig HA, Cummings JL, Fairbanks L, et al: Cerebral blood flow correlates of apathy in Alzheimer's disease. Arch Neurol 1996; 53:1116–1120
72. Knapp MJ, Knopman DS, Solomon PR, et al: A 30-week randomized controlled trial of high-dose tacrine in patients with Alzheimer's disease. JAMA 1994; 271:985–991
73. Thierry AM, Tassin JP, Blanc G, et al: Studies on mesocortical dopamine systems. Adv Biochem Psychopharmacol 1978; 19:205–216
74. Moore RY, Bloom FE: Central catecholamine neuron systems: anatomy and physiology of the dopamine system. Annu Rev Neurosci 1978; 1:129–169
75. Oades RD, Halliday GM: Ventral tegmental (A10) system: neurobi-

ology, I: anatomy and connectivity. Brain Res Brain Res Rev 1987; 12:117–165
76. Scatton B, Javoy-Agid F, Rouquier L, et al: Reduction of cortical dopamine, noradrenaline, serotonin and other metabolites in Parkinson's disease. Brain Res 1983; 275:321–328
77. Agid Y, Rugerg M, Dubois B, Javoy-Agid F: Biochemical substrates of mental disturbances in Parkinson's disease. Adv Neurol 1984; 40:211–218
78. Rinne JO, Rummukainen J, Paljärvi L, et al: Dementia in Parkinson's disease is related to neuronal loss in the medial substantia nigra. Ann Neurol 1989; 26:47–50
79. Alexander MP, Benson DF, Stuss DT: Frontal lobes and language. Brain Lang 1989; 37:656–691
80. Damasio AR, Van Hoesen GW: Focal lesions of the limbic frontal lobe, in Neuropsychology of Human Emotion, edited by Heilman KM, Satz P. New York, Guilford, 1983, pp 85–110
81. Devinsky O, Morrell MJ, Vogt BA: Contributions of anterior cingulate cortex to behavior. Brain 1995; 118:279–306
82. Aggleton JP: The functional effects of amygdala lesions in humans: a comparison with findings from monkeys, in The Amygdala, edited by Aggleton JP. New York, Wiley-Liss, 1992, pp 485–503
83. Bechara A, Tranel D, Damasio H, et al: Double dissociation of conditioning and declarative knowledge relative to the amygdala and hippocampus in humans. Science 1995; 269:1115–1118
84. Cummings JL, Mendez MF: Secondary mania with focal cerebrovascular lesions. Am J Psychiatry 1984; 141:1084–1087
85. Bogousslavsky J, Ferrazzini M, Regli F, et al: Manic delirium and frontal-like syndrome with paramedian infarction of the right thalamus. J Neurol Neurosurg Psychiatry 1988; 51:116–119
86. Jorge RE, Robinson RG, Starkstein SE, et al: Secondary mania following traumatic brain injury. Am J Psychiatry 1993; 150:916–921
87. Starkstein SE, Pearlson GD, Boston J, et al: Mania after brain injury: a controlled study of causative factors. Arch Neurol 1987; 44:1069–1073
88. Rauch SL, Jenike MA, Alpert NM, et al: Regional cerebral blood flow measured during symptom provocation in obsessive-compulsive disorder using oxygen 15-labeled carbon dioxide and positron emission tomography. Arch Gen Psychiatry 1994; 51:62–70
89. McGuire PK, Bench CJ, Firth CD, et al: Functional anatomy of obsessive-compulsive phenomena. Br J Psychiatry 1994; 164:459–468
90. Hay P, Sachdev P, Cumming S, et al: Treatment of obsessive-compulsive disorder by psychosurgery. Acta Psychiatr Scand 1993; 87:197–207
91. Gastaut H, Roger J, Roger A: Sur la signification de certaines fugues épileptiques. A propos d'une observation électroclinique d'état de mal temporal [On the significance of certain features of epilepsy. A proposal on the electroclinical observations of the "dysfunctional temporal state"]. Rev Neurol 1956; 94:298–301
92. Waxman SG, Geschwind N: Hypergraphia in temporal lobe epilepsy. Neurology 1974; 24:629–636
93. Nielsen H, Christensen O: Personality correlates of sphenoidal EEG foci in temporal lobe epilepsy. Acta Neurol Scand 1981; 64:289–300
94. Reeves AG, Plum F: Hyperphagia, rage, and dementia accompanying a ventromedial hypothalamic neoplasm. Arch Neurol 1969; 20:616–624
95. Narabayashi H, Shima F: Which is the better amygdala target, the medial or lateral nucleus for behavioral problems and paroxysms in epileptics? In Surgical Approaches in Psychiatry, edited by Laitinen LV, Livingston KE. Baltimore, University Park Press, 1973, pp 129–134
96. Kiloh LG, Gye RS, Rushworth RG, et al: Stereotactic amygdaloidotomy for aggressive behavior. J Neurol Neurosurg Psychiatry 1974; 37:437–444
97. Schvarcz JR, Driollet R, Rios E, et al: Stereotactic hypothalamotomy for behavior disorders. J Neurol Neurosurg Psychiatry 1972; 35:356–359
98. Lhermitte F, Pillon B, Serdaru M: Human autonomy and the frontal lobes, I: imitation and utilization behavior: a neuropsychological study of 75 patients. Ann Neurol 1986; 19:326–334
99. Raichle M: Exploring the mind with dynamic imaging. Seminars in the Neurosciences 1990; 2:307–315
100. Reiman EM, Raichle ME, Robins E, et al: Neuroanatomical correlates of a lactate-induced anxiety attack. Arch Gen Psychiatry 1989; 46:493–500
101. Yakovlev PI, Lecours A-R: The myelogenetic cycles of regional maturation of the brain, in Regional Development of the Brain in Early Life, edited by Minkowski A. Edinburgh, Blackwell Scientific, 1967, pp 3–70

2

Ventromedial Temporal Lobe Anatomy, With Comments on Alzheimer's Disease and Temporal Injury

Gary W. Van Hoesen, Ph.D.

This chapter reviews the complex surface topography of the ventromedial temporal lobe and recent findings on its cortical connections. The involvement of the ventromedial temporal area has profound effects in Alzheimer's disease, and these, as well as the effects of injury to this area, are also discussed.

TOPOGRAPHY OF THE VENTROMEDIAL TEMPORAL AREA IN THE HUMAN AND NONHUMAN PRIMATES

Component Structures

The ventromedial temporal area is a complicated and functionally diverse collection of neural structures along the innermost margin of the temporal fossa adjacent to the sphenoid bone and the petrous part of the temporal bone. The various structures are progressive in an evolutionary sense, reaching substantial elaboration in higher primates and humans, but all mammals have at least partial copies of the core elements. The largely subcortical amygdala is a key part of the ventromedial temporal area, along with the rolled and enfolded allocortical areas that form the hippocampal formation. The superficial allocortical and periallocortical areas that cover the amygdala and hippocampal formation form the parahippocampal gyrus, the third major structure of the ventromedial temporal area. All structures of the ventromedial temporal area are components of the limbic system, and the parahippocampal gyrus is a prominent part of the limbic lobe.[1-7]

Boundaries

The largest part of the parahippocampal gyrus is formed by Brodmann area 28, the entorhinal cortex.[8] Medially and anteriorly, it abuts the primary olfactory and periamygdaloid allocortices; medially and posteriorly, it abuts the subicular allocortices of the hippocampal formation (Figure 2–1A, B). Posteriorly, the entorhinal cor-

---

This work was supported by Grants NS14944 and PO NS19632 from the National Institutes of Health.

tex nearly reaches the anteriormost part of the lingual gyrus of the occipital lobe in the vicinity of the anterior tip of the calcarine fissure. The lateral borders of the entorhinal cortex, throughout its anterior-posterior extent, is the perirhinal cortex, or Brodmann area 35. This cortex forms a major part of the medial wall of the collateral fissure and intervenes between the highly atypical entorhinal periallocortex and the inferior temporal isocortex, or Brodmann area 36.[9] In lower mammals, the perirhinal cortex may be as narrow as a few cell diameters, whereas in humans it is a sizable (but poorly understood) area.[10,11]

Sulcal Landmarks

The only major sulcal landmark of the ventromedial temporal area in humans is the collateral sulcus, which approximates the lateral boundary of the parahippocampal gyrus (Figure 2–1A, B). Its posterior stem in the occipitotemporal area is nearly invariant from brain to brain, but its anterior parts vary enormously. Unlike the brains of monkeys and great apes, the human brain usually lacks a clear-cut rhinal sulcus.[11] Ironically, this staple of nearly all mammalian brains is nearly absent in the human brain, although our species has a very elaborate entorhinal cortex. If present to any degree, the rhinal sulcus is typically short and/or shallow, resembling a groove more than a fissure or sulcus. Thus, the human entorhinal cortex cannot be viewed as lying "within" the rhinal sulcus as the term would imply and necessitate. Monkeys lack a collateral sulcus and have instead a relatively invariant occipitotemporal sulcus. This entraps a portion of the medial occipitotemporal area into a posterior parahippocampal area posterior to the entorhinal cortex.

Surface Features

Surface bumps and elevations are a prominent feature of the parahippocampal gyrus. They were identified and named near the turn of the century by the Swedish neuroanatomist Retzius.[1] Most conspicuous is the uncus or uncal hippocampal formation. An uncal sulcus, created by the abrupt, hairpin-like medial and upward turn of the anteriormost tip of the hippocampal formation, is an invariant feature of the ventromedial temporal area. Another prominent elevation is the semilunar gyrus, which defines the location of the cortical amygdaloid nuclei.[2,5,7]

Nearly as conspicuous as the uncus, but a few millimeters anterior to it, is the gyrus ambiens, formed by Brodmann area 34 (Figure 2–1A, B). This is an elevation of medial entorhinal cortex that occupies a position medial to the point where the free edge of the tentorium cerebelli grooves the entorhinal cortex before it attaches to the clinoid process. Thus, the gyrus ambiens lies directly in the tentorial aperture and bulges into the space of Bichat and the transverse cerebral fissure. In the gross brain, and in cross-sections of scans and stained tissue, the gyrus ambiens is often misidentified as uncus, although the two are decidedly different structures. The former is entorhinal cortex medial to the tentorial notch, and the latter consists of subicular and hippo-

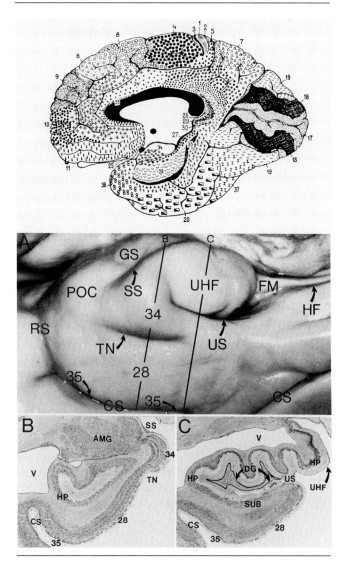

FIGURE 2–1. The human ventromedial temporal area in the gross brain (A) and in cross-section (B and C) with Nissl staining. Brodmann's medial view of the human brain is reproduced at the top for comparison. AMG = amygdala; CS = collateral sulcus; DG = dentate gyrus; FM = fimbria; GS = semilunar gyrus; HF = hippocampal fissure; HP = hippocampal formation; POC = primary olfactory cortex; RS = rhinal sulcus; SS = sulcus semianularis; SUB = subiculum; TN = tentorial notch; UHF = uncal hippocampal formation; US = uncal sulcus; V = inferior horn of lateral ventricle.

campal formation pyramids forced out of hippocampal fissure. The gyrus ambiens is always a few millimeters anterior to the uncus and in a cross-sectional or coronal plane through the posterior amygdala.[7] As discussed later, it is the gyrus ambiens or medial entorhinal cortex that leads herniation of the temporal lobe through the tentorial aperture and that is vulnerable to injury when the brain is forced onto the free edge of the tentorium cerebelli.

A conspicuous topographic feature of the ventromedial temporal area in the human brain is the presence of small wart-like bumps on the surface of the entorhinal cortex. These are visible to the naked eye in both fixed and unfixed specimens. Early neuroanatomists named these *verrucae* to call attention to their resemblance to an epidermal disease of viral origin, and Retzius noted their resemblance to the skin of certain amphibians.[1,2] Verrucae are present in apes and monkeys but are far less notable than in humans, where they cover nearly the entire entorhinal cortex, including its medial parts that form the gyrus ambiens. Histochemical studies have shown that the elevations correspond to modules rich in cytochrome oxidase.[12] From a cellular viewpoint they correlate with the islands of large hyperchromatic multipolar neurons that form layer II of the entorhinal cortex.[12,13] These neurons project powerfully to the hippocampal formation, linking the latter to the cerebral cortex via the perforant pathway.[3] The surface features, cytoarchitecture, and connections of the entorhinal cortex set it apart from all other cortical fields in the hemisphere.

CORTICAL CONNECTIONS OF VENTROMEDIAL TEMPORAL AREA STRUCTURES

Cortical Input

Basic facts about the subcortical projections of ventromedial temporal structures were established early in the century by the use of passive staining methodology and dissection. However, studies using these tools were limited in scope and centered largely on compact bundles like the stria terminalis as an output pathway of the amygdala and the fornix as an output pathway of the hippocampal formation. Ramon y Cajal's detailed Golgi studies established many intrinsic connections within the latter and produced the key observation that the entorhinal cortex of the parahippocampal gyrus projects powerfully to the hippocampal formations via what he labeled the *temporo-ammonic* or *perforant* pathway.[10]

Aside from these observations, it was not until the middle of the century, with the advent of experimental neuroanatomical methodology, that a steady flow of new information became available on ventromedial temporal neuroanatomy, and particularly on the degree and nature of cortical connections that link these structures to the remainder of the hemisphere. Especially noteworthy were two classic reports that took on the challenge of describing all of the major corticocortical connections of the rhesus monkey. For each of these reports, the question of where sensory-related cortices project was central, and the skeleton for all cortical association systems was described.[14,15] A finding common to both of these investigations, and many more that have followed,[16-19] is that one of the targets, or end stations, for multisynaptic cortical sensory association systems is the ventromedial temporal area, and cortical input is the major afferent for the amygdala, hippocampal formation, and parahippocampal gyrus. Thus, the ventromedial temporal area receives cortical input not only from the limbic cortices, but also from the association cortices related to the visual, auditory, and somatic modalities. Along with olfactory input from the primary olfactory cortex and visceral input from the insular cortices, the sum total of their cortical input must be viewed as distinctly multimodal.[3] Collectively, these projections have been labeled *feedforward* systems, denoting their sequential stepwise nature and the manner in which sensory information is disseminated within the cortex.[20-22]

The cortical systems discussed above, although large in size, are by no means diffuse with respect to their ventromedial temporal area targets. For example, corticoamygdaloid projections selectively target only certain amygdaloid nuclei,[23-25] and in some cases only parts of them (Figure 2–2). The lateral amygdaloid nucleus receives at least three cortical projections from the temporal association cortices, and they appear segregated with regard to function. For example, the visual association cortices of the inferior and middle temporal gyri send projections that target the more lateral shell of lateral amygdaloid nucleus and end in a dorsal location. The superior temporal auditory association cortices, in contrast, send axons that end laterally, but ventral to those from the visual association cortices. The more multimodal temporal polar cortex sends corticoamygdaloid axons that end in the more medial parts of the lateral nucleus. In parts of the lateral amygdaloid nucleus, convergence from more than one sensory modality occurs, with putative taste and olfactory association cortical input contributed by the anterior insular, opercular, and posterior orbitofrontal cortices. Although many details remain unknown, it seems fair to state that the amygdala receives powerful input from the association and limbic cortices of the temporal lobe, anterior insula, posterior orbitofrontal, medial frontal, and anterior cingulate cortices. Many of the systems are organized discretely and target only selective nuclei, or parts thereof, in the amygdala.[24]

Direct cortical association input to the hippocampal formation is far less sizable than that to the amygdala, since it relays first in the perirhinal and entorhinal cortices (Figure 2–3).[3,26] However, the indirect nature of the anatomy is deceptive. The entorhinal cortex output to the hippocampal formation forms one of the larger cortical association pathways of the cortex, and certainly the largest in the temporal lobe. Collectively, it is known as the *perforant pathway* because its axons pass through or perforate the subicular cortices en route to synaptic sites on the distal parts of hippocampal and subicular apical dendrites and outer two-thirds of dentate gyrus granule cell dendrites (Figure 2–4). Discrete entorhinal layers provide origin for the perforant pathway.[26,27] For example, layer III projects largely to the subiculum and hippocampal pyramids, whereas layer II contributes the entorhinal-dentate component of the perforant pathway (Figure 2–5). In the dentate gyrus, perforant pathway axons target approximately 80% to 85% of the postsynaptic space and have a powerful excitatory influence on these neurons. In all mammals investigated, and especially primates, perforant pathway entorhinal cortex axons destined for the hippocampal formation form a compact bundle known as the *angular bundle*, located typically just inside the medialmost part of parahippocampal white matter adjacent to the ependymal lining of the unopened portion of the inferior horn of the lateral ventricle. Perforating axon bundles leave the angular bundle throughout the anterior-posterior extent of the hippocampal formation.[27] This arrangement is necessary because in nearly all mammals the entorhinal cortex is not entirely adjacent to the hippocampal formation. In fact, in primates much of the entorhinal cortex is located decidedly anterior to the hippocampal formation.

Cortical Output

Although the older neuroanatomical literature stresses the subcortical projections of the ventromedial tempo-

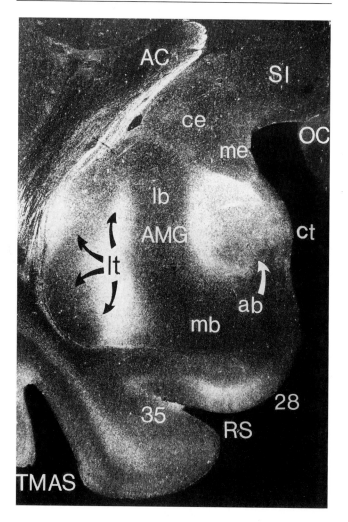

FIGURE 2–2. Darkfield photomicrograph of a cross-section through the amygdala (AMG) of a rhesus monkey after injection of tritiated amino acids in the temporal association cortex, showing terminal axonal labeling *(white)* over various amygdaloid nuclei after autoradiography. ab = accessory basal nucleus; AC = anterior commissure; ce = central nucleus; ct = cortical nuclei; lb = laterobasal nucleus; lt = lateral nucleus; mb = mediobasal nucleus; me = medial nucleus; OC = optic chiasm; RS = rhinal sulcus; SI = substantia innominata; TMAS = temporalis medialis anterior sulcus.

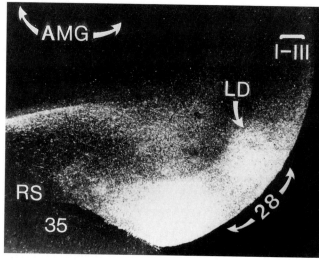

FIGURE 2–3. Darkfield photomicrograph of a cross-section through the entorhinal cortex (area 28) of a rhesus monkey after injection of tritiated amino acids in the posterior parahippocampal area, showing terminal axon labeling *(white)* over entorhinal cortex layers I-III after autoradiography. AMG = amygdala; LD = lamina dissecans; RS = rhinal sulcus.

FIGURE 2–4. Darkfield photomicrograph of a cross-section through the hippocampal formation of a rhesus monkey after injection of tritiated amino acids in the entorhinal cortex and autoradiography showing terminal axon labeling (white) along the perforant pathway (PP) terminal zone. FM = fimbria/fornix; ITG = inferior temporal gyrus; LGN = lateral geniculate nucleus; MTG = middle temporal gyrus; PHG = parahippocampal gyrus; PSUB = presubiculum; SG = stratum granulosum of dentate gyrus; SP = stratum pyramidale of hippocampus; ST = stria terminalis; STG = superior temporal gyrus; SUB = subiculum; TC = tail of caudate nucleus; V = lateral ventricle.

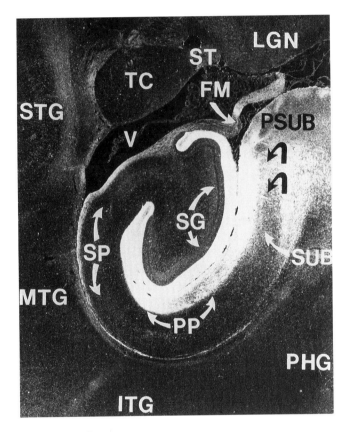

FIGURE 2–5. A Lucifer yellow–filled layer II multipolar neuron from the rhesus monkey rotated in various planes. The neurons give rise to the entorhinal–dentate gyrus part of the perforant pathway.

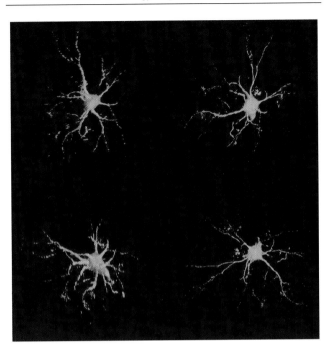

ral structures, newer experimental findings reveal substantial feedback projections to the cortex. The amygdala is particularly impressive in this regard and is now known to send axons to the limbic cortex, association cortex, and in some cases even the primary sensory cortex.[25,28] Some of its output to the cortex of the cingulate sulcus is in a position to influence the cells of origin of corticospinal axons that form part of the cingulate motor cortex. Although some of these cortical axons reciprocate corticoamygdaloid projections, it is clear that many do not. Via subcortical projections to the hypothalamus and parasympathetic centers in the brainstem, the amygdala can influence endocrine and autonomic effectors. Via cortical projections, it can influence somatic effector–related motor areas as well as widespread parts of the limbic, association, and even primary sensory cortices. These outputs suggest a greatly expanded role for the amygdala in diverse behaviors located along much of the neuraxis.

The hippocampal formation also is now known to have extensive feedback projections to the cerebral cortex.[3,26,29,30] These arise directly and selectively from the subicular and CA1 pyramids whose axons end in much of the cortex of the limbic lobe and the orbitofrontal, medial frontal, anterior temporal, and posterior temporal association cortices. A powerful contingent of hippocampal cortical output is directed toward the entorhinal cortex, particularly its deeper layers.[29] This output partially reciprocates the strong entorhino-hippocampal projection carried by the perforant pathway (Figure 2–6A, B). However, the latter arises from the more superficial layers of the entorhinal cortex. Thus, intrinsic cortical axons linking entorhinal layers together complete the reciprocity. Nevertheless, layer IV of the entorhinal cortex, which is the target of subicular and CA1 projections, projects widely to many parts of the limbic cortices and to cortical association areas in the temporal lobe. Together, these direct and indirect hippocampal-cortical projections form strong neural systems disseminating hippocampal output.

In summary, ventromedial temporal structures are now known to be a major target of cortical association feedforward axons, and these form their largest input. The cortical areas in question are the so-called distal association areas, meaning they are several synapses

FIGURE 2–6. A: direct cortical input to hippocampal pyramidal neurons (CA1–3) and dentate gyrus granule cells (DG) via the perforant pathway that arises from the entorhinal cortex (EC). B: hippocampal cortical output to the association and limbic cortices from the hippocampal pyramidal neurons (CA1), subiculum, and layer IV of the entorhinal cortex. AT = anterior thalamus; CS = collateral sulcus; FF = fimbria-fornix; HF = hippocampal fissure; MMB = mammillary bodies; PC = perirhinal cortex.

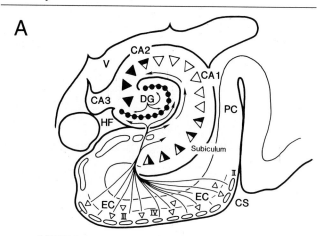

HIPPOCAMPAL CORTICAL INPUT

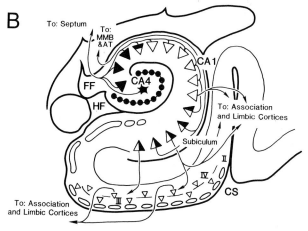

HIPPOCAMPAL CORTICAL OUTPUT

removed from the primary sensory cortices. Some are multimodal association areas, but others retain modality specificity. Functional studies of all types link these cortical areas to higher order processes, including memory. The ventromedial temporal structures reciprocate many of their inputs with feedback axons to the association cortices. However, they are unique in being a source of *non*-reciprocal axons to the early association cortices and even some primary sensory areas. This nonreciprocal projection suggests that ventromedial temporal structures can alter initial sensory processing in the cortex and/or can retroactivate or select percepts unique to, or stored in, early association cortical areas.[31–33]

VENTROMEDIAL TEMPORAL AREAS AND ALZHEIMER'S DISEASE

Hippocampal Formation

It has been known for many years that the ventromedial temporal area contains numerous sites of prediction for the pathological alterations that characterize Alzheimer's disease. The hippocampus has received the greatest attention in this respect, and changes in this structure are nearly an invariant feature of the illness.[34] In terms of neurofibrillary tangles, these intracellular alterations occur with greatest frequency in the subiculum and CA1 parts of the hippocampal formation and are less frequently observed in the CA3 and CA4 parts of the structure. Likewise, the granule cells of the dentate gyrus occasionally contain neurofibrillary tangles but frequently are spared entirely. Neuritic plaques, the second signature change of Alzheimer's disease, have a variable distribution in the hippocampal formation, and when present are most conspicuous in the subiculum and CA1 fields of the structure. Here they are invariably scattered among the pyramidal neurons that form the structures and in the molecular layer through which their apical dendrites and terminal dendritic branches course.[35,36] Although neurofibrillary tangles are less frequent in the granule cells of the dentate gyrus, neuritic plaques are commonly seen in the molecular layer of this structure, typically midway between the granule cell soma and the hippocampal fissure. In the subiculum and CA1 zones neuritic plaques are most abundant at the point where the large radial apical dendrites begin dividing into numerous smaller, terminal secondary and tertiary branches. Curiously, in both the dentate gyrus and the hippocampus neuritic plaques occupy the parts of their respective molecular layers where the largest numbers of terminal axons of the perforant pathway end.[37] Although the hippocampal formation must be regarded as a structure targeted heavily in Alzheimer's disease, it is clear that the distribution of pathology is highly selective, affecting some neurons and not others. No clues exist at this time to explain this selectivity, and hypotheses are equally scarce.

Amygdala

The amygdala is also targeted heavily with pathology in Alzheimer's disease, but only in recent years has its involvement come into sharper perspective.[38,39] Recent studies have focused largely on the presence or absence of pathology in the various subnuclei that form this large and complex structure. Although all amygdaloid subnuclei contain some neurofibrillary tangles and neuritic plaques, a distinct hierarchy exists. For example, the accessory basal, cortical, and mediobasal nuclei all contain large quantities of neurofibrillary tangles and neuritic plaques. The same can be said for the cortical transition area that connects the posteriormost part of the amygdala with the subiculum of the uncal hippocampal formation. In contrast, the medial, central, lateral, and laterobasal nuclei contain less pathology. In fact, the medial and lateral nuclei are very lightly affected, with only occasional evidence of pathology.[38]

The patterns of pathology in the amygdala in Alzheimer's disease are difficult to categorize in either neuroanatomical or functional terms. For example, the accessory basal, laterobasal, and lateral nuclei all have strong input from and project back to the cortex. However, in Alzheimer's disease only the accessory basal nucleus shows substantial pathology. Similarly, the medial and cortical nuclei both receive olfactory bulb input, yet only the latter is targeted heavily in the disease. Even in a phylogenetic sense, correlations are sparse. For example, the more conservative cortical nucleus is damaged heavily, but its partners, the medial and central nuclei, are not. The more progressive lateral nucleus is largely unaffected, but another component of the newer laterobasal complex, the accessory basal nucleus, is nearly destroyed in a high percentage of Alzheimer's disease cases. At this time, the most noteworthy correlation occurs between amygdaloid nuclei connected to the subiculum and CA1 zones of the hippocampal formation. These would include the cortical transition area and the mediobasal and accessory basal nuclei. All receive hippocampal formation afferents, and all contain abundant pathology in Alzheimer's disease.

Parahippocampal Gyrus

There is general agreement that the parahippocampal gyrus is the most heavily damaged part of the cerebral cortex in Alzheimer's disease and the likely focus for the initial appearance of neurofibrillary tangles.[8,9,35,40] The cortical areas in question form Brodmann areas 28 and 35. Changes in these cortical areas are conspicuous even in the gross brain,[41,42] where the cortex of the parahippocampal gyrus appears atrophic, pitted, and discolored (Figure 2–7). A distinct laminar specificity of pathology is observed in Nissl-stained preparations and with pathological stains for neurofibrillary tangles. For example, in the early stages of the disease, layer II contains abundant neurofibrillary tangles, and many of the large multipolar neurons contain this alteration. As the duration of illness progresses, layer III will contain neurofibrillary tangles, followed by layer IV. Layers V and VI appear to be the most resistant to this form of pathology, but they likewise may be affected at end-

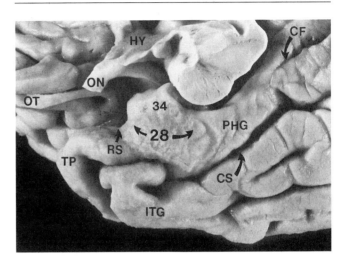

FIGURE 2–7. A ventromedial temporal view of the human brain in Alzheimer's disease showing the atrophic and pitted appearance of area 28, the entorhinal cortex. CF = calcarine fissure; CS = collateral sulcus; HY = hypothalamus; ITG = inferior temporal gyrus; ON = optic nerve; OT = olfactory tract; PHG = parahippocampal gyrus; RS = rhinal sulcus; TP = temporal pole.

stage Alzheimer's disease and with a long duration of illness. A recent report[43] indicates that 90% of layer II neurons can be lost during the course of Alzheimer's disease and that as many as 58% of all entorhinal cortical neurons are lost. These changes devastate the modular organization of entorhinal cortex seen in the normal human brain and eliminate the characteristic bumps or verrucae that are seen in this cortex (Figures 2–8 and 2–9).

Pathological changes in the adjacent perirhinal cortex in Alzheimer's disease are second only to those of the entorhinal cortex, with layers II, III, and V all containing neurofibrillary tangles. In contrast, however, the affected neurons are often seen as distinct columns spanning as many as three layers of cortex.[13] This must be viewed as a modular change as well, but the elements and size of modules differ radically between the entorhinal and perirhinal cortices. In the former, the dominant feature of the modularity is the island of layer II neurons and associated pyramidal neurons of layer III. In the perirhinal cortex, the appearance of modules is sharper and is dominated by columns of neurons, with as few as 5 or 6 affected neurons defining the column width.

As discussed earlier, the entorhinal and perirhinal cortices are key cortical areas for linking the hippocampal formation to the cerebral cortex. In fact, they are the essential conduit through which the visual and auditory association areas communicate with the hippocampal formation. The perforant pathway that arises from layers II and III of the entorhinal cortex is the final link in these neural systems. These neurons are heavily invested by neurofibrillary tangles in Alzheimer's disease. However, hippocampal output back to the cortex is also compromised in this disease because the subicular and CA1 neurons of the hippocampal formation are targeted by neurofibrillary tangles and they are the major source of direct-feedback neural systems. Thus, it is arguable that the selective pathological changes in the hippocampal formation and parahippocampal gyrus in Alzheimer's disease virtually isolate this key ventromedial temporal structure from other cortical neural systems (Figure 2–6A, B).

VENTROMEDIAL TEMPORAL AREAS AND TEMPORAL INJURY

The Tentorium Cerebelli

Although nearly all parts of the ventromedial temporal area are affected by pathology in Alzheimer's disease, other matters related purely to cranial geography jeopardize their integrity. The tight encasement and insertion of the temporal lobe into the irregular bony structure of the temporal fossa creates vulnerability to head injury from direct forces and from forces generated by impact at many points on the skull.[44,45] Likewise, the proximity of the ventromedial temporal area to the inferior horn of the lateral ventricle is also of great consequence with any form of increased intracranial pressure, whether its etiology be tumor, abscess, hematoma, edema, or infarction.[46]

Central to both matters of head trauma and increased intracranial pressure is the fact that the free edge of the tentorium cerebelli cuts across the parahippocampal gyrus before attaching to the petrous apex and the anterior and posterior clinoid processes.[47,48] The collar formed around the brainstem creates an aperture and

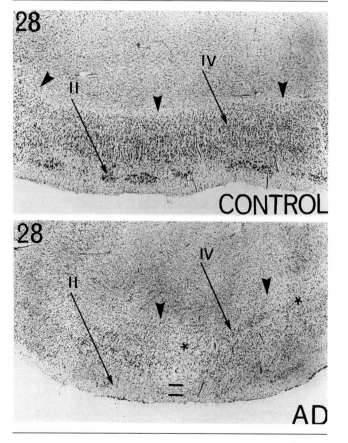

FIGURE 2–8. Nissl-stained cross-sections through the entorhinal cortex (area 28) in a 72-year-old control subject and a 71-year-old Alzheimer's disease (AD) patient, showing the absence of cellular staining in layer II *(crossbars)*. Note that in AD, layer IV is intermittently absent *(asterisks)*. Also note that the normally sharp white matter–gray matter interface *(arrowheads in the control)* is absent in AD because of dense glial cell proliferation.

FIGURE 2–9. Thioflavine S–stained cross-section through the entorhinal cortex at endstage Alzheimer's disease (AD; 12-year duration of illness), showing dense neurofibrillary tangles in all cortical layers.

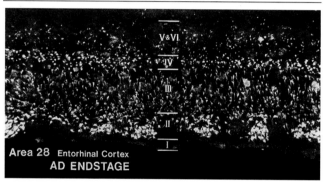

FIGURE 2–10. A reproduction of Figure 1 from Jefferson's classic article "The Tentorial Pressure Cone"[48] showing the course and line of contact of the free edge of the tentorium cerebelli across the parahippocampal gyrus of the human brain. The location of the gyrus ambiens component of the entorhinal cortex is shown.

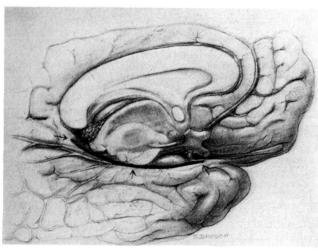

incisura that provides communication between the supratentorial and infratentorial spaces.

The anterior part of the parahippocampal gyrus sits directly in the tentorial aperture, unprotected by dura (Figure 2–10). Although the size of the aperture varies from one individual to another, it has been estimated that the free edge of the tentorium actually contacts and grooves the parahippocampal gyrus in 70% of humans.[49] This tentorial notch or groove ("TN" in Figure 2–1A) approximates the division between Brodmann areas 28 and 34. Retzius[1] labeled the notch the *inferior rhinal sulcus*, and others have followed his lead. However, in the present author's estimation this indentation is nothing more than a surface marking in some human brains relating purely to the disposition of the free edge of the tentorium cerebelli.

Uncal Herniation With Head Injury and Raised Intracranial Pressure

The location of the tentorial incisura and the complications it creates in neurological disease have been long appreciated. Quite simply, spatial compensation is limited in the supratentorial space, and when it is exhausted, the brain will herniate across the free edge of the tentorium into the space of Bichat and the infratentorial compartment.[45] The results necessitate immediate emergency-related measures to preserve life. *Uncal herniation* is an appropriate term to apply to extreme herniation and to pathological specimens where fatality occurred. However, in technical terms, it is not the uncal hippocampus that lies in the tentorial aperture, but in fact the entorhinal cortex of the anterior parahippocampal gyrus. It, and not the uncus, leads the invasion of infratentorial space. Partial herniations that are not fatal may result in entorhinal injury. Behavioral changes in such patients would be of great interest.[44]

In a related chord, a majority of individuals suffering head trauma survive their injuries. However, in such a population it would be expected that many might have injuries to ventromedial temporal areas around the free edge of the tentorium (Figures 2–11A–C), since the temporal lobes are often forced onto a part of it around the incisurum. Posttraumatic behavioral changes in head injury survivors, particularly those with emotion-related and memory-related sequelae, could just as likely be related to ventromedial temporal area injury as to the more obscure causes, such as axon shearing, proposed by some.

CONCLUDING REMARKS

The ventromedial temporal area has long been known to have efferent projections to subcortical structures involved in endocrine and autonomic processes. Less has been known about its connections with the cortex and the manner in which it is related to neural systems operative in memory. The details of a reciprocal interrelationship between ventromedial temporal areas and the remainder of the cortex have been established in the

---

FIGURE 2–11. Three Nissl-stained cross-sections of the ventromedial temporal area showing injury along the tentorial notch (TN) by the free edge of the tentorium cerebelli in a 62-year-old patient who suffered from agitated depression and psychoses following a bicycle fall and head injury 19 years prior to death. Note the deep cut at the tentorial notch separating Brodmann areas 28 and 34 and the abnormal elevation of the unherniated hippocampal formation. AMG = amygdala; CS = collateral sulcus; HP = hippocampus; RS = rhinal sulcus; V = lateral ventricle.

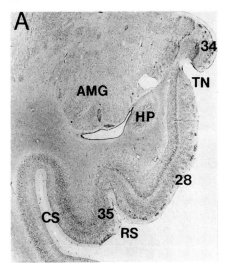
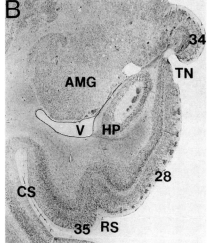
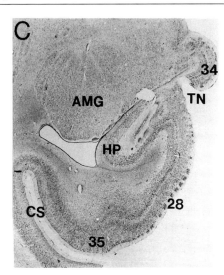

past three decades.[3] A mutually compatible dialogue now exists between anatomy and behavior, and continuing progress and insight have been achieved on both fronts.[50–53]

However, the ventromedial temporal area is involved in a bewildering array of neurological and psychiatric diseases, ranging from Alzheimer's disease to autism[54] and schizophrenia.[40,55] Pathogens such as viruses attack the area, and mechanical injuries around the tentorium cerebelli are well known. Yet the etiologies for many alterations are unknown, and all behavior changes are not clearly in the realm of memory. A foothold has been established with regard to memory, but the four decades that have passed since the studies of the amnestic temporal lobectomy patient H.M. serve only to remind us of the long journey ahead.

REFERENCES

1. Retzius G: Das Menschenhirn. Studien in der makroskopischen Morphologie [The human brain: studies of macroscopic morphology]. Stockholm, Norstedt and Sohne, 1896
2. Klinger J: [Macroscopic anatomy of the hippocampal formation]. Denkschrifter der Schweizerischen Naturforeschenden Gesellschaft [Proceedings of the Swiss Nature Society] 78, Mem. 1, 1–80. Zurich, Bard, 1948
3. Van Hoesen GW: The parahippocampal gyrus: new observations regarding its cortical connections in the monkey. Trends Neurosci 1982; 5:345–350
4. Rosene DL, Van Hoesen GW: The hippocampal formation of the primate brain: a review of some comparative aspects of cytoarchitecture and connections, in Cerebral Cortex: Further Aspects of Cortical Function, Including Hippocampus, vol 6, edited by Jones EG, Peters A. New York, Plenum, 1987, pp 345–456
5. Duvernoy HM (ed): The Human Hippocampus: An Atlas of Applied Anatomy. Munich, Bergmann, 1988
6. Heinsen H, Henn, R, Eisenmenger W, et al: Quantitative investigations on the human entorhinal area: left-right asymmetry and age-related changes. Anat Embryol (Berl) 1994; 190:181–194
7. Van Hoesen GW: Anatomy of the medial temporal lobe. Magn Reson Imaging 1995; 13:1047–1055
8. Braak H, Braak E: The human entorhinal cortex: normal morphology and lamina-specific pathology in various diseases. Neuroscience 1992; 15:6–31
9. Braak H, Braak E: On areas of transition between entorhinal allocortex and temporal isocortex in the human: normal morphology and lamina-specific pathology in Alzheimer's disease. Acta Neuropathol (Berl) 1985; 68:325–332
10. Amaral DG, Insausti R: Hippocampal formation, in The Human Nervous System, edited by Paxinos G. San Diego, CA, Academic Press, 1990, pp 711–755
11. Insausti R: Comparative anatomy of the entorhinal cortex and hippocampus in mammals. Hippocampus 1993; 3:19–26
12. Hevner RF, Wong-Riley MTT: Entorhinal cortex of human, monkey and rat: metabolic map as revealed by cytochrome oxidase. J Comp Neurol 1992; 26:451–469
13. Van Hoesen GW, Solodkin A: Some modular features of temporal cortex in humans as revealed by pathologic changes in Alzheimer's disease. Cereb Cortex 1993, 3:465–475
14. Pandya DN, Kuypers HGJM: Cortico-cortical connections in the rhesus monkey. Brain Res 1969; 13:13–36
15. Jones EG, Powell TPS: An anatomical study of converging sensory pathways within the cerebral cortex of the monkey. Brain 1970; 93:793–820
16. Van Hoesen GW, Pandya DN, Butters, N: Some connections of the entorhinal (area 28) and perirhinal (area 35) of the rhesus monkey, II: frontal lobe afferents. Brain Res 1975; 95:25–38
17. Van Hoesen GW, Pandya DN: Some connections of the entorhinal (area 28) and perirhinal (area 35) cortices of the rhesus monkey, I: temporal lobe afferents. Brain Res 1975; 95:1–24
18. Insausti R, Amaral DG, Cowan WM: The entorhinal cortex of the monkey, II: cortical afferents. J Comp Neurol 1987; 264:326–355
19. Suzuki WA, Amaral DG: Topographic organization of the reciprocal connections between the monkey entorhinal cortex and the perirhinal and parahippocampal cortices. J Neurosci Methods 1994; 14:1856–1877
20. Rockland KS, Pandya DN: Cortical connections of the occipital lobe in the rhesus monkey: interconnections between areas 17, 18, 19 and the superior temporal sulcus. Brain Res 1981; 212:249–270
21. Rockland KS, Pandya DN: Laminar origins and termination of cortical connections of the occipital lobe in the rhesus monkey. Brain Res 1979; 179:3–20
22. Van Essen DC, Felleman DJ, DeYoe EA, et al: Modular and hierarchical organization of extrastriate visual cortex in the macaque monkey. Cold Spring Harbor Symp Quant Biol 1990; 3:679–696
23. Herzog AG, Van Hoesen GW: Temporal neocortical afferent connections to the amygdala in the rhesus monkey. Brain Res 1976; 115:57–69
24. Van Hoesen GW: The differential distribution, diversity and sprouting of cortical projections to the amygdala in the rhesus monkey, in The Amygdaloid Complex, edited by Ben Ari Y. Amsterdam, Elsevier, 1981, pp 77–90
25. Amaral DG, Price JL, Pitkanen A, et al: Anatomical organization of the primate amygdaloid complex, in The Amygdala, edited by Aggleton JP. New York, Wiley-Liss, 1992, pp 1–66
26. Witter MP: Organization of the entorhinal-hippocampal system: a review of current anatomical data. Hippocampus 1993, 3:33–44
27. Witter MP, Van Hoesen GW, Amaral DG: Topographical organization of the entorhinal projection to the dentate gyrus of the monkey. J Neurosci 1989, 9:216–228
28. Amaral DG, Price JL: Amygdalo-cortical projections in the monkey (*Macaca fascicularis*). J Comp Neurol 1984; 230:465–496
29. Rosene DL, Van Hoesen GW: Hippocampal efferents reach widespread areas of cerebral cortex and amygdala in the rhesus monkey. Science 1977; 198:315–317
30. Kosel KC, Van Hoesen GW, Rosene DL: Nonhippocampal cortical projections from the entorhinal cortex in the rat and rhesus monkey. Brain Res 1982; 244:201–214
31. Damasio AR: The brain binds entities and events by multiregional activation from convergence zones. Neural Comp 1988; 1:123–132
32. Damasio AR: The time-locked multiregional retroactivation: a systems level proposal for the neural substrates of recall and recognition. Cognition 1989; 33:25–62
33. Damasio AR: Descartes' Error: Emotion, Reason, and the Human Brain. New York, Grosset/Putnam, 1994
34. Arnold SE, Hyman BT, Flory BT, et al: The topographical and neuroanatomical distribution of neurofibrillary tangles and neuritic plaques in the cerebral cortex of patients with Alzheimer's disease. Cereb Cortex 1991; 1:103–116
35. Hyman BT, Van Hoesen GW, Damasio AR, et al: Alzheimer's disease: cell specific pathology isolates the hippocampal formation. Science 1984; 225:1168–1170
36. Hyman BT, Van Hoesen GW, Kromer LJ, et al: Perforant pathway changes and the memory impairment of Alzheimer's disease. Ann Neurol 1986; 20:472–481
37. Hyman BT, Kromer LJ, Van Hoesen GW: A direct demonstration of

the perforant pathway terminal zone in Alzheimer's disease using the monoclonal antibody Alz-50. Brain Res 1988; 450:392–397
38. Kromer Vogt LJ, Hyman BT, Van Hoesen GW, et al: Pathological alterations in the amygdala in Alzheimer's disease. Neuroscience 1990; 37:377–385
39. Braak H, Braak E: Neuropathological staging of Alzheimer-related changes. Acta Neuropathol (Berl) 1991; 82:239–259
40. Arnold SE, Hyman BT, Van Hoesen GW, et al: Some cytoarchitectural abnormalities of the entorhinal cortex in schizophrenia. Arch Gen Psychiatry 1991; 48:625–632
41. Van Hoesen GW, Damasio AR: Neural correlates of the cognitive impairment in Alzheimer's disease, in The Handbook of Physiology: Higher Functions of the Nervous System, edited by Plum F. Baltimore, Waverly, 1987, pp 871–898
42. Van Hoesen GW, Hyman BT, Damasio AR: Entorhinal cortex pathology in Alzheimer's disease. Hippocampus 1991; 1:1–8
43. Gomez-Isla T, Price J, McKeel DW Jr, et al: Profound loss of layer II entorhinal cortex neurons occurs in very mild Alzheimer's disease. J Neurosci 1996; 16:4491–4500
44. Adams JH, Graham DI, Scott G, et al: Brain damage in non-missile head injury. J Clin Pathol 1980; 33:1132–1145
45. Adams JH: Head injury, in Greenfield's Neuropathology, 4th edition, edited by Adams JH, Corsellis JAN, Dunchen LW. New York, Wiley, 1984, pp 85–124
46. Miller JD, Adams JH: The pathophysiology of raised intracranial pressure, in Greenfield's Neuropathology, 4th edition, edited by Adams JH, Corsellis JAN, Dunchen LW. New York, Wiley, 1984, pp 53–84
47. Meyer A: Herniation of the brain. Archives of Neurology and Psychiatry 1920; 4:387–400
48. Jefferson G: The tentorial pressure cone. Archives of Neurology and Psychiatry 1938; 40:857–876
49. Corsellis JAN: Individual variation in the size of the tentorial opening. J Neurol Neurosurg Psychiatry 1958; 21:279–283
50. Damasio A, Eslinger P, Damasio H, et al: Multimodal amnesic syndrome following bilateral temporal and basal forebrain damage. Arch Neurol 1985; 42:252–259
51. Murray EA, Gaffan D, Mishkin M: Neural substrates of visual stimulus-stimulus association in rhesus monkeys. J Neurosci 1993; 13:4549–4561
52. Suzuki WA, Zola-Morgan S, Squire LR, et al: Lesions of the perirhinal and parahippocampal cortices in the monkey produce long-lasting memory impairment in the visual and tactual modalities. J Neurosci 1993; 13:2430–2451
53. Zola-Morgan S, Squire LR, Clower RP, et al: Damage to the perirhinal cortex exacerbates memory impairment following lesions to the hippocampal formation. J Neurosci 1993; 13:251–265
54. Bauman M, Kemper TL: Histoanatomic observations of the brain in early infantile autism. Neurology 1985; 35:866–874
55. Jakob H, Beckmann H: Prenatal developmental disturbances in the limbic allocortex in schizophrenics. J Neural Transm 1986; 65:303–326

3

The Thalamus and Neuropsychiatric Illness

Arnold B. Scheibel, M.D.

The thalamus is the portal to the cerebral cortex, with which it has concomitantly evolved, reaching its ultimate development in primates and man. Virtually all sensory systems (except olfaction) relay in the thalamus, along with ascending nonspecific (reticular core) and long-loop modulatory systems returning from the basal ganglia and cerebellum. A large number of thalamic nuclei serve as processing centers for these inputs, each idiosyncratic as to histological structure, physiological characteristics, and neurotransmitter content.

This mosaic of subcenters seems to participate in a wide range of functions, including

1. Processing and relaying of information from the external world and the body surfaces, internal and external, to the cerebral cortex.
2. Development and maintenance of the body image.
3. Control over levels of attentiveness, wakefulness, and sleep.
4. Monitoring and modulating of the physiological and chemical milieu of the brain.

It should be noted that these varied roles (and a number of others) can only be performed within the context of the closest functional interaction with overlying cortex. Thalamocorticothalamic connections are the touchstone of all higher level central nervous system function, normal or abnormal. For this reason, it is difficult to conceive of a lesion or dysfunction at one of these levels (cortical or thalamic) that does not affect the other.

In these terms, a uniquely thalamic seat for psychiatric dysfunction might seem especially improbable. However, using the new minimally invasive visualization techniques such as functional magnetic resonance imaging (fMRI) and positron emission tomography (PET), recent studies have pointed to the thalamus as one of the few sites where persistent structural differences may be found in both adult and childhood schizophrenia.[1-3] The degree of resolution of these methods does not yet allow accurate localization of thalamic sites where changes may exist. On theoretical grounds alone, it has been suggested that medial (nonspecific) and midline nuclei might be involved because of their widespread

---

This work was previously presented in shorter form at the American Neuropsychiatric Association annual meeting, Newport, RI, July 21–24, 1994, and the American Psychiatric Association annual meeting, New York, NY, May 5–9, 1996.

connections with multiple neocortical association zones and with limbic centers.[4] Supporting this idea to some degree is the finding that administration of clozapine or haloperidol in rats selectively increased expression of the immediate response gene c-*fos* in a number of medial and midline nuclei, including paraventricular, central medial, and rhomboid nuclei and the nucleus reuniens.[5,6] These nuclei are similar insofar as they all receive strong dopamine innervation from mesencephalic nuclei and express dopamine receptors, thereby placing them in a major ascending dopaminergic pathway. In addition, consistent reduction in thickness of the diencephalic periventricular gray matter has been noted in schizophrenic patients, reiterating the possible role of medial-lying thalamic systems in schizophrenia.[7]

Excepting these data, and persistent reports of diminution of cell number in the medial dorsal nucleus of the thalamus,[8-11] there is still relatively little robust information to support the thalamus as a preferential site of pathogenesis for psychiatric illness. On the other hand, the neurologic literature has an abundance of case reports and reviews evaluating the results of thalamic lesions such as hemorrhage, nonhemorrhagic infarcts, and tumor infiltration. Almost all of these reports suffer from the same weakness: the relative lack of locational specificity of the lesions they document. Naturally occurring lesions are by nature "messy" because multiple nuclei and/or tract systems are usually involved. This limitation notwithstanding, the literature is richly descriptive of thalamic injury–induced changes in consciousness, ranging from coma through persistent drowsiness and obtunded cognitive processes to irritability, hypomania, and manic episodes. Limited attention span, emotional blunting, distractibility, and memory deficits are also frequently noted, especially in the case of medial and dorsal thalamic lesions. Taken collectively, these observations provide a palette of mental signs and symptoms from which recognizable psychiatric syndromology may be constructed.[12-19]

At a more inclusive level, a number of workers have tried to develop the concept of thalamic dysfunction within the more comprehensive matrix of orbitofrontal cortex, basal ganglia, limbic striatum, and thalamus as part of a large-scale filter function failure in psychotic disease. Manic disease, schizophrenia, and obsessive-compulsive disorder (OCD) have all been speculatively attributed to opening of the filter, with resultant increased arousal, or to "aberrant positive feedback loops" between reciprocally excitatory systems.[20-22] There seems little doubt that the central nervous system relies on constant interaction among complex neural ensembles such as these, but concepts of this magnitude about pathogenesis lose predictive rigor.

With these caveats in mind, it may be wiser to approach the problem from the bottom up by examining the inner structure and connections of several representative thalamic nuclei. Accordingly, four thalamic nuclei out of a larger group of 35 to 40 will be used in considering the role of the thalamus in ongoing nervous activity and its possible relationship to psychiatric illness. Special attention will be paid to the input and output connections of these nuclei and, to the extent that data are available, to their internal structural organization (neuropil) and neurotransmitter content. Such information frequently provides insight into normal modes of information processing and transfer and, by extension, into distortions in perception and behavior with which they may be involved.

THE VENTROBASAL COMPLEX

Somesthetic impulses from the spinal cord and medulla ascending in the posterior column–medial lemniscal system terminate in the ventral posterior lateral portion of the ventrobasal complex (Vb). This medial lemniscal system is characterized by a high degree of synaptic security and spatial (and temporal) specificity, and it is usually described as mediating precise (localized) touch, two-point discrimination, and flutter-vibration sense, including sequential changes in muscle stretch and joint position. The spinothalamic (and trigeminothalamic) tracts, which ascend in parallel with the medial lemniscus, terminate in part in the Vb but also in several other receptive zones of the thalamus. Unlike the medial lemniscus, the spinothalamic system is more concerned with affectual components of somesthesis (pain, poorly localized touch and temperature) and much less involved with precise temporospatial aspects of touch.

Thus, the ventrobasal complex receives continuously updated somesthetic information that tends to define the position and role of the body in three-dimensional space and its relationship to objects in its immediate surround.

It has gradually become clear over the past 30 years that body representation is rigorously spelled out in three dimensions at the thalamic level. Medial lemniscal fibers terminate in caudorostral elongated slabs or arbors arranged in concentric lamellae, superimposed on the pools of receptive thalamic cells, both thalamocortical and interneuronal in type.[23-26] A recording microelectrode descending along a vertical track identifies in sequential order the series of contiguous receptive fields on the surface of the body, reported by neurons in that particular track.[27–29] As in the cortex, representation is distorted to reflect patterns of greater receptor density in peripheral parts of the extremities and around the mouth.[4]

Characteristically, processing nuclei along ascending sensory pathways tend to receive far more centrifugal than centripetal information.[30] Sensory thalamic nuclei are similarly organized, and there are at least 10 fibers from more rostral sources (cerebral cortex, nucleus reticularis thalami, etc.) entering the Vb for every afferent fiber ascending from below. The meaning of this distribution is not yet clear, but the terminal patterns generated by these descending afferents provide some clues as to the possible significance behind this skewed distribution of afferents.

When studied with Golgi techniques or by orthograde transport of markers such as horseradish peroxidase or biocytin, descending afferent terminals are found to generate at least three different histologic patterns.[31,32] One fiber type extends caudally through the Vb, repetitively generating small terminal cluster patterns at apparently specific sites along a lengthy anteroposterior trajectory. Some of these enter the midbrain and synapse upon cells of the reticular core (n. cuneiformis). These fibers apparently originate in the nucleus reticularis thalami (nRt) and are known to be inhibitory in function.[33] A second fiber type ends in a progressive series of bifurcations forming an arbor that, in three dimensions, resembles a cone whose base is directed caudally. Attempts at reconstruction of this pattern suggest that these terminal cones are partly embedded in each other, like a stack of ice cream cones turned on its side. A third fiber type generates one or more dense platelike neuropil fields orthogonal to the descending parent fiber. These neuropil plates achieve their maximal dimension in the transverse or coronal plane, appearing to cut across the stacks of cones at right angles, forming what could be likened to a three-dimensional matrix (Figure 3–1). On the assumption that both of these presumably cortically derived neuropil terminal systems extend throughout the entire rostrocaudal extent of the Vb, it has been suggested that each thalamocortical neuron receives innervation from both systems, arranged in gridlike ensemble. As we have noted elsewhere,

imposition of an XY grid upon a spatially regular matrix of discrete elements is an effective way to provide readout capability for any one element in a matrix. Such arrangements are used in the *coincident-current selection technique* for reading out an individual core from a computer memory array. Each member of such an array can be individually selected by convergence of half strength current pulses in two sets of selection lines perpendicular to each other. We can speculate at this time that the two categories of corticothalamic fibers may also represent a cortical readout system playing upon the thalamus. It should be remembered that the Vb field is tridimensional and the presumed readout fibers must also be conceived as organized in three-dimensional space. Packing density and operational reliability due to redundancy of elements are such that this system may function more effectively by several orders of magnitude than any similar system built by man.[31] (p. 75)

Despite the obvious problems in comparing any neurobiological system to a human artifact, it seems reasonable to suggest that the apparently rigorous three-dimensional organization of the ventrobasal complex may provide coding sites for a number of components of the body schema, subject to frequent updating by sensory volleys and to periodic readout and retrieval by the cortex.

A supposition of this sort immediately adds another

FIGURE 3–1. Semischematic drawing of descending axons terminating in the ventrobasal complex. A: Histological structure of two axon types as seen in Golgi impregnations. B: Three-dimensional schematization of these fiber types. C: Schematic drawing indicating how their superposition could impose a grill or matrix on the underlying thalamic neurons (not shown).

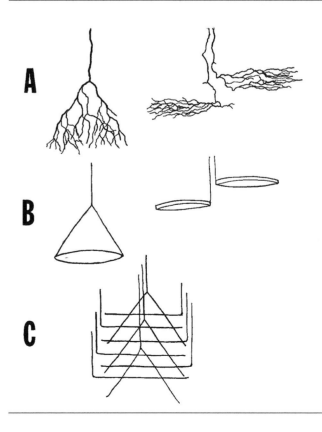

dimension, both literally and figuratively, to the linear, essentially two-dimensional organization of the body on the cortical surface, especially in primary sensory (and primary motor) strips. Although the distortion of the "representation" based on density of sensory receptors rather than size of the part is maintained at the thalamic level, it seems reasonable that the added dimension in thalamic representation provides a richness and depth to the body sense that might be lacking in the extended two-dimensional cortical array.

Clinical lesions entirely localized to the ventrobasal complex are unusual, making clinical correlations difficult. There are occasional reports in the literature of right-left confusion and problems in double simultaneous tactile discrimination, stereognosis, and graphesthesia, which might well represent flaws in experiencing the body image. However, the majority of clinical data point to changes in intensity (hypesthesia or hyperesthesia) of pain, touch, temperature, or proprioceptive senses.[34-36] Less specific but equally disturbing complaints of body distortion, perceived loss of body parts, and so forth, are frequent symptoms of the schizophrenic patient. Disturbing tactile sensations on the body surface, including formication experienced by the psychotic patient or the acutely disturbed alcoholic, may all have their roots in disturbed processing patterns in the Vb.

Body image and sense of self are rich and complex phenomena that are shaped by time, circumstance, and personal emotional state, and yet remain relatively constant. The pubescent individual must refashion the concept of his or her rapidly growing and changing body from its earlier, more familiar contours of the child. The septuagenarian often continues to live with the internal concept of the vigorous adult that matched his or her experiences over the long central period of life, while actually having to deal with visible evidences of the wear and tear of age. How often such individuals say, "I feel that I am still young and vigorous. When I look in the mirror, I can't believe it's really me." This need for relative invariance of image combined with a requirement for continuous reshaping to meet biological reality undoubtedly occurs within the context of thalamocortical interaction. The ascending sensory relays within the Vb have been described as both precise and of high synaptic security.[37] The great descending mass of corticothalamic afferents (more than 10 times as numerous as those from below, as already indicated) must parse and modulate this complex tridimensional representation of the body. Here may lie the representational interface between the structural self and the thoughts and feelings about self. Psychotic delusions about the body and its parts may be played out on the same field.

THE NUCLEUS RETICULARIS THALAMI

The nRt is a thin, sheetlike nucleus of subthalamic derivation that is wrapped around the anterior, ventral, and lateral aspects of the thalamus, separating it from the internal capsule. Its position is such that every thalamocortical and most corticothalamic fibers must traverse its boundaries. In so doing, they make synaptic contact with the large-size but limited number of reticularis cells whose dendrites are arranged at right angles to the perforating fiber systems. Reticularis cells are robustly GABAergic,[38] and their axons project caudally upon thalamus and midbrain.[39,40] The conclusion is inescapable that they serve as an inhibitory feedback system upon subjacent thalamus, and a number of studies support this conclusion.[33,41] Furthermore, the rigorous organization of ascending thalamic fibers as they perforate (and synapse with) elements of reticularis is mirrored by an equally precise arrangement of descending corticothalamic fibers as they traverse reticularis. Thus, fibers from representationally related areas within thalamus and cortex are in register as they cross reticularis and modulate thresholds of delimited reticularis cell groups (Figure 3–2).

This arrangement has led to the "gatelet" concept of reticularis function, wherein the nucleus is conceived of as a continuum of tiny gates, each specific for a small portion of sensory representation (somesthetic, visual, auditory, etc.).[42-44] Research has demonstrated that stimulation of specific sensory thalamic sites (for example, the ventrobasal complex or lateral geniculate nucleus) elicits high-frequency bursts from reticularis neurons located along the trajectory of axons from these sites. These bursts immediately result in inhibition of cell groups close to the original site of stimulation. Effectively, the reticularis gate (to thalamocortical transmission) is temporarily closed by thalamic stimulation.

Stimulation of sites in the mesencephalic reticular formation (n. cuneatus), on the other hand, inhibits reticularis neurons, thereby preventing their inhibitory action on underlying thalamic and mesencephalic sites. This opens the reticularis gate to ascending transmission. In contradistinction to these two sites, the prefrontal cortex interacts with reticularis in a more complex manner through the agency of intercalated neurons in medial thalamic ("nonspecific") nuclei[45] (Figure 3–2). Although the mechanisms of this interaction are not entirely clear, it provides a putative substrate for executive (voluntary) control over reticularis gating of sensory inputs to cortex and a possible mechanism for abilities such as autohypnosis and voluntary pain control.

It now seems fairly well established that interactions of thalamic relay nuclei, nRt, and cerebral cortex result in two modes of thalamocortical activity.[46] The *oscillatory* mode is characterized by slow rhythmic potentials (as in drowsiness, slow-wave sleep, and some anesthetized states). The *transfer* (tonic or relay) mode is marked by EEG desynchronization coincident with REM sleep or attentive wakefulness and appears to enhance the speed and specificity of sensorimotor integration during the waking state. However, the blurred interface between these two states suggests that they might better be thought of as a continuum of thalamic response states. Vestiges of both states are undoubtedly always present as thalamocorticothalamic systems continuously balance the interplay between inner and outer experience through the agency of the nRt.

It has frequently been suggested that schizophrenic patients are overwhelmed by information. A loss of postulated filtering mechanisms results in sensory overload, which is thought to lead, in turn, to shifting attentional focus, distractibility, loss of associations, and so on. A sizable literature has developed on this phenomenon and has recently been summarized by Braff.[47] A number of measures of this gating capacity to screen out irrelevant stimuli have been developed under the general rubric of "proactive interference."[48] As an example, the startle response triggered by a sudden loud tone or bright light is normally at least partially inhibited when preceded at 50 to 500 ms by a weak prestimulus of similar or different mode. This inhibitory and presumably protective response is attenuated or even absent in most patients with schizophrenia and, interestingly, in close nonpsychotic relatives of the patients.[49] It has also been found that prepulse inhibition of an acoustic startle reflex correlates significantly with distractibility scores in chronic psychotic patients, pointing again to faulty sensory gating.[50]

Skinner and Yingling[51] have stressed the importance of the mediothalamic frontocortical system in its capacity to exert variable control over the gating response of the nRt. In their animal model, frontal lobe damage or cryogenic blockade of the inferior thalamic peduncle prevented the animal from directing or maintaining attention and also from suppressing transmission to the cortex of activity evoked by a repetitious and irrelevant stimulus. In this circuit, the medial and intralaminar thalamic nuclei appear to provide the crucial link between frontal granular cortex and the nRt. Both physiologic and anatomic changes in prefrontal cortex of many schizophrenic patients are already well documented.[52-57] This model provides an idea of the complex thalamocorticothalamic circuitry that may be disturbed in the schizophrenic diathesis. Prefrontal cortex may be a significant site of psychosis-related pathology, whereas alterations or interruptions in the timing and phasic activity of nRt control of thalamocorticothalamic systems produced by prefrontal dysfunction could be responsible for problems in cognitive processing seen in schizophrenia.

The thalamic pain syndrome, first described by Dejerine and Roussy,[58] may conceivably serve as a model for a number of sensory distortions that accompany psychiatric syndromes. In reviewing the literature some years

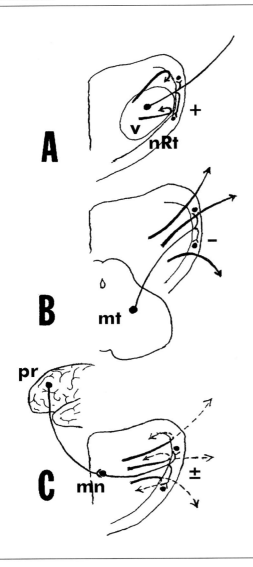

FIGURE 3–2. Schematic drawings showing the results of three types of neural activity on the thalamic nucleus reticularis gate. A: Ventrobasal (v) nuclear stimulation drives (+) the nucleus reticularis (nRt) cells and closes the gate. B: Mesencephalic tegmental (mt) stimulation inhibits (-) nucleus reticularis cells and opens the gate. C: Prefrontal (pr) cortical stimulation operating through synaptic connections in the medial and midline thalamic nuclei (mn) have a variable effect (±), opening or closing the gate depending on the patterns of excitation.

ago,[59] we were struck by the fact that in most individuals with pathological pain who came to autopsy, thalamic nuclear lesions were accompanied by lesions of the adjacent nRt. On this basis, we suggested that reticularis injury might be an essential component of the pain syndrome that developed in the partial or complete absence of the inhibitory feedback usually supplied by reticularis activity. In this regard, it is interesting to note a recent study indicating that microinjections of picrotoxin, a $GABA_A$ receptor antagonist, into the nRt of freely moving rats resulted in painlike behavior of the contralateral extremity.[60]

The thalamic pain syndrome is not usually considered within the purview of the neuropsychiatrist. However, this syndrome might serve as a model for several other distortions of sensory input, especially auditory and visual hallucinations, thereby attaining significance for this overview. I am not familiar with any neuropathological study that has identified geniculate lesions, with or without accompanying reticularis pathology, in hallucinated patients. However, it is possible that functional (biochemical) lesions of such conjoint systems may occur. Positron emission tomography cannot yet provide the spatial resolution necessary to address this problem, but careful histochemical analysis of appropriately harvested postmortem tissue may be useful in such a study.

from the pars reticulata of the substantia nigra. The parvocellular area receives largely from the cortical areas to which it projects (areas 10 and 11), whereas the magnocellular area receives fibers from the basolateral amygdala, the periamygdaloid portion of temporal neocortex, and possibly the substantia innominata.[26]

As with most other thalamic nuclei, it is difficult to identify the precise role of the MD nuclear complex in psychiatric syndromology because thalamocortical relations are close and usually reciprocal (although not in the case of the magnocellular component and the amygdala). With dysfunction or lesion in the amygdaloid/periamygdaloid ensemble, most observers have described hallucinations, delusions, intrusive thoughts, feelings of depersonalization, unusual visceral sensations, and reactions of fear, anxiety, or panic.[61-63] With prefrontal (especially bilateral) dysfunction, the associated clinical picture frequently includes deteriorative personality changes, apathy and negativity, mental lethargy, affective flattening, and memory disturbances. The resemblance of these two clinical pictures

---

FIGURE 3–3. A few connections of the mediodorsal nucleus. Connections between mediodorsal nucleus and cortical stations are reciprocal; connection between amygdala and mediodorsal nucleus is not. a = amygdala; dl = dorsolateral prefrontal cortex; e = frontal eye fields; om = orbitomedial prefrontal cortex; m = magnocellular, p = parvocellular, and l = paralaminar portions of mediodorsal nucleus.

THE MEDIODORSAL NUCLEUS

The mediodorsal nucleus of the thalamus (MD) occupies much of the dorsal and medial portion of the thalamus, medial to the intralaminar complex and lateral to the midline nuclei lying along the ventricular border of the thalamus. In the primate and human thalamus, it is usually divided into three parts: a medial magnocellular division, a somewhat larger lateral parvocellular portion, and, most laterally, a paralaminar portion made up of scattered large cells lying along the edge of the internal medullary lamina.

The efferent projections from these three nuclear zones practically define the scope of the frontal lobe exclusive of primary motor cortex (Figure 3–3). Paralaminar cells project to premotor cortical zones (areas 6 and 8); the parvocellular division projects widely upon dorsolateral prefrontal cortex (particularly areas 9 and 10); and the magnocellular portion projects to medial and orbitofrontal cortex (primarily areas 11, 12, and 32).

Afferent inputs to the three nuclear zones are equally idiosyncratic. The paralaminar zone receives large numbers of fibers from its cortical target areas (6 and 8) and

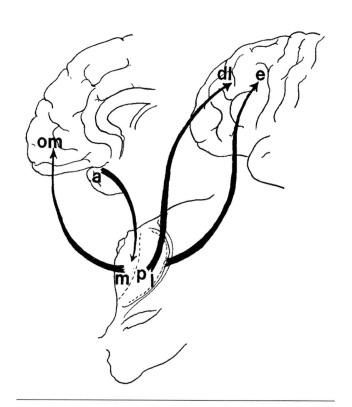

to type I (positive) and type II (negative) schizophrenic reactions, respectively, has been noted.[61,62] It should be mentioned in passing that recent structural studies do not seem to support this correlation. In a recent study,[63] frontal right-left asymmetry was lost in nondeficit (type I) rather than in deficit (type II) patients. Furthermore, the volume of prefrontal cortex was consistently reduced in the former rather than in the latter.

It is rare for naturally occurring lesions selectively to destroy parvocellular or magnocellular portions of the MD. However, there are a number of published reports on the effects of medial thalamic lesions. These lesions affect most of the MD along with varying portions of adjacent structures such as the intralaminar complex and mammillothalamic bundle. Unilateral paramedian infarction of the right (nondominant) thalamus often results in affective disturbances characterized by apathy and lack of initiative punctuated by fits of irritability, sometimes of manic intensity. Bilateral thalamic lesions, either infarcts or tumor infiltrations, tend to produce blunted affect, apathy, fluctuating states of consciousness, dysphoria, and amnesia.[12-19,64] It is not yet entirely clear whether memory disturbances are an intrinsic part of MD pathology or result from the frequently accompanying damage to anterior thalamic nuclei, the mammillothalamic tract and mammillary bodies, and so on. However, stereotaxically placed lesions in the posterior portion of MD in monkeys have been described as producing impairment in memories that in humans would be considered declarative—without altering already learned skill-based tasks (procedural memories).[65]

For almost 40 years, there have been reports of decreased cell density and/or cell numbers in the MD of schizophrenic patients.[8-10] Recent data obtained by modern stereological methods indicate a 40% to 45% decrease in neurons and neuroglia and a significant decrease in volume of the MD. Although these results might conceivably be related to drug treatment, this possibility has been discounted because of the nondopaminergic nature of the nucleus.[11] Once again, the significance of this finding highlights the possible role of the thalamus in the psychotic process without specifying the nature of that role.

Eye-tracking abnormalities are a consistent and robust clinical finding in schizophrenia,[66-68] Although clear-cut anatomical correlations have not been described, the relationship between central thalamus and gaze mechanisms was recognized almost 60 years ago[69] and has been more rigorously evaluated in the past few decades.[70-72] The internal medullary lamina of the thalamus (IML) is usually described as the thalamic source of gaze control; however, it must be remembered that the paralaminar cell component of MD is closely applied to the medial surface of the IML and is often confused, or included, with it. As already indicated, paralaminar cells sustain two-way communication with areas 6 and 8, the latter of which contains the frontal cortical eye fields. Eye movement control circuitry for fast and slow saccades and following movements undoubtedly involves complex circuitry that includes brainstem, cerebellum, colliculi, thalamus, and parietal and frontal cortices, among others. Nonetheless, it seems reasonable to suggest that dysfunction of this slim lateralmost sector of MD may play a significant role in the genesis of abnormal eye-tracking movements in schizophrenia.

Hypometabolism of the overlying prefrontal cortex is a fairly well recognized result of MD dysfunction,[73] although the mechanism is not clear. This phenomenon gains added significance from the description of such alterations described recently in the prefrontal cortices of schizophrenic patients.[74,75]

THE HABENULA AND THE DORSAL THALAMIC PROJECTION

The habenular nuclear complex lies on the medial diencephalic surface at the most posterior and dorsal part of the thalamus, forming small protrusions on either side of the third ventricle. Together with the pineal gland, which lies immediately behind it, this area is often considered a separate thalamic division called *epithalamus*. The habenular nuclei are prominent and very well circumscribed in small mammals but become relatively less distinct in primates and humans. They are composed of a medial nucleus made up of small, densely packed, granule-like cells and a lateral nucleus consisting of larger, more loosely packed multipolar cells. Although their functions have until recently remained enigmatic, their connections and neurotransmitter content, plus recently developed data on physiological effects of stimulation and/or ablation, have begun to reveal significant aspects of their role in brain function and behavior.[76,77]

The largest and most obvious system afferent to the habenular nuclei is the stria medullaris thalami, a dense fiber bundle that runs caudally along the dorsomedial lip of the thalamus. It includes fibers from various sources in the medial forebrain and septum, the limbic lobe, anterior hypothalamus, and medial part of the striatum (Figure 3–4). The major efferent tract for the habenular nuclei is the fasciculus retroflexus of Meynert, which projects predominantly to the interpeduncular nucleus, but also to substantia nigra and ventral tegmental area, the medial raphe complex, and

FIGURE 3-4. Organization of the dorsal thalamic conduction system and its projection on the mesencephalon. f = basal forebrain; hn = habenular nuclei; hy = hypothalamus; i = interpeduncular nuclei; r = midbrain raphe; s = septal area; sm = stria medullaris thalami; st = corpus striatum; v = ventral tegmental area.

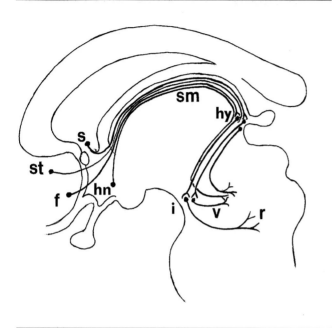

probably the locus coeruleus and central gray. Each of the two habenular nuclei has its own idiosyncratic mix of afferent sources and efferent targets within these groups, with the lateral nucleus adding an extra group of diencephalic targets that includes the mediodorsal and ventral thalamic nuclei and the preoptic hypothalamus. It has been noted that the lateral portion of the lateral habenular nucleus tends to project to dopamine-rich cell ensembles, whereas the more medial part of this nucleus projects to the serotonergic raphe.[77]

The habenular complex, along with the entire dorsal thalamic system of which it is the principal nuclear ensemble, appears to serve as a critical modulatory relay between limbic forebrain structures and the midbrain. Electrical stimulation of the lateral habenula causes inhibition of about 90% of dopamine neurons in the ventral tegmental area and 85% in the substantia nigra. Each nucleus appears to exert bilateral control over dopaminergic nuclei of the midbrain.[78] Lateral habenular electrical stimulation also decreases serotonin release in the dorsal raphe nuclei, the caudate nucleus, and the substantia nigra.[79] Raphe neuronal firing is reduced by habenular stimulation, an effect that is reversed by picrotoxin.[80] These effects appear mediated by both glutamatergic and GABAergic fiber systems.

Conversely, lesions of the habenular nuclei or stria medullaris thalami result in activation of the mesocortical, mesolimbic, and mesostriatal dopaminergic system as measured by increases in dopamine turnover in prefrontal cortex, nucleus accumbens, and striatum.[81,82] In similar fashion, lesions of stria medullaris thalami or the lateral habenula increase cell firing and increase serotonin turnover in the dorsal raphe.[83]

The dorsal thalamic system seems to include a wide range of neurotransmitters in its trajectory from its septal-basal forebrain origins to habenula, interpeduncular nucleus, and midbrain tegmentum. These include glutamate and aspartate,[84] acetylcholine,[85] GABA,[86] opioid peptides,[87,88] and a number of others.

The basic picture that emerges is that of a robust feedback control system regulating levels of dopamine and serotonin utilization in the telencephalon and diencephalon. The dorsal thalamic system seems to be unique in combining and directing striatal and limbic output caudally on major midbrain sources of monoaminergic and cholinergic innervation of forebrain structures.[89] These neurotransmitter systems seem highly significant in the modulation of affect and behavior and have been implicated in schizophrenic and affective psychoses. A recently proposed neurodevelopmental model of schizophrenia[90] suggests alterations in negative feedback control of mesolimbic dopamine circuitry as critical to the disease process. Conjoint problems in feedback control over serotonergic raphe and adjacent acetylcholine-rich brainstem systems are also likely involved. Compared with traditional neuroleptics such as chlorpromazine and haloperidol, the unusual effectiveness of new atypical antipsychotics like clozapine and risperidol has been attributed in part to their broader control over dopaminergic, serotonergic, and perhaps other neurotransmitter and neuromodulator systems. It appears that the habenular nuclei and the dorsal thalamic system normally provide feedback control over those neurotransmitters that have been empirically linked to psychoses and whose modulation by pharmaceutical agents is likely to have therapeutic or ameliorative effects.

Another possible sequel of disturbed (that is, inadequate) habenular feedback may be loss of tonic control over the nRt gatelet system. The importance of mesencephalic reticular control over nRt cells has been described.[45] In view of the demonstrated increased activity of major cell clusters in the midbrain tegmentum after lesions of the stria medullaris thalami or habenula,[81–83] the tegmental cells that project to the n. reticularis may also become more active. Because these cells selectively inhibit the nRt neurons, the effect of their enhanced activity might well be a nonselective opening of the reticularis gate. This would provide another putative mechanism for the sensory flooding phenomenon that has been postulated by many workers.[47]

It has recently been reported that chronic administration of psychotogens such as amphetamine and cocaine to rats results in behavioral abnormalities and degenerative changes in the habenular nuclei and the fasciculus retroflexus.[91] In view of the feedback control exerted by the dorsal thalamic system over brainstem monoaminergic systems, this finding is of obvious interest as a possible model of human psychosis, especially schizophrenia. We are presently examining this system in the upper brainstems of a series of patients diagnosed as schizophrenic, compared with a group of nonpsychotic control subjects.

The relatively small size of the habenula and dorsal thalamic projections makes unlikely the identification of lesions localized to this system. Almost any pathology of the dorsal or dorsomedial thalamus may involve the stria medullaris thalami or habenular nuclei along with underlying thalamic tissue. However, in view of the subtle but widespread histological and cytoarchitectonic changes accompanying psychoses that have recently been described in the limbic system, prefrontal cortex, and striatum[55,56,92–94]—all of which structures send projections back upon habenula and midbrain aminergic centers—it is reasonable to consider this system suspect in the development of some schizophrenic and affective disorders.

COMMENT

A large case-oriented literature already exists linking pathology in various parts of the thalamus to a wide range of sensorimotor, behavioral, and cognitive signs and symptoms. In the human brain, such lesions are at best "natural experiments" subject to a broad range of pathophysiological variables, lack of locational specificity, and ethical constraints, all of which limit what may be learned from the subject. The thalamus is centrally placed across virtually all input lines to the cortex and is a crucial node in the web of connections among the brainstem, cerebellum, corpus striatum, limbic system, and neocortex. With these complications in mind, it has seemed wiser to use a bottom-up approach, citing several thalamic nuclei as models. Patterns of connections, neuropil structure, physiological behavior, and neurochemical attributes have been used wherever possible to conceptualize roles for these centers and the possible results of their dysfunction. These in turn have been related to clinical and pathological data where they exist.

As already stated, the thalamus is a mosaic of idiosyncratic subcenters (Figure 3–5), each linked preferentially to its connecting downstream and upstream targets and all only secondarily to each other. This could be likened to an ensemble of semi-independent modules, clustered together in the diencephalon but concerned primarily with more distant brain systems. Proper interpretation of the flood of fMRI and PET scan data on psychiatric illness now becoming available is still problematic. Do apparently isolated areas of involvement, particularly those obtained by subtraction techniques, really indicate unique areas of pathological involvement, or are more profound, global interpretations necessary? Histological evidence of structural change in schizophrenia, spread widely among prefrontal cortex, cingulate gyrus, hippocampal-dentate complex, entorhinal cortex, and thalamus, would seem to suggest the latter.

The present study provides at best a paradigm effort in this direction. It should be equally useful to examine other thalamic nuclei: for instance, the ventral lateral and ventral anterior nuclei, because of their importance

FIGURE 3–5. Semischematic cross sections through human thalamus showing the nuclei and structures discussed in this chapter. fr = fasciculus retroflexus of Meynert; h = habenular nucleus; md = mediodorsal nucleus; nRt = nucleus reticularis thalami; sm = stria medullaris thalami; vb = ventrobasal nuclear complex.

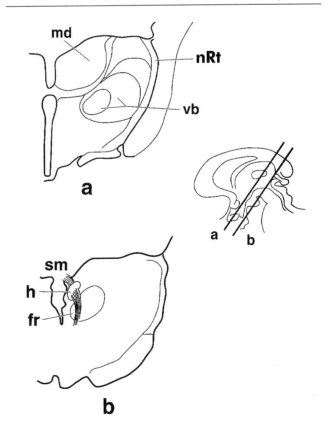

in processing information from the cerebellum and basal ganglia; the medial and midline nuclei, with their broad range of connections with association cortices as well as intrathalamic links; and the anterior thalamic nucleus, placed critically across limbic and hypothalamic circuits. Combinations of the new visualization methods with increasingly high resolution, together with thoughtful clinical correlation and careful postmortem tissue studies, should add greatly to our understanding of these problems.

REFERENCES

1. Andreasen NC, Arndt S, Swayze II V, et al: Thalamic abnormalities in schizophrenia visualized through magnetic resonance image averaging. Science 1994; 266:294–298
2. Buchsbaum MS, Someya T, Teng CY, et al: PET and MRI of the thalamus in never-medicated patients with schizophrenia. Am J Psychiatry 1996; 153:191–199
3. Frazier JA, Giedel JN, Hamburger SD, et al: Brain anatomic resonance imaging in childhood-onset schizophrenia. Arch Gen Psychiatry 1996; 53:617–624
4. Parent A: Carpenter's Human Neuroanatomy, 9th edition. Baltimore, Williams and Wilkins, 1996
5. Cohen BM, Wan W: The thalamus as a site of action of antipsychotic drugs. Am J Psychiatry 1996; 153:104–105
6. Deutch AY, Ongar D, Durnan RS: Antipsychotic drugs induce Fos protein in the thalamic paraventricular nucleus: a novel locus of antipsychotic drug action. Neuroscience 1995; 66:337–346
7. Lesch A, Bogerts B: The diencephalon in schizophrenia: evidence for reduced thickness of the periventricular grey matter. European Archives of Psychiatry and Neurological Sciences 1984; 234:212–219
8. Funfgeld E: Pathological-anatomische Untersuchungen bei Dementia praecox mit besonderen Beruchsichtigung des Thalamus opticus [Pathological-anatomical examination of dementia praecox with special attention to the optic thalamus]. Zeitschrift für die gesamte Neurologie und Psychiatrie 1925; 95:411–463
9. Baumer H: Veranderungen des Thalamus bei Schizophrenie [On the nature of the thalamus in schizophrenia]. J Hirnforsch 1954; 1:157–172
10. Treff U, Hempel K: Die Zelldichte bei Schizophrenen und Klinische Gesunden [Cell density patterns in schizophrenics and in clinically healthy controls]. J Hirnforsch 1958; 4:314–369
11. Pakkenberg B: Pronounced reduction of total neuron number in mediodorsal thalamic nucleus and nucleus accumbens in schizophrenia. Arch Gen Psychiatry 1991; 47:1023–1025
12. Gentilini M, de Renzi E, Grisi E: Bilateral paramedian thalamic artery infarcts: report of eight cases. J Neurol Neurosurg Psychiatry 1987; 50:900–912
13. Sandson TA, Daffner KR, Carvalho PA, et al: Frontal lobe dysfunction following infarction of the left-sided medial thalamus. Arch Neurol 1991; 48:1300–1303
14. Starkstein SE, Boston JD, Robinson RG: Mechanisms of mania after brain injury: 12 case reports and review of the literature. J Nerv Ment Dis 1988; 176:87–100
15. Stuss DT, Guberman A, Nelson R, et al: The neuropsychology of paramedian infarction. Brain Cogn 1988; 8:348–378
16. Mennemeier M, Fennell E, Valenstin E, et al: Contributions of the left intralaminar and medial thalamic nuclei to memory. Arch Neurol 1992; 49:1050–1058
17. Gutmann DH, Grossman RE, Mollinan JE: Personality changes associated with thalamic infiltration. J Neurooncol 1990; 8:263–267
18. Graff-Radford NG, Eslinger PJ, Damasio AR, et al: Nonhemorrhagic infarction of the thalamus: behavioral, anatomic and physiologic correlates. Neurology 1984; 34:14–23
19. Sener RN, Alper H, Yunten N, et al: Bilateral acute thalamic infarcts causing thalamic dementia. American Journal of Roentgenology 1993; 161:678–679
20. Carlsson A: The current status of the dopamine hypothesis of schizophrenia. Neuropsychopharmacology 1988; 1:179–186
21. Modell JG, Mowitz JM, Curtis GC, et al: Neurophysiologic dysfunction in basal ganglia, limbic striatal, and thalamocortical circuits as a pathogenetic mechanism of obsessive-compulsive disease. J Neuropsychiatry Clin Neurosci 1989: 1:27–36
22. Swerdlo N, Koob G: Dopamine, schizophrenia, mania and depression: toward a unified hypothesis of cortico-striato-pallido-thalamic functions. Behav Brain Sci 1987; 10:197–245
23. Ramon y Cajal S: Histologie du Système Nerveux de l'Homme et des Vertebres [Histology of the nervous system of man and vertebrates], vol 2. Paris, A Maloine, 1911, pp 394–402
24. Scheibel ME, Scheibel AB: Patterns of organization in specific and nonspecific thalamic fields, in The Thalamus, edited by Purpura D, Yahr M. New York, Columbia University Press, 1966, pp 13–46
25. Jones EG, Friedman DP, Hendry SHC: Thalamic basis of place- and modality-specific columns in monkey somato-sensory cortex: a correlative anatomical and physiological study. J Neurophysiol 1982; 48:545–568
26. Jones EG: The Thalamus. New York, Plenum, 1985
27. Mountcastle VB, Henneman E: Patterns of tactile representation in thalamus of the cat. J Neurophysiol 1949; 12:85–100
28. Mountcastle VB, Henneman E: The representation of tactile sensibility in the thalamus of the monkey. J Comp Neurol 1952; 97:409–440
29. Jones EG, Friedman DP: Projection pattern of functional components of the thalamic ventrobasal complex on monkey somatosensory cortex. J Neurophysiol 1982; 48:521–544
30. Sherman SM, Koch C: Thalamus, in The Synaptic Organization of the Brain, edited by Shepherd GM. New York, Oxford University Press, 1990, pp 246–278
31. Scheibel ME, Scheibel AB: Thalamus and body image: a model. Biol Psychiatry 1971; 3:71–76
32. Scheibel ME, Scheibel AB, Davies TH: Some substrates for centrifugal control over thalamic cell ensembles, in Corticothalamic Projections and Sensorimotor Activities, edited by Frigyesi T, Rinvik E, Yahr MD. New York, Raven, 1972, pp 131–160
33. Schlag J, Waszak M: Characteristics of unit responses in nucleus reticularis thalami. Brain Res 1970; 21:286–288
34. Bogouslavsky J, Regli F, Uske A: Thalamic infarcts: clinical syndromes, etiology and progress. Neurology 1988; 38:837–848
35. Gutrecht JA, Zavari AA, Pandya DN: Lacunar thalamic stroke with pure cerebellar and proprioceptive deficits. J Neurol Neurosurg Psychiatry 1992; 55:854–856
36. Melo TP, Bogouslavsky J: Hemiataxia-hypesthesia: a thalamic stroke syndrome J Neurol Neurosurg Psychiatry 1992; 55:581–584
37. Poggio GF, Mountcastle V: A study of the functional contributions of the lemniscal and spinothalamic systems to somatic sensibility. Bulletin of the Johns Hopkins Hospital 1960; 106:266–316
38. Houser CR, Vaughn JE, Barber RP, et al: GABA neurons are the major cell type of the nucleus reticularis thalami. Brain Res 1980; 200:341–354
39. Scheibel ME, Scheibel AB: The organization of the nucleus reticularis thalami: a Golgi study. Brain Res 1966; 1:43–62
40. Yen C-T, Jones EG: Intracellular staining of physiologically identified neurons and axons in the somatosensory thalamus of the cat. Brain Res 1983; 280:148–154
41. Steriade M, Domich L, Oakson G: Reticularis thalami neurons revisited: activity changes during shifts in states of vigilance. J Neurosci 1986; 6:68–81

42. Scheibel AB: Anatomical and physiological substrates of arousal: a view from the bridge, in The Reticular Formation Revisited, edited by Hobson, JA, Brazier MAB, New York, Raven, 1980, pp 55–66
43. Scheibel ME, Scheibel AB: Input-output relations of the thalamic nonspecific system. Brain Behav Evol 1972; 6:332–358
44. Crick F: Function of the thalamic reticular complex: the searchlight hypothesis. Proc Natl Acad Sci USA 1984; 81:4586–4590
45. Yingling CD, Skinner JE: Gating of thalamic input to cerebral cortex by nucleus reticularis thalami, in Progress in Clinical Neurophysiology, vol l, edited by Desmedt JE. Basel, Karger 1977, pp 70–96
46. Steriade M, Jones EG, Llinas R: Thalamic Oscillations and Signaling. New York, Wiley, 1990
47. Braff DL: Information processing and attention dysfunctions in schizophrenia. Schizophr Bull 1993; 19:233–259
48. Merriam AE, Kay SR, Oplir LA, et al: Information processing deficits in schizophrenia: a frontal lobe sign? (letter). Arch Gen Psychiatry 1989: 46:760
49. Braff DL, Stone C, Calloway E, et al: Prestimulus effects on human startle reflex in normals and schizophrenics. Psychopathology 1978; 14:339–343
50. Karper LP, Freeman GK, Grillon C, et al: Preliminary evidence for an association between sensorimotor gating and distractibility in psychosis. J Neuropsychiatry Clin Neurosci 1996; 8:60–66
51. Skinner JE, Yingling CD: Central gating mechanisms that regulate event-related potentials and behavior: a neural model for attention, in Progress in Clinical Neurophysiology, vol l, edited by Desmedt JE. Basel, Karger, 1977, pp 30–69
52. Weinberger DR, Torrey EF, Zec RF: Physiologic dysfunction of dorsolateral prefrontal cortex in schizophrenia, I: regional cerebral blood flow evidence. Arch Gen Psychiatry 1986; 43:114–124
53. Weinberger DR, Berman K, Illowsky BP: Physiologic dysfunction of dorsolateral prefrontal cortex in schizophrenia, IV: a new cohort and evidence for a monoaminergic mechanism. Arch Gen Psychiatry 1988; 45:609–615
54. Benes FM, McSparren J, Bird ED, et al: Deficits in small interneurons in prefrontal and cingulate cortices of schizophrenic and schizoaffective patients. Arch Gen Psychiatry 1991; 48:996–1001
55. Benes FM, Davidson J, Bird ED: Quantitative cytoarchitectural studies of the cerebral cortex of schizophrenics. Arch Gen Psychiatry 1986; 43:31–35
56. Akbarian S, Kiru JJ, Potkin SG, et al: Maldistribution of interstitial neurons in prefrontal white matter of the brains of schizophrenic patients Arch. Gen Psychiatry 1996; 53:425–436
57. Akbarian S, Sucher NJ, Bradley D, et al: Selective alterations in gene expression for NMDA receptor subunits in prefrontal cortex of schizophrenics. J Neurosci 1996; 16:19–30
58. Dejerine J, Roussy G: Le syndrome thalamique [The thalamic syndrome]. Revue Neurologique 1906; 14:521–532
59. Scheibel ME, Scheibel AB: The basis of thalamic pain: a structurofunctional hypothesis. Transactions of the American Neurological Association 1969; 94:149–152
60. Oliveras JC, Montagne-Clavel J: The GABA-A receptor antagonist picrotoxin induces a "pain-like" behavior when administered into the thalamic reticular nucleus of the behaving rat: a possible model for "central" pain. Neurosci Lett 1994; 179:21–24
61. Crow TJ: Molecular pathology of schizophrenia; more than one process? British Medical Journal 1980; 280:66–68
62. Pantelis C, Barnes T, Nelson H: Is the concept of frontal subcortical dementia related to schizophrenia? Br J Psychiatry 1992; 160:442–460
63. Buchanan RN, Breier A, Kirkpatrick B, et al: Structural abnormalities in deficit and nondeficit schizophrenia. Am J Psychiatry 1993; 150:59–65
64. Partlow GD, del Carpio-O'Donovan R, Melenson D, et al: Bilateral thalamic glioma: review of eight cases with personality change and mental deterioration. Am J Neuroradiol 1992; 13:1225–1230
65. Zola-Morgan S, Squire LR: Amnesia in monkeys after lesions of the mediodorsal nucleus of the thalamus. Ann Neurol 1985; 17:558–564
66. Holzman PS, Proctor LR, Levy DL, et al: Eye-tracking dysfunctions in schizophrenic patients and their relatives. Arch Gen Psychiatry 1992; 31:143–151
67. Abel LA, Friedman L, Jesberger J, et al: Quantitative assessment of smooth pursuit gain and catch-up saccades in schizophrenia and affective disorders. Biol Psychiatry 1991; 29.1063–1072
68. Cegalis JA, Sweeney JA: Eye movements in schizophrenia: a quantitative analysis. Biol Psychiatry 1979; 14:13–26
69. Godlowski W: Experimentelle Untersuchungen über die durch Reizung des Zwischen—und Mittelhirns herrorgerufenen assoziierten Augenbewegungen [Experimental study of the effects of stimulation of midbrain and diencephalon on associated eye movements]. Zeitschrift für die gesamte Neurologie und Psychiatrie 1938; 162:160–182
70. Schlag J, Schlag-Rey M: Induction of oculomotor responses from thalamic internal medullary lamina in the cat. Exp Neurol 1971; 33:498–505
71. Schlag J, Schlag-Rey M: Visuomotor functions of central thalamus in monkey, II: unit activity related to visual events, targeting and fixation. J Neurophysiol 1984; 51:1175–1195
72. Schlag-Rey M, Schlag J: Visuomotor functions of central thalamus in monkey, I: unit activity related to spontaneous eye movements. J Neurophysiol 1984; 51:1149–1173
73. Hennerici M, Halsband W, Kuwert T, et al: PET and neuropsychology in thalamic infarction: evidence for associated cortical dysfunction. Psychiatry Res 1989; 29:363–365
74. Pappeta S, Mazoyer B, Tren Dinh S, et al: Effects of capsular or thalamic stroke on metabolism in the cortex and cerebellum: a positron tomography study. Stroke 1990; 21:519–524
75. McGilchrist I, Goldstein L, Jadresic D, et al: Thalamo-frontal psychosis. Br J Psychiatry 1993; 163:113–115
76. Sutherland RJ: The dorsal diencephalic conduction system: a review of the anatomy and functions of the habenular complex. Neurosci Biobehav Rev 1982; 6:1–13
77. Sandyk R: Relevance of the habenular complex to neuropsychiatry: a review and hypothesis. Int J Neurosci 1991; 61:189–219
78. Christoph GR, Leonzio RJ, Wilcox KS: Stimulation of the lateral habenula inhibits dopamine-containing neurons in the substantia nigra and ventral tegmental area of the rat. J Neurosci 1986; 6:613–619
79. Reisine TD, Soubrie P, Artand F, et al: Involvement of the lateral habenula-dorsal raphe neurons in the differential regulation of striatal and nigral serotonergic transmission in cats. J Neurosci 1982; 2:1062–1071
80. Wang R, Aghajanian G: Physiological evidence for habenula as a major link between forebrain and midbrain raphe. Science 1977; 157:89–91
81. Lisoprawski A, Herve D, Blanc G, et al: Selective activation of the mesocortico-frontal dopaminergic neurons induced by lesions of the habenula in the rat. Brain Res 1980; 183:229–234
82. Nishikawa T, Fage D, Scatton B: Evidence for and nature of the tonic inhibitory influence of the habenulointerpeduncular pathways upon cerebral dopaminergic transmission in the rat. Brain Res 1986; 373:324–336
83. Speciale S, Neckers L, Wyatt R: Habenular modulation of raphe indoleamine metabolism. Life Sci 1980; 27:2367–2372
84. Matsuda Y, Fujimura K: Action of habenular efferents on ventral tegmental area neurons studied in vitro. Brain Res Bull 1992; 743–749
85. Woolf NJ, Butcher LL: Cholinergic systems in the rat brain, IV: descending projections of the pontomesencephalic tegmentum. Brain Res Bull 1989; 23:519–540
86. Gottesfeld Z, Massari V, Muth E, et al: Stria medullaris: a possible pathway containing GABAergic afferents to the lateral habenula. Brain Res 1977; 130:184–189
87. Artymyshyn R, Eckenrode TC, Murray M: Opiate binding sites in

the interpeduncular nucleus of the rat: normal distribution and the effects of fasciculus retroflexus lesions. Neuroscience 1987; 23:159–172
88. Atweh SF, Kuhar MJ: Autoradiographic localization of opiate receptors in rat brain, II: the brain stem. Brain Res 1977; 129:1–12
89. Nauta WJH: Evidence of a pallidohabenular pathway in the cat. J Comp Neurol 1974; 156:19–28
90. Weinberger D: Implications of normal development for the pathogenesis of schizophrenia. Arch Gen Psychiatry 1987; 44:660–669
91. Ellison G: Stimulant-induced psychosis, the dopamine theory of schizophrenia and the habenula. Brain Res Rev 1994; 19:223–239
92. Benes FM: Neurobiological investigations in cingulate cortex of schizophrenic brain. Schizophr Bull 1993; 19:537–549
93. Bogerts B, Meertz E, Schonfeldt-Bausch R: Basal ganglia and limbic system pathology in schizophrenia: a morphometric study of brain volume and shrinkage. Arch Gen Psychiatry 1985; 42:784–791
94. Falkai P, Bogerts B, Rozumak M: Cell loss and volume reduction in the entorhinal cortex of schizophrenics. Biol Psychiatry 1988; 24:515–521

4

The Accumbens

Beyond the Core–Shell Dichotomy

Lennart Heimer, M.D., George F. Alheid, Ph.D., José S. de Olmos, Ph.D.,
Henk J. Groenewegen, M.D., Ph.D., Suzanne N. Haber, Ph.D., Richard E. Harlan, Ph.D.,
Daniel S. Zahm, Ph.D.

The purpose of this chapter is to highlight some major advances of the past several years that have profoundly changed our view not only of the accumbens, but of the anatomical organization of the basal forebrain in general. Discoveries related to the accumbens have played a significant role in this changing perspective. Since most progress emerged from experiments in the rat, this will be the immediate focus of our discussion. Translation of these insights to the primate accumbens, however, is proceeding apace, as is briefly discussed later in the chapter.

1. GENERAL OVERVIEW

Ever since Matthysse and Stevens focused attention on the mesolimbic dopaminergic (DA) system and its relation to schizophrenia in the early 1970s,[1,2] the accumbens has been a central structure in theories exploring the anatomical substrates of schizophrenia. However, clinical interest in the accumbens is based on more than its possible role in schizophrenia and other affective disorders. As a prominent part of the ventral striatum, and as a main target of the mesotelencephalic dopamine system, the accumbens is a natural focus for theories of reward and motivation, including aspects of drug abuse.[3-8]

Particularly in the work of Mogenson et al.,[9] the accumbens was envisioned as the "limbic-motor interface." Although this catchy phrase is sometimes invoked to describe the functions of the accumbens, the phrase is somewhat unsettling in the sense that the term *limbic* is rather murky; it lacks an adequate operational definition. Mogenson and his co-workers referred to the motivational processes that result in action or movements. In the context of this chapter it may be more important to emphasize that, whatever meaning one assigns to the term *limbic* and whatever role the accumbens plays in

---

The authors thank Dr. Michael Forbes for his excellent editorial work and Ms. Vickie Loeser for secretarial assistance. Dr. David Song prepared the material used in section 2, and Drs. Chris I. Wright and Arno V. J. Beijer produced the tracing material presented in section 3. The authors also acknowledge Dr. John Jane, Department of Neurological Surgery, University of Virginia, who sponsored a meeting that provided the incentive and initial draft for this essay.

This work was supported by United States Public Health Service Grants NS17743 (L.H. and G.F.A) and DA06194 and NS24148 (R.E.H.); Consejo Nacional de Investigaciones Científicas y Técnicas of Argentina (J. de O.); National Institute of Mental Health Grant MH45573; a grant from the Lucille P. Markey Charitable Trust Award (S.N.H.); and Grant NS23805 from the American Parkinson Disease Association and United Parkinson Foundation (D.S.Z.).

linking motivation to action, the accumbens is likely to share this role with several other structures of the basal forebrain, including especially the "extended amygdala" (see section 6).

Although the physiological and behavioral studies of the last 20 years have served to establish the accumbens as a pivotal structure in motivation and reward mechanisms, the anatomical discoveries related to the accumbens have provided the structural framework for this progress. The term *nucleus accumbens septi* was introduced by Ziehen[10] 100 years ago (see also introductory article by Chronister and DeFrance[11]), and for many years the accumbens remained in this uncertain relationship with the septum. In the mid-1970s, however, it became evident that the accumbens is an integral part of the striatal complex.[12,13] This assertion is reinforced by its developmental pattern (see section 2) and by the extrinsic connections of accumbens, which are reminiscent of those in the rest of the striatum.[14,15]

Based in part on the cortical input that the rat accumbens receives from allocortical and periallocortical areas (hippocampal formation, entorhinal area, and olfactory cortex), as well as from medial (anterior cingulate, prelimbic, infralimbic) and lateral (sulcal, agranular insular, and perirhinal) proisocortical areas, it is now customary to include the accumbens, together with the ventromedial part of caudate-putamen and the olfactory tubercle, in a larger entity referred to as the *ventral striatum* (see reviews by Alheid et al.[14] for primate and Heimer et al.[15] for rat). The identification of the ventral striatopallidal system,[12] in which the accumbens and its projections to a ventral extension of the pallidum (referred to as *ventral pallidum*) occupy a prominent position, added an important channel to the striatopallidal circuitry; in short, it demonstrated that the whole cortical mantle, and not only the neocortex, is interlinked with the basal ganglia. This added several corticostriatopallidothalamic circuits to the well-known corticostriatopallidothalamic loop by which information from various neocortical regions reaches the motor regions in the frontal lobe via synaptic relays in basal ganglia and thalamus. The corticosubcortical reentrant circuits related to the ventral striatopallidal system relay allocortical and periallocortical afferents via primarily the mediodorsal thalamic nucleus to the prefrontal cortex. This circuitry is in contrast to the ventral anterior–ventral lateral thalamic complex, with its output to the premotor cortex, which characterizes the channels through the dorsal portions of the striatopallidal system.[16]

The identification of the corticosubcortical reentrant circuits that originate in various allocortical, periallocortical, and proisocortical areas and pass through the ventral striatopallidal system and the mediodorsal thalamus back to the prefrontal cortex has inspired several theoretical schemes attempting to explain various symptoms of neuropsychiatric disorders. Because these circuits have received considerable attention both in the preclinical and the clinical literature,[17–28] they will not be the focus of further discussion here. Suffice it to say that the nature of these circuits—that is, the extent to which they are "closed" (parallel and independent of each other) or "open" (interrelated with each other)—is still being debated (see reviews;[15,20,29,30] see also section 5).

Although the corticosubcortical reentrant circuits just mentioned are undoubtedly of great significance in the context of neuropsychiatric disorders, it should be emphasized that the basal ganglia, including the ventral striatopallidal system, establish several other interneuronal relations that are equally important. Among these are output channels via the internal pallidal segment and the reticular part of substantia nigra to various brainstem areas (reviewed[14,15,31]). Other axons interconnect the striatal complex with the mesencephalic dopamine neurons. Since the dopamine system occupies a central role in the etiology and pharmacotherapy of some of the most common brain disorders, including schizophrenia and Parkinson's disease, we will briefly outline in section 5 the neuronal circuits that the midbrain dopamine neurons establish with the striatal complex, including the accumbens. An important concept highlighted in that section, the modulation of the dorsal striatal complex by the ventral striatum via dopamine circuits, was first suggested on the basis of studies in the rat by Nauta et al.[32] As we shall see in section 5, this intriguing way of integrating information from ventral and dorsal parts of the basal ganglia via the ascending mesotelencephalic dopamine system is only one of several possible systems for integrating information from different sources within the basal ganglia.

Once the "striatal" nature of the accumbens was established, its compartmental infrastructure became the focus of interest for a number of research groups, analogous with the intense interest in the compartmental organization of the dorsal striatum. Armed with a variety of immunohistochemical and in situ hybridization methods, receptor binding techniques, and increasingly more sensitive tract-tracing methods, workers gradually defined the compartmental or "patch-matrix" organization of the accumbens as similar, but not exactly parallel, to that elucidated for the dorsal striatum (the caudate-putamen).[28,33,34] In attempting to encompass the compartmental organization of ventral striatum in a structural theory of accumbens function, one hypothesis focuses on the existence of segregated clusters of neurons as neuronal ensembles that are characterized by different combinations of inputs and outputs that determine specific components of various behaviors.[35] The anatomical substrates for this theory will be discussed in section 3.

The accumbens is directly continuous with the main dorsal part of the striatum, the caudate-putamen (Figure 4–1). It is generally agreed that the accumbens is an integral part of the striatal complex. It is obvious, however, that the accumbens has unique features compared with the rest of the striatal complex. For example, it can be divided into a central core that is surrounded on its medial, ventral, and lateral sides by a shell.[36] In most respects, the core cannot be easily distinguished from the rest of the caudate-putamen (Figure 4–1). The shell, on the other hand, has in addition to its striatal characteristics a number of features that are atypical for a striatal structure. Without directly dividing the accumbens into two different parts, some papers that appeared in the 1970s foreshadowed this important point (see, for example, Nauta et al.[32]). In addition to projections to ventral pallidum, the accumbens, especially its medial part, was characterized by projections to regions such as the bed nucleus of stria terminalis and lateral hypothalamus and to mesopontine targets caudal to the mesencephalic dopamine neurons.[33,37] It was later revealed that these projections, atypical for the striatum, originate primarily in the shell rather than in the core of the accumbens.[38]

Aside from having distinctive efferents, the core and the shell of the accumbens appear to receive distinctive cortical afferents from generally different cortical areas, suggesting functionally separate circuits.[21,22,28,39] In fact, the last few years have seen a rapidly increasing number of pharmacological and physiological studies in which the functional differences between the core and shell of the accumbens have been explored. These studies are especially relevant in the field of neuropsychiatry.

Ever since it was proposed, almost 20 years ago, that the accumbens plays an important role in reward mechanisms,[40] the function of dopamine and its interaction with neuronal elements in the accumbens have been central themes in drug abuse studies.[4,6,41,42] In the last few years it has been reported that many of the abused drugs (including cocaine, amphetamine, and morphine) preferentially stimulate the release of dopamine and increase the energy metabolism in the shell rather than the core.[43–45] Even nicotine seems to derive much of its addictive effect from interference with neuronal mechanisms in the shell of the accumbens.[46] Another drug that has received a lot of attention in recent years is methylenedioxymethamphetamine, better known as "ecstacy." The physiological and pharmacological effects of "ecstasy" on the monoaminergic mechanisms in the brain, and in particular on the accumbens, have recently been reviewed in great detail.[47] As expected, there are significant differences between the core and the shell in their response to "ecstacy." The core–shell dichotomy also appears to be relevant in the context of antidepressant medication.[48]

Stress-induced activation of the dopamine systems in the brain[49,50] is another subject of considerable importance in neuropsychiatry. Several recent reports indicate that it is the neuronal circuits related to the shell rather than the core that are sensitive to stress[30,51–54] and that therefore may be related to symptoms of schizophrenia that are influenced by stress. In this context, it

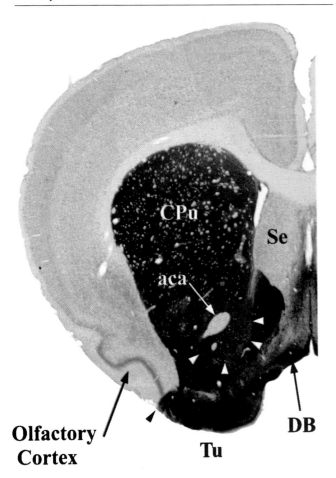

FIGURE 4–1. Acetylcholinesterase-stained coronal section through the rat brain at the level of the olfactory tubercle. The section was counterstained with thionine. Note how the striatal complex (darkly stained with acetylcholinesterase) extends toward the ventral surface of the brain and includes not only the accumbens, but also large parts of the olfactory tubercle (Tu). The subdivision of the accumbens into a central core surrounded on its medial, ventral, and lateral sides by a shell is also evident. The border between the core and the shell is marked with *arrowheads*. The core extends into the dorsal part of the striatal complex (caudate-putamen; CPu), without sharp boundary. aca = anterior limb of the anterior commissure; DB = diagonal band; Se = Septum.

is especially interesting to realize that the shell rather than the core seems to be a primary target for antipsychotic drug action.[52,55,56] Because of the functional differences between the core and the shell and their apparent clinical relevance, we discuss in section 4 some of the many histochemical differences that exist between the core and the shell, including histochemical patterns altered by the administration of antipsychotic drugs.

Relevant to understanding the functional segmentation of the nucleus accumbens is the concept of the extended amygdala (see section 6). As originally proposed by Johnston[57] and elaborated in modern form by de Olmos, Alheid, and co-workers,[58–60] this structure stretches from the temporal lobe to the caudal portions of the accumbens. Portions of the ventral striatum, including part of the accumbens, are intimately related to the extended amygdala. In particular, we will explore the extent to which the accumbens, and especially parts of its shell, can be considered a transitional area between the striatal complex and extended amygdala. The identification of the core and shell of the accumbens and the continuing elaboration of the theory of the extended amygdala have significantly improved our understanding of basal forebrain anatomical organization. Because the core–shell dichotomy and the extended amygdala appear to be as relevant in the primate, including the human,[58,61–65] as they are in the rat, the continuing exploration of these structures is likely to be especially relevant to understanding the etiology and treatment of neuropsychiatric disorders.[58,66] As indicated above, many studies have focused attention on the shell of the accumbens as being of special relevance in drug abuse. The shell of the accumbens and the extended amygdala have many characteristics in common, so it is not surprising to learn that results obtained in recent experiments indicate that the extended amygdala is also an important structure in the context of drug abuse.[67–70]

2. SOME DEVELOPMENTAL ASPECTS

Developmental studies of specific brain structures can often yield insights into overall organizational features. For instance, early developmental and connectional studies[13] contributed to the recognition that the accumbens is more like the striatum than the septum. These studies, however, were performed prior to the recognition that the striatum is divided into "patch and matrix" compartments with different developmental histories and that the accumbens is divided into core and shell regions.

In the caudate, and to a lesser degree in the putamen, a striking feature of intrinsic organization is the presence of acetylcholinesterase-poor striosomes or opiate receptor–rich patches surrounded by a complementary stained "matrix."[71–73] These compartments appear to bear a regular relationship to the input-output structure of the dorsal striatum; for example, there are patches that target tegmental dopamine neurons. Studies examining the ontogeny of opiate receptor expression provide a means of tracking the development of striatal patches (striosomes) and help explain the adult configuration of these important structures.

The recent development of antibodies to the mu opiate receptor has facilitated immunohistochemical demonstration of the patch compartment (Figure 4–2) and helps to shed some light on the arrangement of these structures in the accumbens. Within the adult dorsal striatum, the patch compartment consists of two parts: elongated groups of neurons stained intensely but diffusely with mu receptor antibody, and a collection of similarly labeled neurons adjacent to the corpus callosum and external capsule, often termed the *subcallosal streak*. Within the ventral striatum, the core of the accumbens is partitioned into moderately stained and poorly stained regions, while much of the shell of the accumbens is diffusely immunoreactive. Along the most medial boundary of the shell is found a band of intense immunoreactivity somewhat reminiscent of the subcallosal streak in the dorsal striatum.

During development of the telencephalon, the lateral walls of the neural tube evaginate, forming mirror-image tubes straddling the diencephalon. Within each telencephalic tube, three broad regions of neuroepithelium, which surround a somewhat triangular-shaped lateral ventricle, can be delineated as precursors of large brain regions. The medial wall will give rise to the septum, hippocampus, and interconnecting fiber systems. The dorsal and dorsolateral walls will form the cerebral cortex, and the ventrolateral wall is thrown into two longitudinal ridges, the medial and lateral ganglionic eminences. It is generally agreed that the lateral ridge, also called the striatal ridge or eminence, will develop into the dorsal striatum, that is, the caudate-putamen (CPu), in the rat.[74]

However, the derivatives of the medial ganglionic eminence are still subject to debate. Recent atlases of neural development have designated the pallidum as a major derivative of the medial ganglionic eminence.[75,76] However, developmental studies of patterns of two major neuropeptide genes in the forebrain, preproenkephalin and preprotachykinin A, are consistent with the possibility that the medial ganglionic eminence gives rise to much of the extended amygdala,[77] as origi-

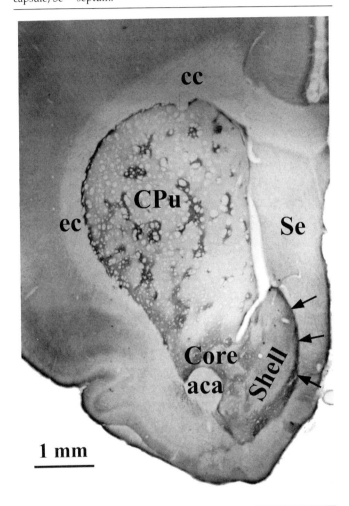

FIGURE 4–2. Photomicrograph of a section through the rat striatum stained immunocytochemically with an antibody to the mu opiate receptor. Note the intensely stained patches and subcallosal streak in the caudate-putamen (CPu), the moderately stained areas in the core and portions of the shell of the nucleus accumbens, and the intensely stained band along the medial edge of the shell *(arrows)*. aca = anterior limb of anterior commissure; cc = corpus callosum; ec = external capsule; Se = septum.

nally suggested by Johnston in 1923 (reviewed in Heimer et al.[66]).

These three broad regions of neuroepithelium differ in basic strategies of development. The rostral end of the medial wall undergoes an outside-to-inside pattern of migration; that is, the oldest septal neurons are found furthest from the ventricle, with younger neurons occupying positions progressively nearer the ventricle.[78] In the rat, the caudal end of the medial wall develops into the hippocampal formation through a complicated process of layering and infolding, and most neurons of the dentate gyrus are formed after birth.[79] Indeed, more recent studies have shown that, in the rat, dentate granule cells are continually born along the inner edge of the granule cell layer and migrate to the outer edge throughout life.[80] The dorsal and dorsolateral wall develop through an unusual inside-out process, whereby the oldest neurons occupy layers closest to the ventricle, forcing younger neurons to migrate through the older layer to reach more superficial layers of the cortex. As reviewed below, the lateral ridge of the ventrolateral wall (lateral ganglionic eminence) develops through a double migration, with early arising neurons first migrating to a position far away from the ventricle, then migrating back toward the ventricle. Migratory patterns of the derivatives of the medial ganglionic eminence are still being investigated.

On the basis of previous studies with ^{3}H thymidine that demonstrated that neurons of the patch compartment in general arise at an earlier time than the neurons of the matrix,[81] the origins of the patch and matrix compartments in striatum were reexamined by use of the thymidine analog bromodeoxyuridine (Brdu). This was injected into pregnant rats on either embryonic day 14 (E14) or E19 (the date of birth usually being E22). The earlier injection labels neurons in the patch compartment; matrix neurons are labeled by the later injection at E19. The E14-injected embryos were sacrificed either at 2 hours after injection or on E16, E19, or the day of birth (P0). E19-injected embryos were sacrificed either 2 hours after injection or on P0. Brains were removed, frozen, and processed for immunocytochemical localization of Brdu, as described previously.[82]

From these studies, our current model for the formation of the patch compartment in the dorsal striatum (Figure 4–3) has emerged. Specifically, the neurons that are the first to be formed migrate away from the neuroepithelium to occupy the developing field of the striatum. By E19, these neurons occupy much of the lateral and ventrolateral portion of the developing caudate-putamen. Neurons arising on E19 are restricted initially to the neuroepithelium but then begin to migrate ventrolaterally. By P0, these neurons have occupied essentially the entire caudate-putamen. In so doing, they appear to surround and push the early-arising neurons back toward the lateral ventricle. The earlier-arising neurons clump together to form the distinct striosomes of the patch compartment. Presumably blocked by the boundary formed by the external capsule, the later-arising E19 neurons are unable to surround and displace early-arising neurons lying adjacent to this developing fiber tract. Thus, these early-arising neurons form the subcallosal streak. The early-arising neurons undergo two migrations, one away from the ventricle and a second, probably passive, migration back toward the ventricle. This results in an actual decrease in the distance between the ventricle and the closest E14-labeled cells between days E19 and P0.[82]

As indicated above, the accumbens is also partitioned into mu receptor–rich and –poor areas, bounded medially by a mu receptor–rich stripe, perhaps analogous to the subcallosal streak. At E16, the lateral ventricle produces a deep crevice between the striatum and the septum (Figure 4–4A). The neuroepithelium on the lateral side of this slit is proposed to form the core of the accumbens, and the neuroepithelium on the medial and ventral sides likely produces much of the shell. At E16, neurons labeled at E14 fill all of the presumptive accumbens territory between the ventricle and the external surface of the brain (Figure 4–4B). By E19, the early-arising neurons have migrated out of the neuroepithelium but continue to occupy the entire region of the presumptive accumbens developing field (Figure 4–5C). The neuroepithelium at this age is filled with newly labeled cells (Figure 4–5B). Thus, neurons originating at

FIGURE 4–3. Schematic model of the development of the patch-matrix organization of the rat dorsal striatum. Left-side sections at embryonic day 14 (E14), E19, and the day of birth (P0) are shown. *Filled circles* represent cells labeled by injection of bromodeoxyuridine at E14; *open circles* represent cells labeled by injection at E19. The *dashed lines* indicate the ventrolateral border of the neuroepithelium, as determined by counterstaining with basic fuchsin. Originating in the neuroepithelium adjacent to the lateral ventricle (LV), neurons initially migrate ventrolaterally into the developing field. Between E19 and P0, later-arising neurons invade the territory of the earlier-arising cells, surrounding and pushing them back in groups toward the ventricle; these groups become the patches, characterized in adults by several markers, including mu opiate receptors. Some early-originating neurons remain adjacent to the external capsule (ec) and corpus callosum (cc), presumably because the later-arising neurons are unable to surround them owing to the presence of these fiber tracts; these early-arising neurons form the subcallosal streak underneath the external capsule.

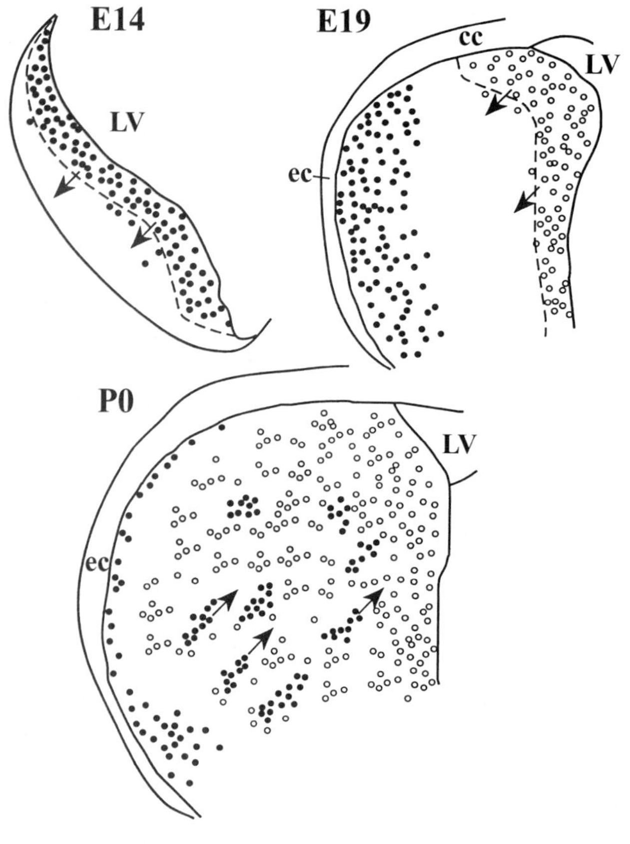

FIGURE 4–4. Camera lucida drawings of a section through the left side of an E16 rat brain, following injection of bromodeoxyuridine (Brdu) at E14. The rectangle in the low-power view (A) indicates the region drawn in panel B. The *dotted lines* in A indicate the approximate boundaries that divide the telencephalic tube into three major regions: the medial wall, which will produce the septum, hippocampus, and related fiber systems; the dorsal and dorsal and dorsolateral wall, which will produce the neocortex; and the ventrolateral wall, which will develop into the striatum. Notice the deep ventricular groove between the developing striatum and septum; the wall of this groove will form the nucleus accumbens, with the lateral wall forming the core and the medial wall forming much of the shell. The *dashed lines* represent the borders of the neuroepithelium. In panel B, Brdu-labeled cells are plotted. *Solid dots* represent cells with intense, uniform labeling throughout the nucleus; *open circles* represent cells with intense spots of immunoreactivity in the nucleus. Note that E14-labeled cells fill all of the neuroepithelium and the developing field out to the external surface of the brain. CPu = caudate-putamen; LV = lateral ventricle; Se = septum.

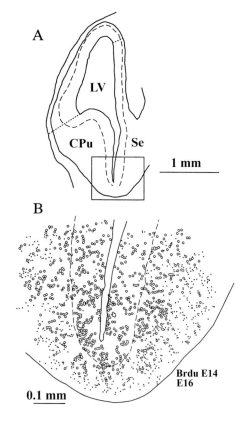

E14 and at E19 occupy exclusive but adjoining fields in the E19 accumbens rather than being separated by a substantial gap such as that found in the caudate-putamen at this age (Figure 4–3). By P0, the later-arising neurons have migrated out of the neuroepithelium to occupy most of the developing accumbens (Figure 4–6B). Within the core of the accumbens are groups of early-arising neurons, reminiscent of the patches of the caudate-putamen (Figure 4–6C); fewer such groups are found in the shell of the accumbens. Along the medial edge of the shell is a discontinuous band of early-arising neurons, although it is not as distinct as the subcallosal streak.

These results suggest that similar developmental pro-

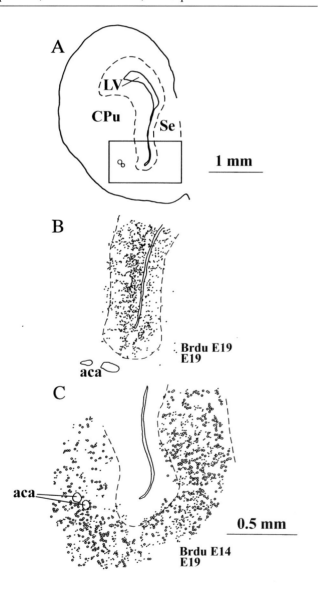

FIGURE 4–5. Camera lucida drawings of sections through the left side of E19 rat brains following injection of bromodeoxyuridine (Brdu) at either E19 (B) or E14 (C). The low-power drawing (A) is from the same section shown in panel C; the rectangle in A shows the area plotted in C. Symbols for labeled cells are as in Figure 4–3. Note that the E19-labeled cells are restricted to the neuroepithelium and the E14-labeled cells occupy the adjacent developing field.
aca = anterior limb of the anterior commissure; CPu = caudate-putamen; LV = lateral ventricle; Se = septum.

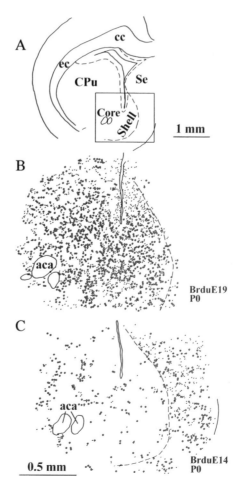

FIGURE 4–6. Camera lucida drawings of sections through the left side of P0 rat brains, following injection of bromodeoxyuridine (Brdu) at either E19 (B) or E14 (C). The low-power drawing (A) is from the same section shown in panel C; the rectangle represents the area plotted in C. Symbols for labeled cells are as in Figure 4–3. Note that the E19-labeled cells occupy the entire core of the accumbens but are rarely found in the septum (Se) or diagonal band of Broca. Far fewer E14-labeled cells are found in the accumbens. Within the core, note the groups of labeled cells, which may be similar to the groups of early developing neurons that form the patches in the caudate-putamen. Within the shell, labeled cells tend to be more isolated, although some accumulation along the medial boundary of the shell is evident. The solid line along the medial edge of the drawing in C represents the midline. aca = anterior limb of the anterior commissure; cc = corpus callosum; CPu = caudate-putamen; ec = external capsule.

cesses occur in the dorsal and ventral striatum, at least in the accumbens. In both regions, neurons originate in the neuroepithelium and migrate away from the ventricle into the developing field. In the caudate-putamen and, to some extent, in the core of the accumbens, later-developing neurons surround and push earlier-arising neurons in groups back toward the ventricle. Early-developing neurons form lateral (subcallosal streak) and, to some extent, medial boundaries of the striatum (medial edge of the accumbens shell). The lateral boundary, however, is delimited by the external capsule, whereas the medial boundary appears to be formed primarily by early-arising septal neurons, which may exert repulsive forces on the migration of the later-developing accumbens cells.

3. ENSEMBLES OF NEURONS IN THE ACCUMBENS: DIFFERENT COMBINATIONS OF INPUT AND A VARIETY OF TARGETS

Most, if not all, of the afferent fibers to the accumbens show heterogeneous patterns of distribution, and the accumbens' projection neurons that reach particular targets are also heterogeneously distributed. These inhomogeneous patterns can, to a large degree, be related to the distinction between the accumbens shell and core regions, as well as to patch-matrix patterns in the core of the accumbens. Furthermore, the nucleus accumbens is an integral part of a larger unit, the ventral striatum, and so there are few, if any, afferent projections that are restricted to the accumbens; most of the terminal fields are continuous dorsally into the ventral and medial parts of the caudate-putamen complex or ventrally into the striatal elements of the olfactory tubercle. In a caudal direction, terminal fields of several afferent systems are continuous into various parts of the extended amygdala, including the bed nucleus of the stria terminalis, the subpallidal portions of the extended amygdala, and the interstitial nucleus of the posterior limb of the anterior commissure (see section 6). Likewise, populations of projection neurons in the accumbens in most instances blend into neighboring regions of ventral striatum and/or extended amygdala.

Functional Aspects of Accumbens Neurons

Medium-sized spiny neurons form the major class of neurons (around 95% in the rat and 70% in the primate) of the striatum, and, as in the dorsal striatum, these cells constitute the output neurons of the accumbens. The remaining accumbens neurons are larger and aspiny or sparsely spined interneurons. They include the cholinergic interneurons and a variety of peptidergic interneurons,[83] as well as a less well studied complex of smaller neurons that generally occur more frequently medially, ventrally, and along the margins of the striatum.[14,84-86]

The medium-sized projection cells are among the most densely spined neurons in the brain, which suggests that a major function of these cells is the integration of information from different sources. Physiologically, medium-sized spiny neurons tend to be "silent," and it has been shown that a large proportion of these cells, like their counterparts in the dorsal striatum,[87] exhibit two distinct states in their resting membrane potential: an "up state" with a resting membrane potential of approximately –55 mV, close to the threshold for generating action potentials, and a hyperpolarized "down state" with a resting membrane potential of –85 to –90 mV.[35,88]

When these neurons are in their up state, they may generate bursts of spikes. Neurons that are in the hyperpolarized down state must first be brought into the up state, as for example by the activity of one of the excitatory inputs of the accumbens, before firing of these neurons can be elicited, as for example by a second excitatory input. Lesions of the cerebral cortex may cause medium-sized striatal neurons to be in a permanent down state,[87] whereas lesions of the fornix-fimbria, which interfere with input from the hippocampus formation, may have similar effects on accumbens neurons.[88] When the latter effects occur, excitation of prefrontal afferents of the accumbens does not lead to firing of accumbens output neurons, as is normally the case. This has led O'Donnell and Grace[88] to suggest that the hippocampal afferents "gate" the prefrontal throughput in the accumbens.

Similar gating mechanisms have recently been suggested for the amygdala and prefrontal afferents at the level of the accumbens.[89] Interactions between hippocampal and amygdaloid inputs of the accumbens, with respect to both short-term (paired pulse facilitation) and long-term (long-term potentiation and long-term depression) plastic processes, have also been described.[90]

Specific Relationships of the Thalamic and Amygdaloid Afferents With the Prefrontal Cortex–Ventral Striatal System

Considering these interactions, it may be assumed that the activity in a combination of distinct afferent sources will to some degree determine the output of the accumbens. As indicated earlier, excitatory inputs to the rat accumbens originate in allocortical, periallocortical, and proisocortical areas, as well as in the midline thalamic nuclei and basal amygdaloid complex.[91] In the present context, the organization of the midline thalamic, basal amygdaloid, and hippocampal inputs will be described in some detail. For all three sources of input to the accumbens, it has been shown that they also project, in a topographical fashion, to proisocortical prefrontal areas.[92-96] Furthermore, the midline thalamic and the basal amygdaloid projections have been clearly shown to be organized so that specific parts of these nuclei project to both an area in the prefrontal cortex and a region in the ventral striatum (including the accumbens) that are, in turn, connected by way of prefrontal corticostriatal projections (Figure 4–7A and B).[21,95,96] Through such an arrangement, a particular region of the ventral striatum may be reached by excitatory inputs from either of two sources (see Figure 4–7B): a specific midline thalamic nucleus, such as the paraventricular nucleus, or a specific part of the basal amygdaloid complex, such as the caudal parvicellular basal nucleus. These inputs can occur both directly and indirectly through thalamocortical or amygdalocortical and subsequent prefrontal corticostriatal projections, as for instance from the ventral prelimbic and infralimbic areas (see Figure 4–7B). The midline paraventricular thalamic nucleus, in addition, projects to the caudal basal amygdaloid complex, which provides this nucleus yet another indirect route to reach the same region in the ventral striatum (Figure 4–7B).

These arrangements of connections between the midline thalamus, the basal amygdaloid complex, and the prefrontal cortex–ventral striatal system indicate

FIGURE 4–7. Schematic representation, in two sagittal views of the rat brain, of the relationships of the ventral striatum. In A, the relationships of the intermediodorsal thalamic nucleus (IMD) with the lateral prefrontal cortex (agranular insular area; AI) and the lateral part of the nucleus accumbens (Acb lat) are shown. These parts of the prefrontal cortex and the nucleus accumbens are connected to each other by way of corticostriatal projections. The IMD projects also to the rostral part of the basal amygdaloid complex, in particular to the rostral magnocellular basal nucleus (Bmg), which in turn projects to the lateral parts of the prefrontal cortex and the nucleus accumbens. In B, similar arrangements are depicted for the paraventricular thalamic nucleus (PV), the medial part of the prefrontal cortex, the medial part of the nucleus accumbens (Acb med), and the caudal parvicellular basal amygdaloid nucleus (Bpc). CPu = caudate-putamen complex, dorsal striatum; MD = mediodorsal thalamic nucleus; PL/IL = prelimbic/intralimbic; SNr = substantia nigra pars reticulata; VP = ventral pallidum.

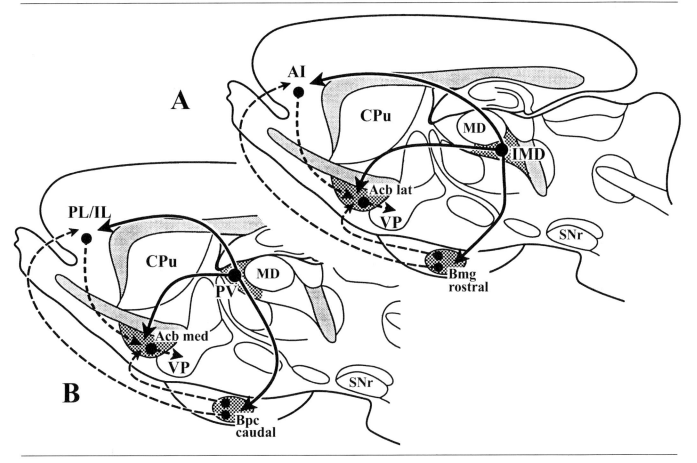

FIGURE 4–8. Color photomicrographs of double-immunostained sections, showing the relationships of afferent fiber systems with each other or with the output neurons of the accumbens. A: Transverse section through the caudal accumbens showing fibers from the caudal basal amygdaloid complex *(black)* with those from the midline paraventricular thalamic nucleus *(brown)*. In B, a higher magnification of part of this section is shown; *asterisk* and *arrow* indicate similar positions in A and B. Note that in the dorsomedial part of the accumbens, close to the septum (Se), fibers from the amygdala and the thalamus overlap, whereas more ventrally there exists an almost complete segregation between the two fiber systems, most clearly illustrated in B. Areas avoided by the brown thalamic fibers are specifically innervated by the black amygdaloid afferents *(asterisk and arrow)*. C: Photomicrograph shows the relationship of the afferents from the caudal basal amygdaloid complex *(black fibers)* with the distribution of immunoreactivity for the calcium-binding protein calbindin D_{28K}, which marks the shell/core division of the accumbens *(arrowheads)*. D: Transverse section through the rostral part of the accumbens, double-stained for enkephalin immunoreactivity (to show the accumbens compartments) and neurons retrogradely labeled from a small injection of a retrograde tracer in the ventral pallidum. Note the clustering of retrogradely labeled neurons (in this case outside the enkephalin-positive patches *[asterisk]*). E: Transverse section from the ventromedial accumbens, double-stained for anterogradely labeled fibers from the basal amygdaloid complex *(black)* and neurons retrogradely labeled from an injection in the mesopontine tegmentum. Note the close association of the labeled fibers and the clusters of projection neurons. aca = anterior limb of the anterior commissure; CPu = caudate-putamen; LV = lateral ventricle.

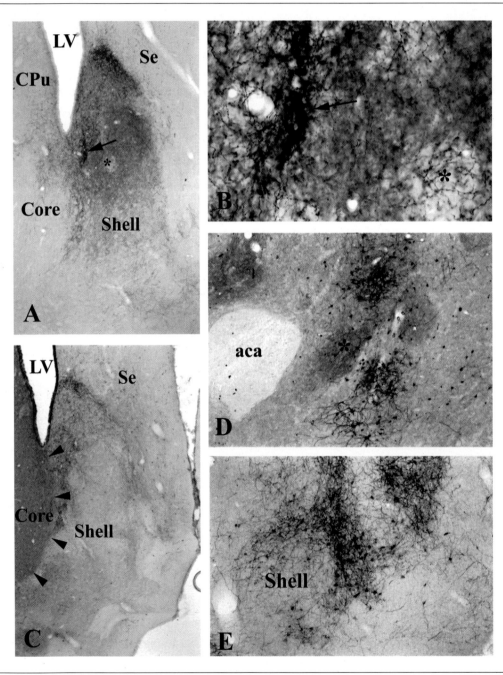

that these structures are interconnected in a distributed way.[25] Activity in the midline thalamic and basal amygdaloid nuclei may in this way have a profound influence on the activity of the prefrontal cortex–ventral striatal system. Determining whether the hippocampus is involved in a similar way in this distributed network will require further detailed study.

Convergence and Segregation of Afferents of the Accumbens

The previous subsection describes some of the general "rules" in the connections between different structures that project to the accumbens; the precise patterns of overlap and segregation of these afferents appear to be rather complicated. In general, the projections to the nucleus accumbens from the cerebral cortex, the hippocampal formation, the basal amygdaloid complex, and the midline thalamus are topographically organized.[39,97–100] For example, the ventral subiculum of the hippocampal formation projects to the medial part of the accumbens, most dominantly in the caudomedial shell, whereas progressively more dorsal parts of the subiculum project to successively more lateral and rostral parts of the accumbens.[99] Likewise, the projections from the basal amygdaloid complex show a mediolateral topography: the caudal part of the parvicellular basal nucleus projects to the caudomedial accumbens (Figure 4–8C), whereas, at the other extreme, the rostral part of the magnocellular basal nucleus projects primarily laterally. The hippocampal projections are more restricted (largely to the shell and to the medial and rostrolateral parts of the core[99,101]) than those of the basal amygdaloid complex, which involve the shell, the entire core, and extensive parts of the caudate-putamen complex.[100,102] Therefore, only in a rather restricted part of the accumbens—predominantly in the medial and rostral shell and in the medial core—may interactions between hippocampal and amygdaloid inputs take place. Furthermore, in view of the topography in both afferent systems, only specific subsystems of these afferents may converge; for example, the ventral subiculum and the caudal part of the basal amygdaloid complex may converge in the caudomedial accumbens, whereas the dorsal subiculum and the rostral basal amygdaloid complex both target the rostrolateral part of the accumbens. Even within these regions of the accumbens, which receive terminations from both the hippocampus and the amygdala, the terminal fields do not necessarily completely overlap, such that areas with particular cytoarchitectural or immunohistochemical characteristics may receive inputs either from the hippocampus or from the amygdala, but not from both.[103]

Similar patterns of overlap and segregation have been found for other sets of afferents of the accumbens. For example, in the caudomedial part of the shell, the afferents from the ventral prelimbic area and the caudal basal amygdaloid nucleus overlap extensively, but the terminal fields of these two afferents are largely segregated from the projection area occupied by the fibers from the paraventricular thalamic nucleus (Figure 4–8A and B). However, all three afferents converge in the patches of the medial part of the core (Figure 4–9).[104] In part, the specific areas of convergence and segregation can be recognized in the cytoarchitecture or chemoarchitecture of the accumbens, for example in cell-dense regions (within which appear clear cell clusters) or cell-poor regions in the shell.[104]

Neuronal Ensembles in the Accumbens: Projections to a Multitude of Targets

The above data on the topographical organization of different afferent fiber systems suggest that within the accumbens a patchwork of heterogeneously distributed inputs from various sources exists. The results of injections with retrograde tracers in target areas of the accumbens have shown that the output neurons are inhomogeneously distributed and, to a certain degree, organized in clusters (Figure 4–8D and E).

FIGURE 4–9. Convergence and segregation of accumbens afferents. Schematic representation of the relationships of three fiber systems of the medial part of the accumbens: the prelimbic area (PL) of the prefrontal cortex, the midline paraventricular thalamic nucleus (PV), and the caudal part of the basal amygdaloid complex (BA). *Black dots* mark a cell cluster zone as seen in Nissl-stained sections; *gray shades* indicate the immunoreactivity patterns of the calcium-binding protein calbindin D_{28K} in the shell and core. Prelimbic cortical and caudal basal amygdaloid fibers project to the cell cluster zone. Paraventricular thalamic fibers project to the moderately cellular zone and show a light, homogeneous calbindin immunoreactivity. All three afferent systems converge in the patches of the core, which are characterized by light calbindin staining. Note that the deep and superficial layers of the prelimbic area project to different areas in the accumbens. aca = anterior limb of the anterior commissure. Modified from Wright and Groenewegen.[104]

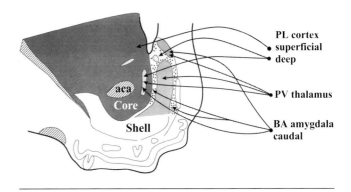

An important question is to what extent these clusters of output neurons receive specific sets of inputs; that is, to what extent are there specific input/output channels through the accumbens? Preliminary results from experiments in which injections of anterograde tracers in one of the input structures of the accumbens were combined with injections of retrograde tracers in one of its targets indicate that, indeed, clusters of output neurons may receive rather specific sets of inputs (Figure 4–8E).[103] These anatomical results support the idea that the accumbens consists of a collection of neuronal ensembles that may be differentially recruited under different functional or behavioral circumstances.[35]

The output from the neuronal ensembles in the accumbens has the potential to directly engage a number of functional-anatomical circuits. Via output to the ventral pallidum, which in turn projects to the mediodorsal thalamus, the accumbal ensembles have direct access to the thalamo-prefrontal circuitry.[21,28] Projections from, especially, the shell of the accumbens to the central division of the extended amygdala and the lateral hypothalamus[31,38,105] provide access to a multitude of autonomic and somatomotor targets, which are part of the anatomical substrate for goal-directed behaviors. Projections from the accumbens to the basal forebrain corticopetal cholinergic projection neurons[106,107] may provide a means for accumbens to affect cortical arousal, attention, and cognitive functions.[58,108] As discussed in more detail in section 5, the accumbens also projects to the dopaminergic neurons in the ventral tegmental area and the pars compacta of the substantia nigra, and it may also have input to the pars reticulata, either directly or indirectly via a relay in the ventral pallidum.[109–112]

This short review of the accumbens' most prominent output channels gives an idea of the many functions and behavioral acts that are likely to be controlled or activated by the neuronal ensembles in the accumbens. To what extent such neuronal ensembles indeed exist as distinct anatomical and functional units remains to be seen. How and under what circumstances are these neuronal ensembles activated, and can they influence each other?

More detailed anatomical studies of the type described in this section would facilitate the solution to these questions, and such investigations must also be carried to the ultrastructural level in order to investigate potential convergence of various afferents on single projection neurons in the accumbens.[113–115] In fact, the elucidation of the microanatomy of the accumbens and of the rest of striatum is an essential component in efforts to understand how information is being processed in the basal ganglia.[116,117]

On the functional level, a multidisciplinary approach is needed, as indicated by Pennartz et al.,[35] in which pharmacological, electrophysiological, and behavioral experiments are used in different combinations and under different conditions (for example, in live, freely moving animals, or in brain slices). When determining the actual placement for electrodes or pipettes used in these experiments, it is no longer enough to refer to rostral versus caudal, or lateral versus medial accumbens; the placement of the electrode or pipette tip will have to be described in relation to the various subdivisions of the accumbens as revealed in histological or histochemical stains.

4. THE CHAMELEON-LIKE NATURE OF THE SHELL OF THE ACCUMBENS

A conventional Nissl-stained preparation (Figure 4–10A) demonstrates that it is difficult to distinguish boundaries between any of the striatal components. As a chameleon blends into its background, the accumbens appears to blend into the rest of the striatum. Careful inspection, however, reveals some subterritorial patterns of organization even in Nissl-stained material. For example, the lateral part of the shell is characterized by lower cell density (arrows in Figure 4–10A) than the adjacent core. On the other hand, by examining preparations stained with substance P immunoreactivity (Figure 4–10B), one can make the point that the shell is a rather uniform structure having rather well-defined boundaries, except perhaps towards parts of the olfactory tubercle and the extended amygdala (see section 6).

Although the histochemical differences between the core and the shell, and within the shell itself, will be emphasized in this section, it is also important to point out that in some histochemical stains it is difficult or even impossible to detect a difference between core and shell. Although the case is somewhat exceptional, this is for instance true in regard to binding sites for amylin.[118] The histochemical stain for this peptide, incidentally, tends to show that the central division of the extended amygdala, or at least its IPAC component, is closely related to the accumbens. (IPAC abbreviates "interstitial nucleus of the posterior limb of the anterior commissure"; see section 6.)

That certain neurochemical consistencies exist throughout the shell, distinguishing it from most surrounding structures, is also shown in a preparation where the monoamine-selective neurotoxin 6-hydroxydopamine (6-OHDA) was injected into the ventral teg-

FIGURE 4–10. Coronal sections through the rat brain that illustrate organizational features of the accumbens at a level where both core and shell are present. A: Nissl-stained preparation. Note that striatal structures, including the accumbens core and shell, caudate-putamen (CPu), and olfactory tubercle (Tu), exhibit similar cell densities and are not clearly delimited from each other. Careful observation distinguishes subterritories: a particularly conspicuous boundary is evident between the core and lateral shell, which has a noticeably lesser cell density *(arrows)*. B: Substance P immunoreactivity (SP-ir). The shell exhibits moderately dense immunoperoxidase product that extends uniformly from its most dorsomedial to the most lateral parts. Whereas a sharp boundary exists between the core and shell *(arrowheads)*, it is difficult to identify a boundary between the core and the dorsally adjoining caudate-putamen. Note that a module lacking SP-ir is present dorsomedially *(asterisk)*. C: Tyrosine hydroxylase immuno- reactivity (TH-ir) 2 days following injection of 6-hydroxydopamine into the ventral mesencephalon. Note that immunoreactivity is preserved in the shell *(arrows)* and the medial part of the olfactory tubercle (Tu). D: Following the same lesion, the TH-ir innervation is preserved also in the lateral bed nucleus of stria terminalis and in the interstitial nucleus of the posterior limb of the anterior commissure (BSTL and IPAC) as well as in the central nucleus of the amygdala (not shown), in contrast to adjacent parts of the caudate-putamen. E: Calbindin D_{28K} immunoreactivity (CBP-ir). Staining is robust in the core, whereas the shell is characterized by relatively weak CBP-ir. The boundary between core and shell is indicated by *arrowheads*. CBP-ir intensity can be seen to be greater in the lateral *(arrow)* than in the medial part of the shell. F: Acetylcholinesterase (AChe) reaction, which demonstrates that the shell has a stronger reaction than the core. The staining is very strong in the dorsomedial and lateral parts of the shell, which have between them an intermediate sector with a reaction of moderate intensity. aca = anterior limb of the anterior commissure; lot = lateral olfactory tract; Se = septum; V = ventricle. The scale bar shown in A applies also to B–E.

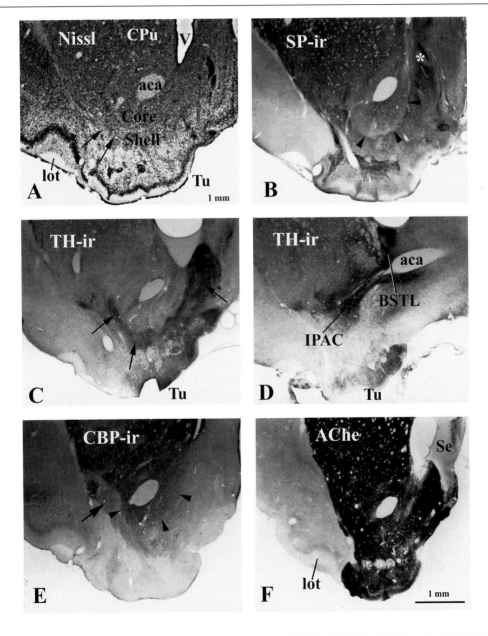

mental area and revealed a differential vulnerability of the tyrosine hydroxylase immunoreactive (dopaminergic) innervation in the core and shell, in the sense that the shell is more resistant than the core and the adjoining caudate-putamen[119] (Figure 4–10C). The same is true as well for the medial part of the olfactory tubercle (Figure 4–10C and D) and parts of the extended amygdala, including especially the lateral bed nucleus of stria terminalis and the interstitial nucleus of the posterior limb of the anterior commissure (BSTL and IPAC in Figure 4–10D), as well as the central amygdaloid nucleus (not shown). In other words, the caudomedial part of the shell blends into some other part of the basal forebrain without forming a clear border. The extended amygdala (including the IPAC) and its relation to the accumbens are discussed in section 6.

A similarly transient retention of shell dopamine innervation is observed following 1-methyl-4-phenyl-1,2,3,6-tetrahydropyridine (MPTP) administration,[120] ventral mesencephalic ibotenic acid injections,[119] and methamphetamine administration.[121] In the accumbens core and caudate-putamen, dopamine innervation is rapidly eliminated by the toxin, thus indicating a fundamental difference between the dopamine innervation in the shell of the accumbens and that of the core and caudate-putamen. Following ventral mesencephalic 6-OHDA lesions, the differential resistance of the dopamine innervation is not the same throughout the entire shell: the loss of fibers in the lateral part of the shell is more pronounced than in the medial part (Figure 4–10C). This suggests that the fundamental difference between the dopamine innervation in the shell and core may also to some extent characterize the dopamine innervations of the medial and lateral parts of the shell. It is noteworthy that the dopamine innervation of core and caudate-putamen in the vicinity of the lateral shell is rapidly lost following 6-OHDA administration.

As indicated above, few histochemical preparations actually reveal uniform neurochemical features throughout the shell.[122–124] Rather, most neurochemical markers of the shell tend in varying degrees to distinguish medial, lateral, and, occasionally, intermediate parts of the shell. Calbindin immunoreactivity, for example, is very light in the dorsomedial shell, but in more ventral and lateral parts of the shell that happen to be adjacent to other structures that stain very strongly for calbindin, namely the accumbens core and lateral parts of the caudate-putamen, the shell calbindin immunoreactivity also is somewhat stronger (Figure 4–10E). Thus, one observes neurochemical gradients in the shell, indicating heterogeneity among the different parts of the shell.

This point is nicely shown with acetylcholinesterase histochemistry, particularly if the reaction has been carried out with a light touch (Figure 4–10F). In a section that was incubated for a short time in substrate, acetylcholinesterase staining is exceedingly strong in the dorsomedial and lateral parts of the shell but much less prominent in the intermediate shell. In other words, acetylcholinesterase staining serves to identify the dorsomedial part of the shell as an area that is likely quite different from the remainder of the shell. In this regard, it is interesting to note that the caudomedial part of the shell is also the part that has many features in common with the extended amygdala. However, except in the IPAC component of the extended amygdala, which is directly continuous with the shell, the two structures do not share strong acetylcholinesterase reactivity (see section 6).

Response of the Shell to Antipsychotic and Psychoactive Drugs

The contrast between neurotensin negative and positive patches in the dorsomedial shell is enhanced by treatments with a variety of drugs, including antipsychotics. Figure 4–11A shows neurotensin immunoreactivity in the accumbens after administration of the antipsychotic drug haloperidol, which gives a clear image of the entire shell. In this kind of preparation, a dorsomedial part of the shell is observed in which the organization is conspicuously modular. Because the microenvironments seem to be so mutable under these circumstances, the studies by Groenewegen and his colleagues, which were discussed in the previous section, become especially relevant. The neurochemical environments of these specialized districts are to some extent distinct, while the boundaries between them are likely quite vague. Nevertheless, one can easily generate hypotheses regarding ways by which the different neurochemical environments in the subterritories might affect the functions of neurons. For example, the propensity of neurons to shift into the "up state" from the hyperpolarized state and hence increase the probability of firing (as reviewed in the previous section) could very well differ between these regions, which would certainly differentially regulate cortical throughput in these regions.

Laterally in the shell (Figure 4–11A), one also sees a clear, albeit more homogeneous, enhancement of the neurotensin staining pattern. Between the dorsomedial and lateral shell, there is an intermediate region within which input-output relationships are unclear. It is interesting that neurotensin immunoreactivity is enhanced considerably by typical antipsychotic drugs medially and laterally in the shell, but minimally in this intermediate region. This pattern is similar to the pattern observed in material exhibiting strong acetylcholinesterase activity (Figure 4–10F). It was shown that this intermediate region may correspond to an area of heightened D_3 dopamine receptor mRNA[125] and that haloperidol, putatively acting through the D_3 receptor, tends to blunt the synthesis of proneurotensin mRNA specifically in this region.[126] Thus, the possibility becomes real that the relatively weak neurotensin immunoreactivity in this intermediate part of the shell is related to a specific receptor interaction (see also Figure 11B in Zahm[127]).

The response of the immediate-early gene, c-*fos*, and its protein product, Fos, following administration of antipsychotic drugs[52,128–130] is a useful tool for identifying neurons that have been stimulated by the drug. The diagram in Figure 4–11B shows a large amount of the immunoperoxidase product for Fos in a rat sacrificed 2 hours after administration of haloperidol. Whereas Fos is virtually absent from the striatum of control brains, the haloperidol-treated animal exhibits intense Fos reaction in the dorsomedial shell as well as throughout the core and caudate-putamen, particularly dorsolaterally. Considering the recent advances in the treatment of schizophrenia based on the use of so-called atypical antipsychotic drugs, it is especially interesting to note that such drugs, as exemplified by clozapine, produce an equally robust response in the dorsomedial shell but fail to elicit significant Fos immunoreactivity in the caudate-putamen.[52,56,129]

A complementary picture emerges from preparations showing perikaryal neurotensin immunoreactivity, which is of considerable interest in view of the proposed transcriptional regulation of neurotensin by Fos[131] and a considerable literature suggesting that neurotensin and dopamine coregulate each other in the basal ganglia (recently reviewed in Lambert et al.[132]). Figure 4–11C shows that following administration of haloperidol, the distribution of neurotensin-immunolabeled neurons is very weak in the dorsomedial shell but is stronger laterally in the shell, which is the converse of what is observed for Fos immunoperoxidase product following the same drug treatment (Figure 4–11B). Inasmuch as Fos and neurotensin exhibit the most robust histochemically detectable responses so far observed in striatum following antipsychotic drug treatments, it is noteworthy that a complementarity characterizes their distributions and serves to further distinguish the dorsomedial and lateral part of the accumbens shell in functional-anatomical terms.

Following administration of amphetamine, which results in increased stimulation of striatal dopamine receptors, one observes moderate numbers of very strongly labeled neurotensin immunoreactive neurons in the lateral, but not the dorsomedial part of the shell (Figure 4–12A and B). This response can be blocked by a dopamine D_1 antagonist (Figure 4–12C). Yet again, something quite different appears to be happening in the medial and lateral parts of the shell. Although this response is not observed at all in the medial shell, it is seen in the rostral part of the olfactory tubercle and caudodorsal parts of the caudate-putamen (D. S. Zahm and M. A. Welch, unpublished data).

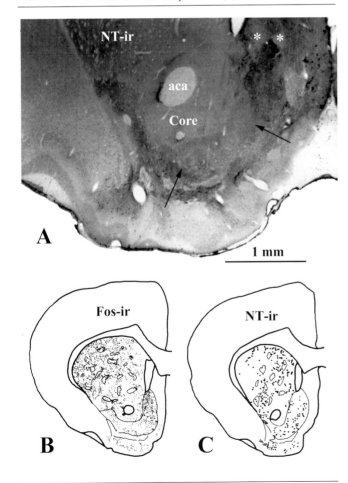

FIGURE 4–11. A: Neurotensin immunoreactivity (NT-ir) following administration of haloperidol. NT-ir is strong dorso-medially and laterally in the shell and weak to moderate in an intermediate zone of the shell *(between arrows)*. The dorsomedial sector contains modules with weak NT-ir *(asterisks)*. B and C: Diagrams of frontal sections through the rat brain showing the distributions of Fos (Fos-ir; B) and neurotensin (C) immunoreactive neurons at 2 and 24 hours, respectively, following administration of haloperidol (2 mg/kg). Neither marker is present in appreciable quantity in control brains. Note that haloperidol induces robust Fos immunoreactivity in the dorsomedial shell and dorsolateral caudate-putamen (B). Following haloperidol administration, NT-ir neurons are more numerous in lateral parts of the shell and in a band extending from the dorsomedial to the ventrolateral caudate-putamen (C).

FIGURE 4–12. A: Neurotensin immunoreactivity (NT-ir) following administration of *d*-amphetamine (4 mg/kg at 4, 14, and 24 hours prior to sacrifice). The shell exhibits dorsomedial and lateral sectors in which the immunoreactivity is strong and an intermediate sector in which it is moderate. Also note the numerous immunoreactive cell bodies in the lateral shell, which are absent from control brains. The dorsomedial shell exhibits modules with weak immunoreactivity *(asterisks)*. B: Enlargement of the lateral shell, showing NT-ir neurons *(arrows)*. C: Section through the lateral shell (corresponding approximately to the area shown in B) from brain treated with *d*-amphetamine and SCH 39166, a dopamine D_1 receptor antagonist. Note relative paucity of NT-ir neurons. aca = anterior limb of the anterior commissure.

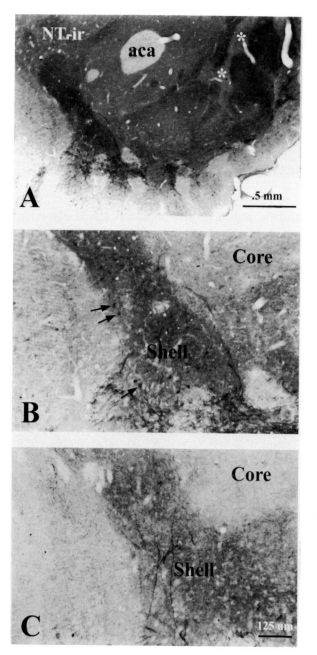

5. ORGANIZATION OF THE PRIMATE STRIATUM AND ITS RELATION TO THE MESENCEPHALIC DOPAMINE NEURONS

The Ventral Striatum in the Primate

In the primate brain, including that of the human, the accumbens is more or less synonymous with the fundus striati, which appears as a broad continuum between the caudate nucleus and the putamen underneath the rostral part of the internal capsule (Figure 4–13). The accumbens of the primate, like that of the rat, is a prominent part of the ventral striatum, which in addition includes the rather ill-defined primate olfactory tubercle, as well as the rostroventral parts of the caudate nucleus and putamen.[22] The primate striatum, like the striatal complex in the rat, may be divided into separate domains based on cortical input (Figure 4–13). Cortical afferents to the ventral striatum (light gray in Figure 4–13; see also Haber et al.[22]) originate in allocortical and periallocortical regions (hippocampal formation, entorhinal area, and olfactory cortex), proisocortical regions (anterior cingulate gyrus, orbitofrontal and insular regions), the cortical-like basal amygdaloid complex, and some isocortical orbitofrontal and inferior temporal gyrus association areas. The accumbens in

FIGURE 4–13. Schematic representation of the "functional" map of the striatum based on cortical input. Levels of overlap are indicated by intermediate shades of gray. *Light gray:* input from allocortical and periallocortical regions (hippocampus formation, entorhinal area, and olfactory cortex), proisocortical regions (anterior cingulate gyrus, orbitofrontal and insular regions), and some isocortical orbitofrontal and inferior temporal gyrus association areas. All of these areas are sometimes referred to as "limbic-related cortex." *Medium gray:* input from a wide range of association cortices. *Dark gray:* input from the sensorimotor cortex, the supplementary motor area, and the frontal eye field.

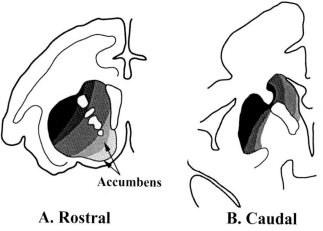

primates, like that in rats, is divided into core and shell subterritories. The core is continuous with the rest of the striatum, whereas the shell, which can be distinguished from the rest of the ventral striatum by its relatively low levels of calbindin (CaBP) immunoreactivity,[64] surrounds the core at its medial and ventral borders.

The cingulate and prefrontal-orbitofrontal cortical input, as well as the thalamic afferents to the core and shell of the accumbens and to the rest of ventral striatum, in the primate have recently been described in considerable detail in a series of papers by Haber and her collaborators,[22,133,134] which should be consulted for further details. The dorsolateral part of the striatum receives inputs from sensorimotor cortex, supplementary motor area, and the frontal eye fields (dark gray in Figure 4–13). The large central area of the rostral striatum and much of the caudate nucleus posterior to the commissure receive inputs from a wide range of association cortices, as well as direct afferents from primary and secondary visual and auditory cortices. For a discussion of the primate corticosubcortical reentrant circuits via the basal ganglia and the thalamus back to cortex, the reader is referred to Alexander et al.[16,17] and to Hoover and Strick.[135] In the following segments we will turn our attention to the mesotelencephalic dopamine system and its interaction with the accumbens and the rest of the striatum.

Mesencephalic Dopamine Neurons

As indicated in Figure 4–14, the dopamine cells of the substantia nigra pars compacta (SNc) are closely associated and almost imperceptibly merge with the immediately adjacent dopamine cell groups of the ventral tegmental area. Based on the phenotypic characteristics, the midbrain dopamine neurons can be divided into a dorsal tier and a ventral tier. The dorsal tier, whose cells are CaBP positive, includes both the dorsal substantia nigra pars compacta and the contiguous ventral tegmental area. The ventral tier includes cells in both the densocellular and the ventral group that form cell columns penetrating deep into pars reticulata. The ventral tier cells are CaBP negative and have relatively high levels of expression for dopamine transporter and the D_2 receptor. The main projections of both groups of dopamine neurons are to the striatum and to cortex. The dorsal group of the SNc is composed of loosely arranged cells, extending dorsolaterally to circumvent the ventral and lateral aspects of the superior cerebellar peduncle and the red nucleus. These dorsal neurons are oriented horizontally, just dorsal to a dense cluster of neurons referred to as the densocellular region, and form a continuous band with the ventral tegmental area. The den-

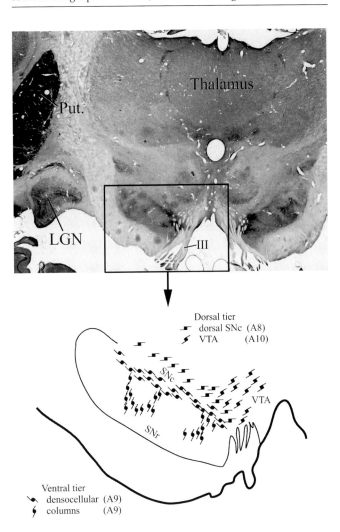

FIGURE 4–14. Schematic drawing of the base of the left cerebral peduncle to show the distribution of dopaminergic cells in the substantia nigra and ventral tegmental area. Put. = putamen; LGN = lateral geniculate nucleus; SNc = Subtantia nigra pars compacta; SNr = substantia nigra pars reticulata; VTA = ventral tegmental area.

drites of this dorsal group stretch in a mediolateral direction, and they do not extend into the ventral parts of the pars compacta or into the substantia nigra pars reticulata. In contrast, the dendritic arborizations of the densocellular region are oriented ventrally and penetrate the major portion of the pars reticulata in primates.

The Mesotelencephalic Dopamine Projections

The projections of the dopamine neurons to the striatum are more loosely organized than the corticostriatal projections with respect to their functional domains. The dorsal tier projects to the ventral striatum.[62]

FIGURE 4–15. Composite drawing of the midbrain projection to the striatum in two rostrocaudal views. Whereas the dorsal tier of dopaminergic neurons projects to the ventral striatum, the densocellular part of the ventral tier projects throughout the striatum. ac = anterior commissure; C = caudate; ic = internal capsule; Pu = putamen; SNc = substantia nigra pars compacta; VP = ventral pallidum; VTA = ventral tegmental area.

FIGURE 4–16. Schematic figure to show how the ventral striatum might be able to modulate the dorsal striatum by projecting to substantia nigra neurons *(diamonds)* that in turn project to dorsal striatum. ac = anterior commissure; C = caudate; Pu = putamen; SN = substantia nigra; VP = ventral pallidum.

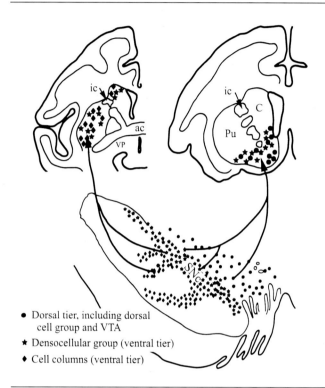

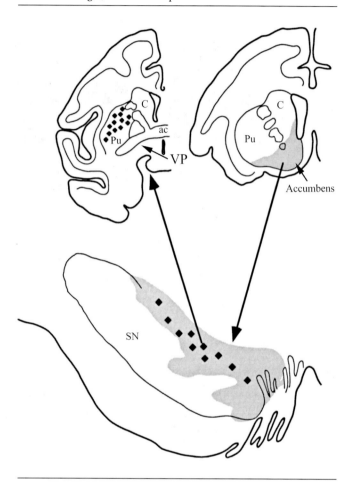

The cell columns that penetrate into pars reticulata project primarily to the dorsolateral (sensorimotor) striatum. The densocellular part of the ventral tier, however, projects throughout the striatum (Figure 4–15). In particular, the central part of the densocellular region of the ventral tier projects to both the ventral and dorsal parts of the striatum. Within the center of this midbrain region, there is an intermingling of cells that project to different striatal territories, thereby allowing a large area of the striatum to be modulated by inputs from this midbrain region. Therefore, in contrast to cortical projections to the striatum, there is not a simple point-to-point relationship between different groups of substantia nigra neurons and the different functional domains of the striatum. The majority of midbrain dopamine cells projecting to the cortex arise from both the ventral tegmental area and the dorsal group of the pars compacta throughout its rostrocaudal extent.[136,137] The densocellular region and the ventral groups of the substantia nigra do not appear to project heavily to cortex. Thus the dorsal tier, but not the ventral tier, of midbrain neurons gives rise to the dopamine-cortical projection.

In summary, the dorsal tier neurons project to a limited region of striatum—in particular, to the ventral striatum. However, the dorsal tier neurons project widely throughout cortex. In contrast, the ventral tier cells project widely throughout the striatum but do not project to cortex.

Projections to the Midbrain Dopamine Neurons

The striatum projects massively back to the substantia nigra[138] (see also the review by Alheid et al.[14]). Ventral striatonigral inputs terminate in the medial pars reticulata and much of the densocellular part of the pars compacta. It is important to note that the ventral striatal fibers terminate throughout a wide mediolateral extent of the densocellular region of the ventral tier.[139,140] Thus, the ventral striatal input to the substantia nigra is in a position to influence a large population of dopamine neurons that in turn project to the dorsal striatum (Figure 4–16). This important concept was first developed on the basis of experimental studies in the rat.[32,141] The dorsolateral striatonigral inputs to the substantia nigra, on the other hand, terminate ventrally in the pars reticulata, and therefore are not in a position to influence a large population of dopamine neurons.[32,142] Descending projections from the central nucleus of the amygdala also terminate in a wide mediolateral region of dopamine cells, primarily in the dorsal pars compacta and in the densocellular regions.[143] In addition, cells from the bed nucleus of the stria terminalis and the sublenticular part of the extended amygdala project to the dorsal tier (S. N. Haber, unpublished observations). Thus, like the ventral striatum, the extended amygdala may therefore be able to influence a wide range of dopamine neurons, and it may thus have direct access to striatal and cortical structures by way of the mesencephalic dopamine neurons (see also Gonzales and Chesselet[144]).

The simple scheme shown in Figure 4–16 illustrates a way whereby pieces of information that originally come from different parts of the cerebral cortex can be integrated via a nigrostriatal feedback loop. However, there are other possibilities whereby functionally diverse information might be integrated in the basal ganglia, such as by local circuit neurons within the striatum or via "open" corticosubcortical reentrant circuits through the basal ganglia.[109,145] Yet another possibility has recently been proposed by Bevan and his collaborators,[109] who traced descending projections from dorsal and ventral pallidal territories to individual neurons in substantia nigra in a combined light-electron microscopic study. They found that individual neurons in both the pars compacta and the pars reticulata do receive input from both the dorsal and the ventral pallidum. In other words, stimuli from different parts of the striatopallidal system may converge on single neurons in the substantia nigra. However, some caution is needed, since it is difficult to make tracer injections in the ventral pallidum without involving neurons in the extended amygdala. (See Figure 6 in Heimer et al.[31])

6. ACCUMBENS AND EXTENDED AMYGDALA

Although the amygdaloid complex is often treated as a monolithic structure in summary behavioral assessments of forebrain function, it is a highly differentiated region with many nuclei (reviewed by Alheid et al.[59]), and ample evidence suggests distinct functions for subgroups of these nuclei. The nuclei of the *basolateral complex* (lateral, basolateral, and basomedial nuclei), for example, may function to some degree as cortical areas. Although lacking a clear laminated appearance, they appear to be made up of generally pyramidal-like cells with a neurohistochemical makeup similar to the nearby areas of cortex. The basolateral complex shares reciprocal connections with a variety of cortical areas and projects to dorsal and ventral striatum.[102] This projection might heuristically be considered as a kind of corticostriatal projection. Also contributing to the innervation of ventral striatum are various other cortical-like *superficial nuclei* such as the anterior cortical nucleus, the nucleus of the lateral olfactory tract, and the posterior cortical nuclei. Distinct from the superficial nuclei and the basolateral complex are central and medial amygdaloid nuclei (the *centromedial amygdala*); these nuclei seem to be the main relays for amygdaloid information routed to the hypothalamus, although they do not have exclusive rights to this role. The central nucleus, moreover, is the main source of amygdaloid projections to autonomic and somatomotor areas of the brainstem.

Extended Amygdala

Over the past two decades, it has become increasingly clear that neuronal cell groups composing central and medial amygdaloid nuclei are not bounded by the arbitrary borders that are generally used to depict the limit of the amygdaloid body. Rather, more or less continuous columns of cells extend from these two nuclei through the sublenticular areas (just below the globus pallidus) and within the fascicles of the stria terminalis in order to merge with the bed nucleus of the stria terminalis. In fact, the symmetries between the cells, connections, and neurochemistry of the bed nucleus of the stria terminalis and the centromedial amygdala are so strong that it has become convenient to talk of the interconnecting cells and the bed nucleus of the stria terminalis as an extension of the amygdala into the basomedial forebrain,[60,146,147] an "extended amygdala."[58,59] This situation is illustrated in a schematic fashion in Figure 4–17. Insofar as the portions of this structure related to the central amygdala and those most closely related to the medial amygdala have distinctive efferent targets, it is practical to partition the extended amygdala into "central" and "medial" divisions (shown in yellow and green, respectively, in Figure 4–17), with the central division possessing the most direct projections to the lateral hypothalamus and brainstem (repre-

62   THE NEUROPSYCHIATRY OF LIMBIC AND SUBCORTICAL DISORDERS

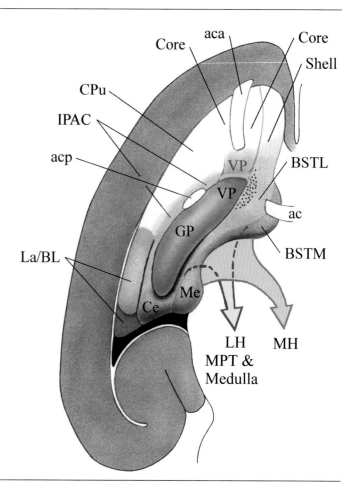

FIGURE 4–17. Schematic, partly three-dimensional drawing of the basal forebrain of the rat in horizontal view. The major subcortical forebrain structures—the striatum (*white*), represented by the caudate-putamen (CPu) and the core and shell of the accumbens; the pallidum (*pink*), represented by the globus pallidus (GP) and the ventral pallidum (VP); and the extended amygdala with its two subdivisions, the central (*yellow*) and medial (*green*) divisions—form large diagonally oriented "columns." Note that the core of the accumbens can generally be easily distinguished from the shell; there is a distinct border between the core and shell of the accumbens, whereas the core is directly continuous with the rest of the striatum without any clear border dividing the two parts. The *yellow*-colored central division of the extended amygdala continues without sharp boundary into the caudomedial part of the shell, which can be considered a transitional area between the striatum and the extended amygdala. The IPAC (interstitial nucleus of the posterior limb of the anterior commissure) component of the extended amygdala (shown in *yellow* since it appears to constitute a subdivision of the central extended amygdala) has traditionally been included in the caudate-putamen or striatum. The ventral extension of the globus pallidus (the ventral pallidum) is sandwiched between the striatum, the main part of the extended amygdala, and the rostral arm of the extended amygdala (the IPAC). In order not to compromise the rendition of the core–shell dichotomy of the accumbens, the rostroventral extension of the ventral pallidum into the deep part of the olfactory tubercle has only been implied with a gradually disappearing *pink* color underneath the core of the accumbens. The *broad arrows* illustrate how the two divisions of the extended amygdala are characterized by, in general, different efferent targets. Whereas the medial extended amygdala is closely related to the medial hypothalamus (MH), the central extended amygdala provides the main output to the lateral hypothalamus (LH) and to somatomotor and autonomic centers in the brainstem. aca = anterior limb of the anterior commissure; acp = posterior limb of the anterior commissure; BSTL = lateral bed nucleus of stria terminalis; BSTM = medial bed nucleus of stria terminalis; Ce = central amygdaloid nucleus; La/BL = lateral and basolateral amygdaloid nuclei; Me = medial amygdaloid nucleus; MPT = mesopontine tegmentum.

sented by the light yellow arrow in Figure 4–17) and with efferents from the medial division of the extended amygdala massively favoring the medial hypothalamus (represented by the light green arrow in Figure 4–17; the stria terminalis and its accompanying cell groups are not indicated in the schematic illustrations in Figure 4–17).

In the context of a review of the accumbens, it is the central division that is the most immediately relevant. Particularly similar to the shell of the accumbens,[38] the central division of extended amygdala makes extensive projections to the lateral hypothalamus and to the mesopontine tegmentum (see review by Alheid and Heimer[105]). Another similarity to the accumbens is that the central division of extended amygdala receives a dense innervation from the basolateral complex of the amygdala; however, as with the shell of the accumbens, this appears to preferentially originate in the posterior (parvicellular) part of the basolateral amygdala. The anterior (magnocellular) part of the basolateral amygdala seems to have a more exclusive relation to the dorsal striatum and to lateral portions of the ventral striatum.[102]

Transitions Between Accumbens and Extended Amygdala

One territory where the histochemistry of the accumbens and central division of extended amygdala seem to blend is in the caudal aspects of the accumbens, particularly its shell division. This zone, which is adjacent to the rostral face of the lateral bed nucleus of the stria terminalis (BSTL), is indicated by a gradually fading yellow color in Figure 4–17. The caudal aspects of the accumbens are also separated from the sublenticular portions of the extended amygdala by the ventral extension of the pallidum (VP in Figure 4–17). Histochemically, these areas are rather rich in acetylcholinesterase but somewhat lighter than the dorsal striatum generally or the more rostral portions of the shell of the accumbens; these caudomedial zones of accumbens are also immunoreactive for angiotensin II (Figure 4–18) and enriched in a number of neuropeptides (such as neurotensin, cholecystokinin, and opioid peptides), which also distinguish extended amygdala from nearby dorsal or ventral striatum. The presence of vasopressin/oxyto-

FIGURE 4–18. A series of coronal sections through the rat forebrain stained for angiotensin II to show the strong immunoreactivity in the caudomedial shell of the accumbens (A and B) and its continuity with interstitial nucleus of the posterior limb of the anterior commissure (IPAC in C) and the bed nucleus of stria terminalis (BST in D) at the level of the crossing of the anterior commissure (ac). aca = anterior limb of the anterior commissure; acp = posterior limb of the anterior commissure; cc = corpus callosum; Cpu = caudate-putamen; GP = globus pallidus; LV = lateral ventricle; Se = septum; Tu = olfactory tubercle.

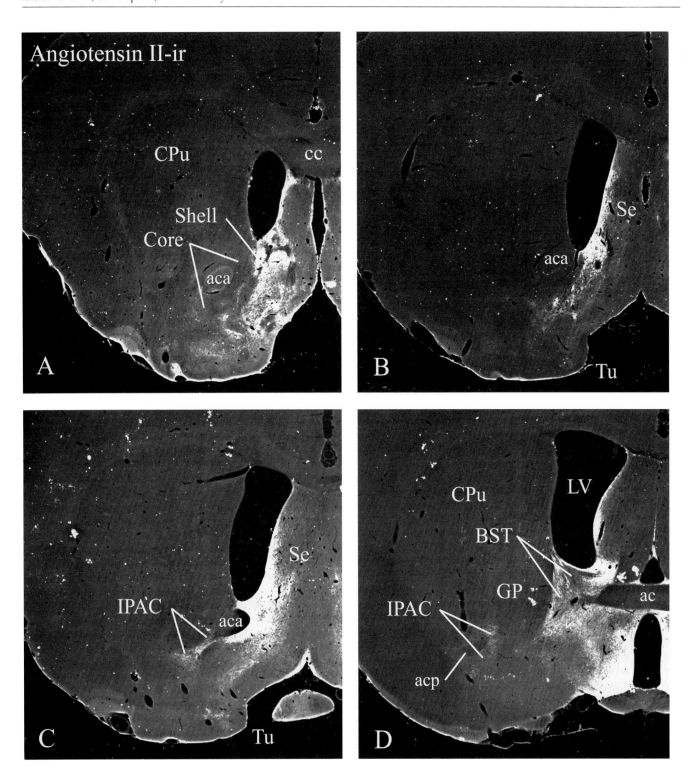

FIGURE 4–19. Anterograde labeling of the accumbens after phyto-hemagglutin-stimulated lymphocyte (PHA-L) injections in the extended amygdala. A: PHA-L injection in the lateral bed nucleus of the stria terminalis (BSTL). The injection site is shown in the lower inset; the upper inset is a wider view of the area of termination. B: PHA-L injection in the interstitial nucleus of the posterior limb of the anterior commissure (IPAC). The injection site is again shown in the lower inset, with a wider view of the section above. aca = anterior limb of the anterior commissure.

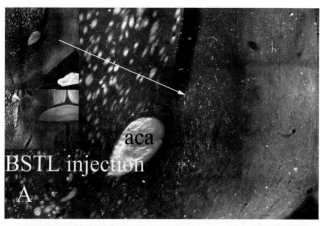

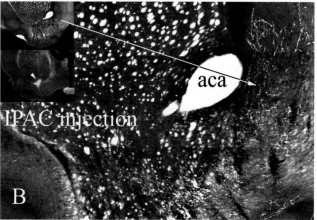

cin receptors[148,149] also distinguishes the caudomedial accumbens and the extended amygdala from the rest of striatum.

In terms of connections, these caudal shell areas are closely interconnected with the central division of the extended amygdala. Projections from the caudal shell regions reach the lateral bed nucleus of the stria terminalis, the sublenticular portions of extended amygdala, and the central nucleus of the amygdala.[32,38] The extended amygdala appears to reciprocate these projections: the lateral bed nucleus of stria terminalis and the IPAC component of the extended amygdala project to the caudal shell of the accumbens medially and ventrally (Figure 4–19), as well as laterally across the rear face of the accumbens, and the central amygdaloid nucleus projects with a less dense projection medially in the shell.[39] Because of the relay to the amygdala, these caudal areas of the accumbens are in a position to funnel information from medial prefrontal cortex, insular cortex, and hippocampal formation to the central division of the extended amygdala. It also is notable that afferents relayed to the central division of the extended amygdala through these caudal areas of the accumbens shell should be strongly affected by the dense dopaminergic innervation of these areas, in addition to the conspicuous, but less dense, dopaminergic terminations seen in the lateral bed nucleus of the stria terminalis, in the central division of the sublenticular extended amygdala, and in the central amygdaloid nucleus.

Given the many histochemical similarities and strong interrelations between the caudal shell areas of the accumbens and the extended amygdala, it may seem tempting to include the caudomedial shell of the accumbens within the boundaries of the extended amygdala. On the other hand, the medial shell, like the rest of the accumbens, appears to have a well-developed projection to the ventral pallidum, which in turn projects to the mediodorsal thalamus (see review by Heimer et al.[15]). The caudomedial shell, therefore, is an integral part of a significant ventral striatopallidothalamocortical loop, and from this point of view it is more similar to other striatal or basal ganglia structures than to the extended amygdala. One approach to this problem is to consider the caudal areas of accumbens—and in particular its caudomedial shell area, which has the closest neurochemical and hodological relations with the extended amygdala—as "transitional" with this latter structure.[58,59] That is to say, for the most part we cannot simply separate or define borders between neurons that seem to participate in striatopallidal circuits and those that participate in circuits typical of the central division of the extended amygdala.

The Interstitial Nucleus of the Posterior Limb of the Anterior Commissure (IPAC)

IPAC, which was illustrated in coronal section in Figures 4–10D, 4–18, and 4–19 and in the schematic drawing in Figure 4–17, is represented by collections of neurons that, in the rat, accompany the posterior limb of the anterior commissure across the face of the pallidum. It has traditionally been included in the caudate-putamen (striatum), as reflected by terms such as *fundus striati*[150] or *subcommissural striatal pocket*, when viewed in coronal rat brain sections. As explained below, however, the term *interstitial nucleus of the posterior limb of the anterior commissure*[146] seems more appropriate. This term reflects its general association with the posterior limb of the anterior commissure and has the virtue of being neutral with respect to its association with the striatum or amygdala. It is also in keeping with the rather reasonable tradition of accepting the nomenclature proposed by the first individual to identify a novel anatomical structure.

The reasons for introducing a short discussion of IPAC in the context of the accumbens are manifold. As indicated in Figures 4–17 and 4–18, IPAC, like the rest of the central division of the extended amygdala, is directly continuous with the caudal aspects of the accumbens, and it shares with the caudal medial shell area of the accumbens reciprocal interconnections with other parts of the central extended amygdala (Figure 4–19).[32,59,151–153] It also appears to contain the same variety of histochemical markers (including angiotensin II, neurotensin, secretoneurin, opioid peptides, vasopressin, and oxytocin receptors), as well as androgen and estrogen receptors (for references, see Alheid et al.[59] and Heimer et al.[66]), that serve to distinguish the extended amygdala from the adjacent striatal territories. However, like the rest of the accumbens and ventral striatum, IPAC is also characterized by significant acetylcholinesterase staining and by an especially great density of dopamine terminals, which, like the dopamine terminals in the caudomedial shell (see section 4, Figure 4–10C and D), are relatively more resistant to toxic or surgical manipulation than the dopamine terminals in the rest of the accumbens. In contrast to other parts of the ventral striatum, including the caudomedial shell of the accumbens, IPAC does not appear to have any significant projections to ventral or dorsal pallidum,[154] and it thus may not be part of the corticostriatopallidothalamic circuitry. Therefore, even though IPAC has stronger acetylcholinesterase reaction and a greater density of dopamine terminals than other parts of the extended amygdala, it nonetheless appears to be an integral part of the extended amygdala, and from this point of view it is a little different from the caudomedial shell of the accumbens, which we defined as a transition area between the basal ganglia and extended amygdala.

IPAC, like the rest of the continuum formed by the central extended amygdala, and the shell of the accumbens for that matter, has a number of features that are highly relevant in the context of neuropsychiatric disorders. This subject has been discussed at some length in a recent article by several of the present authors[66] and will not be dwelt upon here. Suffice it to say that the IPAC component of the extended amygdala and its continuation into the shell of the accumbens belong to some of the most densely dopamine-innervated parts of the basal forebrain. This particular corridor, therefore, is likely to be of special interest in evaluating therapeutic agents targeting dopamine receptors or drugs of abuse that are known to interact strongly with dopaminergic mechanisms. Although the theory of the extended amygdala appears to be as relevant in the primate brain, including the human,[61,63] as it is in the rat, the special segment represented by IPAC still remains to be adequately described in the primate.

7. CONCLUDING REMARKS

It appears that the term *nucleus accumbens septi* does not adequately describe this important part of the basal forebrain; in fact, the accumbens can hardly be referred to as a nucleus, since, with the major exception of clear boundaries towards the septum and the anterior olfactory areas, it cannot be clearly demarcated either from nearby parts of ventral striatum or from parts of the extended amygdala. When it was acknowledged that the accumbens is an integral and major part of the ventral striatum, it was only a matter of time before the corticosubcortical reentrant circuits through the accumbens were described and quickly adapted to theories intended to help explain major symptoms in various neuropsychiatric disorders. Gradually the patch-matrix organization of the accumbens has been clarified, and, as has been reviewed to some extent in this chapter, additional discoveries are now being made that make it possible to propose hypotheses based on the anatomical connections of ensembles of neurons within different parts of the accumbens and to analyze subpopulations of neurons as to their potential involvement in the drug treatment of neuropsychiatric disorders and in some aspects of drug abuse.

The recent anatomical discoveries related to the accumbens, including the identification of the core–shell dichotomy and its relevance in the context of connectional arrangements and immunohistochemical staining patterns, have added another level of sophistication to our understanding of the organization of basal forebrain functional-anatomical systems in general. Of special note was the realization that many of the features that distinguish the shell, especially its caudomedial part, from the core and the rest of striatum are also characteristic of the central division of the extended amygdala.[58] As a way of summarizing our understanding of the nature of the accumbens and its position in the overall organizational scheme of the basal forebrain, we would like to refer again to the schematic drawing in Figure 4–17. Although such schemes carry the risk of oversimplification, they nonetheless provide for overall perspectives, which are otherwise difficult to obtain.

Since a sharp border does not exist between the core of the accumbens and the rest of the striatal complex (caudate-putamen), and since the core parallels in almost every aspect the rest of the striatum, it is difficult to escape the conclusion that the accumbens is an integral part of the striatal complex. However, as indicated by a fine line in Figure 4–17, and as illustrated in many of the previous figures, there is a rather distinct border between the core and the shell of the accumbens. Although the shell has many striatal features that justify its inclusion in the notion of the ventral striatum, the shell is also endowed with very specific characteristics, atypical of the striatum, that justify the subdivision of the accumbens into two major subterritories as defined by the core–shell dichotomy.

The accumbens and its relevance to neuropsychiatric disorders cannot be brought into perspective without some appreciation of the theory of the extended amygdala. As indicated by the gradually fading yellow color in Figure 4–17, parts of the accumbens, especially its caudomedial shell, serve as areas of transition to the central division of the extended amygdala (in yellow). Together, the shell of the accumbens and the extended amygdala form an extensive forebrain continuum of apparent importance in neuropsychiatry (for further discussion of this subject, see Alheid and Heimer[58] and Heimer and co-workers[66]). The shell of the accumbens is similar, although not identical, to the central division of the extended amygdala, with which it is directly continuous and with which it appears to share histochemical and connectional features. These shared features include an especially high density of dopamine terminals and many similar inputs from areas in the prefrontal-orbitofrontal cortex and medial temporal lobe, including the hippocampal formation and the large "cortical-like" basolateral amygdaloid complex. These cortical regions are precisely the ones that appear to be prominently involved in many neuropsychiatric disorders, including especially those characterized by emotional derangements and associated viscero-endocrine and behavioral symptoms, such as Alzheimer's disease, schizophrenia, depressive disorders, and obsessive-compulsive disorder.

REFERENCES

1. Matthysse S: Antipsychotic drug actions: a clue to the neuropathology of schizophrenia? Federation Proceedings 1973; 32:200–205
2. Stevens JR: An anatomy of schizophrenia? Arch Gen Psychiatry 1973; 29:177–189
3. Fibiger HC, Phillips AG: Reward, motivation, cognition: psychobiology of mesotelencephalic dopamine systems, in Handbook of Physiology: The Nervous System IV, edited by Mountcastle VB, Bloom FE, Geyer SR. Bethesda, MD, American Physiological Society, 1986, pp 647–675
4. Koob GF: Drugs of abuse: anatomy, pharmacology and function of reward pathways. Trends Neurosci 1992; 13:177–184
5. Robbins TW, Everitt BJ: Neurobehavioural mechanisms of reward and motivation. Curr Opin Neurobiol 1996; 6:228–236
6. Wise RA: Neurobiology of addiction. Curr Opin Neurobiol 1996; 6:243–251
7. Chang JY, Paris JM, Sawyer SF, et al: Neuronal spike activity in rat nucleus accumbens during cocaine self-administration under different fixed-ratio schedules. Neuroscience 1996; 74:483–497
8. Fibiger HC: Mesolimbic dopamine: an analysis of its role in motivated behavior. Semin Neurosci 1993; 5:321–327
9. Mogenson GJ, Jones DL, Yim CY: From motivation to action: functional interface between the limbic system and the motor system. Prog Neurobiol 1980; 14:69–97
10. Ziehen T: Das Zentralnervensystem der Monotremen und Marsupialier [The central nervous systems of monotremes and marsupials], I, II. Denkschriften der Medizinischnaturwissenschaftlichen Gesellschaft zu Jena 1897; Bd 6
11. Chronister RB, DeFrance JF: The Neurobiology of the Nucleus Accumbens. Sebasco Estates, ME, Haer Institute for Electrophysiological Research, 1981
12. Heimer L, Wilson RD: The subcortical projections of allocortex: similarities in the neuronal associations of the hippocampus, the piriform cortex and the neocortex, in Golgi Centennial Symposium Proceedings, edited by Santini M. New York, Raven, 1975, pp 173–193
13. Swanson LW, Cowan WM: A note on the connections and development of the nucleus accumbens. Brain Res 1975; 92:324–330
14. Alheid GF, Heimer L, Switzer RC: Basal ganglia, in The Human Nervous System, edited by Paxinos G. San Diego, CA, Academic Press, 1990, pp 483–582
15. Heimer L, Zahm DS, Alheid GF: Basal ganglia, in The Rat Nervous System, 2nd edition, edited by Paxinos G. San Diego, CA, Academic Press, 1995, pp 579–628
16. Alexander GE, Crutcher MD, DeLong MR: Basal ganglia–thalamocortical circuits: parallel substrates for motor, oculomotor, "prefrontal" and "limbic" functions, in The Prefrontal Cortex: Its Structure, Function and Pathology (Progress in Brain Research, vol 85), edited by Uylings HBM, Van Eden CG, De Bruin JPC, et al. Amsterdam, Elsevier, 1990, pp 119–146
17. Alexander GE, DeLong MR, Strick PL: Parallel organization of functionally segregated circuits linking basal ganglia and cortex. Annu Rev Neurosci 1986; 9:357–381
18. Cummings JL: Frontal-subcortical circuits and human behavior. Arch Neurol 1993; 50:873–880
19. Deutch AY, Bourdelais AJ, Zahm DS: The nucleus accumbens core and shell: accumbal compartments and their functional attributes, in Limbic Motor Circuits and Neuropsychiatry, edited by Kalivas PW, Barnes CD. Boca Raton, FL, CRC Press, 1993, pp 45–88
20. Groenewegen HJ: Cortical-subcortical relationships and the limbic forebrain. Behav Neurol 1996; 9:29–48
21. Groenewegen HJ, Berendse HW: Anatomical relationships between the prefrontal cortex and the basal ganglia in the rat, in Motor and Cognitive Functions of the Prefrontal Cortex, edited by Thierry AM, Glowinski J, Goldman-Rakic PS, et al. Berlin and Heidelberg, Springer-Verlag, 1994, pp 31–77
22. Haber SN, Kunishio K, Mizobuchi M, et al: The orbital and medial prefrontal circuit through the primate basal ganglia. J Neurosci 1995; 15:4851–4867
23. Mega MS, Cummings JL: Frontal-subcortical circuits and neuropsychiatric disorders. J Neuropsychiatry Clin Neurosci 1994; 6:358–370
24. Modell JG, Mountz JM, Curtis GC, et al: Neurophysiologic dysfunction in basal ganglia/limbic striatal and thalamocortical circuits as a pathogenetic mechanism of obsessive-compulsive disorder. J

Neuropsychiatry Clin Neurosci 1989; 1:27–36
25. Price JL, Carmichael ST, Drevets WC: Networks related to the orbital and medial prefrontal cortex: a substrate for emotional behavior? In The Emotional Motor System (Progress in Brain Research, vol 107), edited by Holstege G, Bandler R, Saper CB. Amsterdam, Elsevier, 1996, pp 461–484
26. Salloway S, Cummings J: Subcortical disease and neuropsychiatric illness. J Neuropsychiatry Clin Neurosci 1994; 6:93–99
27. Swerdlow NR, Koob GF: Dopamine, schizophrenia, mania and depression: toward a unified hypothesis of cortico-striato-pallido-thalamic function. Behav Brain Sci 1987; 10:197–245
28. Zahm DS, Brog JS: On the significance of subterritories in the "accumbens" part of the rat ventral striatum. Neuroscience 1992; 50:751–767
29. Joel D, Weiner I: The organization of the basal ganglia-thalamocortical circuits: open interconnected rather than closed segregated. Neuroscience 1994; 63:363–379
30. Deutch AY: Prefrontal cortical dopamine systems and the elaboration of functional corticostriatal circuits: implications for schizophrenia and Parkinson's disease. J Neural Transm 1993; 91:197–221
31. Heimer L, Alheid GF, Zahm DS: Basal forebrain organization: an anatomical framework for motor aspects of drive and motivation, in Limbic Motor Circuits and Neuropsychiatry, edited by Kalivas PW, Barnes CD. Boca Raton, FL, CRC Press, 1993, pp 1–43
32. Nauta WJH, Smith GP, Faull RLM, et al: Efferent connections and nigral afferents of the nucleus accumbens septi in the rat. Neuroscience 1978; 3:385–401
33. Groenewegen HJ, Meredith GE, Berendse HW, et al: The compartmental organization of the ventral striatum in the rat, in Neural Mechanisms in Disorders of Movement, edited by Crossman AR, Sambrook MA. London, John Libbey, 1989, pp 45–54
34. Herkenham S, Moon-Edley S, Stuart J: Cell clusters in the nucleus accumbens of the rat, and the mosaic relationship of opiate receptors, acetylcholinesterase and subcortical afferent terminations. Neuroscience 1984; 11:561–593
35. Pennartz CMA, Groenewegen HJ, Lopes da Silva FH: The nucleus accumbens as a complex of functionally distinct neuronal ensembles: an integration of behavioural, electrophysiological and anatomical data. Prog Neurobiol 1994; 42:719–761
36. Záborszky L, Alheid GF, Beinfeld MC, et al: Cholecystokinin innervation of the ventral striatum: a morphological and radioimmunological study. Neuroscience 1985; 14:427–453
37. Groenewegen HJ, Russchen FT: Organization of the efferent projections of the nucleus accumbens to pallidal, hypothalamic, and mesencephalic structures: a tracing and immunohistochemical study in the cat. J Comp Neurol 1984; 223:347–367
38. Heimer L, Zahm DS, Churchill L, et al: Specificity in the projection patterns of accumbal core and shell in the rat. Neuroscience 1991; 41:89–125
39. Brog JS, Salyapongse A, Deutch AY, et al: The patterns of afferent innervation of the core and shell in the "accumbens" part of the rat ventral striatum: immunohistochemical detection of retrogradely transported fluoro-gold. J Comp Neurol 1993; 338:255–278
40. Roberts DCS, Koob GF, Klonoff P, et al: Extinction and recovery of cocaine self-administration following 6-hydroxydopamine lesions of the nucleus accumbens. Pharmacol Biochem Behav 1980; 12:781–787
41. Di Chiara G: The role of dopamine in drug abuse viewed from the perspective of its role in motivation. Drug Alcohol Depend 1995; 38:95–137
42. Altman J, Everitt BJ, Glautier S, et al: The biological, social and clinical bases of drug addiction: commentary and debate. Psychopharmacol (Berl) 1996; 125:285–345
43. Pontieri FE, Colangelo V, La Riccia M, et al: Psychostimulant drugs increase glucose utilization in the shell of the rat nucleus accumbens. Neuroreport 1994; 5:2561–2564
44. Pontieri FE, Tanda G, Di Chiara G: Intravenous cocaine, morphine, and amphetamine preferentially increase extracellular dopamine in the "shell" as compared with the "core" of the rat nucleus accumbens. Proc Natl Acad Sci USA 1995; 92:12304–12308
45. Pierce RC, Kalivas PW: Amphetamine produces sensitized increases in locomotion and extracellular dopamine preferentially in the nucleus accumbens shell of rats administered repeated cocaine. J Pharmacol Exp Ther 1995; 275:1019–1029
46. Pontieri FE, Tanda G, Orzi F, et al: Effects of nicotine on the nucleus accumbens and similarity to those of addictive drugs. Nature 1996; 382:255–257
47. White SR, Obradovic T, Imel KM, et al: The effects of methylenedioxymethamphetamine (MDMA, ecstasy) on monoaminergic neurotransmission in the central nervous system. Prog Neurobiol 1996; 49:455–479
48. Zachrisson O, Mathe AA, Stenfors C, et al: Region-specific effects of chronic lithium administration on neuropeptide Y and somatostatin mRNA expression in the rat brain. Neurosci Lett 1995; 194:89–92
49. Thierry AM, Tassin JP, Blanc G, et al: Selective activation of the mesocortical dopamine system by stress. Nature 1976; 263:242–244
50. Abercrombie ED, Keefe KA, DiFrischia DS, et al: Differential effect of stress on in vivo dopamine release in striatum, nucleus accumbens, and medial frontal cortex. J Neurochem 1989; 52:1655–1657
51. Kalivas PW, Duffy P: Selective activation of dopamine transmission in the shell of the nucleus accumbens by stress. Brain Res 1995; 675:325–328
52. Deutch AY, Lee MC, Iadarola M: Regionally specific effects of atypical antipsychotic drugs on striatal Fos expression: the nucleus accumbens shell as a locus of antipsychotic action. Mol Cell Neurosci 1992; 3:332–341
53. King D, Zigmond MJ, Finlay JM: Effects of dopamine depletion in the medial prefrontal cortex on the stress-induced increase in extracellular dopamine in the nucleus accumbens core and shell. Neuroscience 1997; 77:141–153
54. Horger BA, Elsworth JD, Roth RH: Selective increase in dopamine utilization in the shell subdivision of the nucleus accumbens by the benzodiazepine inverse agonist FG 7142. J Neurochem 1995; 65:770–774
55. Graybiel AM, Mortalla R, Robertson HA: Amphetamine and cocaine induce drug-specific activation of the c-*fos* gene in striosome-matrix compartments and limbic subdivisions of the striatum. Proc Natl Acad Sci USA 1990; 87:3915–3934
56. Merchant KM, Dorsa DM: Differential induction of neurotensin and c-*fos* gene expression by typical versus atypical antipsychotics. Proc Natl Acad Sci 1993; 90:3447–3451
57. Johnston JB: Further contribution to the study of the evolution of the forebrain. J Comp Neurol 1923; 35:337–481
58. Alheid GF, Heimer L: New perspectives in basal forebrain organization of special relevance for neuropsychiatric disorders: the striatopallidal, amygdaloid, and corticopetal components of substantia innominata. Neuroscience 1988; 27:1–39
59. Alheid GF, de Olmos JS, Beltramino CA: Amygdala and extended amygdala, in The Rat Nervous System, 2nd edition, edited by Paxinos G. San Diego, CA, Academic Press, 1995, pp 495–578
60. de Olmos JS, Alheid GF, Beltramino CA: Amygdala, in The Rat Nervous System, edited by Paxinos G. Sydney, Academic Press, 1985, pp 223–334
61. de Olmos JS: Amygdala, in The Human Nervous System, edited by Paxinos G. San Diego, CA, Academic Press, 1990, pp 583–710
62. Lynd-Balta E, Haber SN: The organization of midbrain projections to the ventral striatum in the primate. Neuroscience 1994; 59:609–623
63. Martin LJ, Powers RE, Dellovade TL, et al: The bed nucleus-amygdala continuum in human and monkey. J Comp Neurol 1991; 309:445–485
64. Meredith GE, Pattiselanno A, Groenewegen HJ, et al: The shell and core in monkey and human nucleus accumbens identified with

antibodies to calbindin-D_{28K}. J Comp Neurol 1996; 365:628–639
65. Voorn P, Brady LS, Berendse HW, et al: Densitometrical analysis of opioid receptor ligand binding in the human striatum, I: distribution of µ opioid receptor defines shell and core of the ventral striatum. Neuroscience 1996; 75:777–792
66. Heimer L, Harlan RE, Alheid GF, et al: Substantia innominata: a notion which impedes clinical-anatomical correlations in neuropsychiatric disorders. Neuroscience 1997; 76:957–1006
67. Caine SB, Heinrichs SC, Coffin VL, et al: Effects of the dopamine D-1 antagonist SCH 23390 microinjected into the accumbens, amygdala or striatum on cocaine self-administration in the rat. Brain Res 1995; 692:47–56
68. McGregor A, Roberts DCS: Dopaminergic antagonism within the nucleus accumbens or the amygdala produces differential effects on intravenous cocaine self-administration under fixed and progressive ratio schedules of reinforcement. Brain Res 1993; 624:245–252
69. Koob GF, Robledo P, Markou A, Caine SB: The mesocorticolimbic circuit in drug dependence and reward: a role for the extended amygdala? In Limbic Motor Circuits and Neuropsychiatry, edited by Kalivas PW, Barnes CD. Boca Raton, FL, CRC Press, 1993, pp 289–309
70. Brown VJ, Latimer MP, Winn P: Memory for the changing cost of a reward is mediated by the sublenticular extended amygdala. Brain Res Bull 1996; 39:163–170
71. Gerfen CR: The neostriatal mosaic: striatal patch-matrix organization is related to cortical lamination. Science 1989; 246:385–388
72. Graybiel AM, Ragsdale CW: Histochemically distinct compartments in the striatum of human, monkey and cat demonstrated by acetylcholinesterase staining. Proc Natl Acad Sci USA 1978; 75:5723–5726
73. Herkenham M, Pert C: Mosaic distribution of opiate receptors, parafasicular projections and acetylcholinesterase in the rat striatum. Nature (London) 1981; 291:415–418
74. Deacon TW, Pakzaban P, Isacson O: The lateral ganglionic eminence is the origin of cells committed to striatal phenotypes: neuronal transplantation and developmental evidence. Brain Res 1994; 668:211–219
75. Altman J, Bayer SA: Atlas of Prenatal Rat Brain Development. Boca Raton, FL, CRC Press, 1995
76. Alvarez-Bolado G, Swanson LS: Developmental Brain Maps. Amsterdam, Elsevier, 1996
77. Song DD, Harlan RE: The development of enkephalin and substance P neurons in the basal ganglia: insights into neostriatal compartments and the extended amygdala. Brain Res Dev Brain Res 1994; 83:247–261
78. Bayer SA, Altman J: Neurogenesis and neuronal migration, in The Rat Nervous System, 2nd edition, edited by Paxinos G. San Diego, CA, Academic Press, 1995, pp 1041–1078
79. Bayer A: Development of the hippocampal region in the rat, I: neurogenesis examined with [^{3}H]thymidine autoradiography. J Comp Neurol 1980; 190:87–114
80. Cameron HA, McEwen BS, Gould E: Regulation of adult neurogenesis by excitatory input and NMDA receptor activation in the dentate gyrus. J Neurosci 1995; 15:4687–4692
81. van der Kooy D, Fishell G: Neuronal birthdate underlies the development of striatal compartments. Brain Res 1987; 401:155–161
82. Song DD, Harlan RE: Genesis and migration patterns of neurons forming the patch and matrix compartments of the rat striatum. Brain Res Dev Brain Res 1994; 83:233–246
83. Meredith GE, Pennartz CMA, Groenewegen HJ: The cellular framework for chemical signalling in the nucleus accumbens, in Chemical Signalling in the Basal Ganglia (Progress in Brain Research, vol 99), edited by Arbuthnott GW, Emson PC. Amsterdam, Elsevier, 1993, pp 3–24
84. Hartz-Schütt CG, Mai JK: Cholinesterase-Aktivität im menschlichen Striatum unter besonderer Berücksichtigung der Insulae terminales [Cholinesterase activity in the human striatum with special reference to the "insulae terminales"]. J Hirnforsch 1991; 32:317–342
85. Meyer G, Gonzalez-Hernandez T, Carrillo-Padilia F, et al: Aggregations of granule cells in the basal forebrain (islands of Calleja): a Golgi and cytoarchitectonic study in different mammals including man. J Comp Neurol 1989; 284:405–478
86. Sanides F: Die Insulae terminales des Erwachsenen Gehirns des Menschen ["Insulae terminales" in the brains of adult humans]. J Hirnforsch 1957; 3:243–273
87. Wilson CJ: The generation of natural firing patterns in neostriatal neurons, in Chemical Signalling in the Basal Ganglia (Progress in Brain Research, vol 99), edited by Arbuthnott GW, Emson PC. Amsterdam, Elsevier, 1993, pp 277–297
88. O'Donnell P, Grace AAG: Synaptic interactions among excitatory afferents to nucleus accumbens neurons: hippocampal gating of prefrontal cortical input. J Neurosci 1995; 15:3622–3639
89. Moore H, Grace AA: Interactions between amygdala and prefrontal cortical afferents in the nucleus accumbens and their modulation by dopamine receptor activation (abstract). Society for Neuroscience Abstracts 1996; 22:431.20
90. Mulder AB, Lopes da Silva H: Interactions of the hippocampal and basolateral amygdala inputs to the nucleus accumbens of the rat. Society for Neuroscience Abstracts 1996; 22:358.4
91. Fuller TA, Russchen FT, Price JL: Sources of presumptive glutamatergic afferents to the rat ventral striatopallidal region. J Comp Neurol 1987; 258:317–338
92. Berendse HW, Groenewegen HJ: Restricted cortical termination fields of the midline and intralaminar thalamic nuclei. Neuroscience 1991; 42:73–102
93. Jay T, Witter MP: Distribution of hippocampal CA1 and subicular efferents in the prefrontal cortex of the rat studied by means of anterograde transport of *Phaseolus vulgaris*-leucoagglutinin. J Comp Neurol 1991; 313:574–586
94. Krettek JE, Price JL: Projections from the amygdaloid complex to the cerebral cortex and thalamus in the rat and cat. J Comp Neurol 1978; 172:687–722
95. McDonald AJ: Organization of amygdaloid projections to the prefrontal cortex and associated striatum in the rat. Neuroscience 1991; 44:1–14
96. McDonald AJ: Topographic organization of amygdaloid projections to the caudate-putamen, nucleus accumbens, and related striatal-like areas of the rat brain. Neuroscience 1991; 44:15–33
97. Berendse HW, Galis-de Graaf Y, Groenewegen HJ: Topographical organization and relationship with ventral striatal compartments of prefrontal corticostriatal projections in the rat. J Comp Neurol 1992; 316:314–347
98. Berendse HW, Groenewegen HJ: The organization of the thalamostriatal projections in the rat, with special emphasis on the ventral striatum. J Comp Neurol 1990; 299:187–228
99. Groenewegen HJ, Vermeulen-Van der Zee E, te Kortschot A: Organization of the projections from the subiculum to the ventral striatum in the rat: a study using anterograde transport of *Phaseolus vulgaris*-leucoagglutinin. Neuroscience 1987; 23:103–120
100. Wright CI, Beijer AVJ, Groenewegen HJ: Basal amygdaloid complex afferents to the rat nucleus accumbens are compartmentally organized. J Neurosci 1996; 16:1877–1893
101. Kelley AE, Domesick VB: The distribution of the projections from the hippocampal formation to the nucleus accumbens in the rat: an anterograde and retrograde horseradish peroxidase study. Neuroscience 1982; 7:2321–2335
102. Kelley AE, Domesick VB, Nauta WJH: The amygdalostriatal projection in the rat: an anatomical study by anterograde and retrograde tracing methods. Neuroscience 1982; 7:615–630
103. Beijer AVJ, Groenewegen HJ: Specific anatomical relationships between hippocampal and basal amygdaloid afferents and different populations of projection neurons in the nucleus accumbens of rats (abstract). Society for Neuroscience Abstracts 1996; 22:164.11
104. Wright CI, Groenewegen HJ: Patterns of convergence and segrega-

tion in the medial nucleus accumbens of the rat: relationships of prefrontal cortical, midline thalamic, and basal amygdaloid afferents. J Comp Neurol 1995; 361:383–403

105. Alheid GF, Heimer L: Theories of basal forebrain organization and the "emotional motor system," in The Emotional Motor System (Progress in Brain Research, vol 107), edited by Holstege G, Bandler R, Saper CB. Amsterdam, Elsevier, 1996, pp 461–484

106. Haber SN: Anatomical relationship between the basal ganglia and the basal nucleus of Meynert in human and monkey forebrain. Proc Natl Acad Sci USA 1987; 84:1408–1412

107. Záborszky L, Cullinan WE: Projections from the nucleus accumbens to cholinergic neurons of the ventral pallidum: a correlated light and electron microscopic double immunostaining study in the rat. Brain Res 1992; 750:92–101

108. Muir JL, Everitt BJ, Robbins TW: AMPA-induced excitotoxic lesions of the basal forebrain: a significant role for the cortical cholinergic system in attentional function. J Neurosci 1994; 14:2313–2326

109. Bevan MD, Smith AD, Bolam JP: The substantia nigra as a site of synaptic integration of functionally diverse information arising from the ventral pallidum and the globus pallidus in the rat. Neuroscience 1996; 75:5–12

110. Zahm DS, Williams E, Wohltmann C: Ventral striatopallidothalamic projection, IV: relative involvements of neurochemically distinct subterritories in the ventral pallidum and adjacent parts of the rostroventral forebrain. J Comp Neurol 1996; 364:340–362

111. Deniau JM, Menetrey A, Thierry AM: Indirect nucleus accumbens input to the prefrontal cortex via the substantia nigra pars reticulata: a combined anatomical and electrophysiological study in the rat. Neuroscience 1994; 61:533–545

112. Thierry AM, Maurice N, Deniau JM, et al: Anatomical and functional relationships between the prefrontal cortex and the basal ganglia in the rat (abstract). Society for Neuroscience Abstracts 1996; 22:164.10

113. Smith AD, Bolam JP: The neural network of the basal ganglia as revealed by the study of synaptic connections of identified neurones. Trends Neurosci 1990; 13:259–265

114. Johnson LR, Aylward RLM, Hussain Z, et al: Input from the amygdala to the rat nucleus accumbens: its relationship with tyrosine hydroxylase immunoreactivity and identified neurons. Neuroscience 1994; 61:851–865

115. Van Bockstaele EJ, Sesack SR, Pickel VM: Dynorphin-immunoreactive terminals in the rat nucleus accumbens: cellular sites for modulation of target neurons and interactions with catecholamine afferents. J Comp Neurol 1994; 341:1–15

116. Groves PM, Garcia-Munoz M, Linder JC, et al: Elements of the intrinsic organization and information processing in the neostriatum, in Models of Information Processing in the Basal Ganglia, edited by Houk JC, Davis JL, Beiser DG. Cambridge, MA, MIT Press, 1994, pp 51–96

117. Woodward DJ, Kirillov AB, Myre CD, et al: Neostriatal circuitry as a scalar memory: modeling and ensemble neuron recording, in Models of Information Processing in the Basal Ganglia, edited by Houk JC, Davis JL, Beiser DG. Cambridge, MA, MIT Press, 1994, pp 315–336

118. Sexton PM, Paxinos G, Kenney MA, et al: In vitro autoradiographic localization of amylin binding sites in rat brain. Neuroscience 1994; 62:553–567

119. Zahm DS: Compartments in rat dorsal and ventral striatum revealed following injection of 6-hydroxydopamine into the ventral mesencephalon. Brain Res 1991; 552:164–169

120. Turner BH, Wilson JS, McKenzie JC, et al: MPTP produces a pattern of nigrostriatal degeneration which coincides with the mosaic organization of the caudate nucleus. Brain Res 1988; 473:60–64

121. Broening HW, Pu C, Vorhees CV: Dopaminergic innervation to the nucleus accumbens core and shell is differentially vulnerable to methamphetamine-induced neurotoxicity (abstract). Society for Neuroscience Abstracts 1996; 22:1915

122. Curran EJ, Watson SJ Jr: Dopamine receptor mRNA expression patterns by opioid peptide cells in the nucleus accumbens of the rat: a double in situ hybridization study. J Comp Neurol 1995; 361:57–76

123. Mathieu AM, Caboche J, Besson MJ: Distribution of preproenkephalin, preprotachykinin A, and preprodynorphin mRNAs in the rat nucleus accumbens: effect of repeated administration of nicotine. Synapse 1996; 23:94–106

124. Le Moine C, Bloch B: Expression of the D_3 dopamine receptor in peptidergic neurons of the nucleus accumbens: comparison with the D_1 and D_2 dopamine receptors. Neuroscience 1996; 73:131–143

125. Diaz J, Lévesque D, Lammers CH, et al: Phenotypical characterization of neurons expressing the dopamine D_3 receptor in the rat brain. Neuroscience 1995; 65:731–745

126. Diaz J, Levesque D, Griffon N, et al: Opposing roles for dopamine D_2 and D_3 receptors on neurotensin mRNA expression in nucleus accumbens. Eur J Neurosci 1994; 6:1384–1387

127. Zahm DS: Subsets of immunoreactive neurons revealed following antagonism of the dopamine-mediated suppression of neurotensin immunoreactivity in the rat striatum. Neuroscience 1992; 46:335–350

128. Brog JS, Zahm DS: Two subpopulations of striatal neurons revealed by neurotensin and Fos immunoreactivities following acute 6-hydroxydopamine lesions and reserpine administration. Neuroscience 1995; 65:71–86

129. Robertson GS, Fibiger HC: Neuroleptics increase c-*fos* expression in the forebrain: contrasting effects of haloperidol and clozapine. Neuroscience 1992; 46:315–328

130. Senger B, Brog JS, Zahm DS: Subsets of neurotensin immunoreactive neurons in the rat striatal complex following antagonism of the dopamine D-2 receptor: an immunohistochemical double-labeling study using antibodies against Fos. Neuroscience 1993; 57:649–660

131. Merchant KM, Miller MA, Ashleigh A, et al: Haloperidol rapidly increases the number of neurotensin mRNA-expressing neurons in neostriatum of the rat brain. Brain Res 1991; 540:311–314

132. Lambert PD, Gross R, Nemeroff CB, et al: Anatomy and mechanisms of neurotensin–dopamine interactions in the central nervous system. Ann NY Acad Sci 1995; 757:377–389

133. Giménez-Amaya JM, McFarland NR, de las Heras S, et al: Organization of thalamic projections to the ventral striatum in the primate. J Comp Neurol 1995; 354:127–149

134. Kunishio K, Haber SN: Primate cingulostriatal projection: limbic striatal versus sensorimotor striatal input. J Comp Neurol 1994; 350:337–356

135. Hoover JE, Strick PL: Multiple output channels in the basal ganglia. Science 1993; 259:819–821

136. Gaspar P, Stepniewska I, Kaas JH: Topography and collateralization of the dopaminergic projections to motor and lateral prefrontal cortex in owl monkeys. J Comp Neurol 1992; 325:1–21

137. Porrino LJ, Goldman-Rakic PS: Brainstem innervation of prefrontal and anterior cingulate cortex in the rhesus monkey revealed by retrograde transport of HRP. J Comp Neurol 1982; 205:63–76

138. Parent A, Hazrati L-N: Multiple striatal representation in primate substantia nigra. J Comp Neurol 1994; 344:305–320

139. Haber SN, Lynd E, Klein C, et al: Topographic organization of the ventral striatal efferent projections in the rhesus monkey: an anterograde tracing study. J Comp Neurol 1990; 293:282–298

140. Hedreen JC, DeLong MR: Organization of striatopallidal, striatonigral, and nigrostriatal projections in the macaque. J Comp Neurol 1991; 304:569–595

141. Somogyi P, Bolam JP, Totterdell S, et al: Monosynaptic input from the nucleus accumbens-ventral striatum region to retrogradely labelled nigrostriatal neurones. Brain Res 1981; 217:245–263

142. Haber SN, Lynd-Balta E, Spooren WPTM: Integrative aspects of basal ganglia circuitry, in The Basal Ganglia IV, edited by Percheron G, McKenzie JS, Féger J. New York, Plenum, 1994, pp 71–80

143. Price JL, Amaral DJ: An autoradiographic study of the projections of the central nucleus of the monkey amygdala. J Neurosci 1981; 1:1242–1259

144. Gonzales C, Chesselet M-F: Amygdalonigral pathway: an anterograde study in the rat with *Phaseolus vulgaris* leucoagglutinin (PHA-L). J Comp Neurol 1990; 297:182–200
145. Heimer L, Zahm DS, Alkeid GF: Basal ganglia, in The Rat Nervous System, 2nd edition, edited by Paxinos G. San Diego, CA, Academic Press, 1995, pp 579–628
146. de Olmos JS: The amygdaloid projection field in the rat as studied with the cupric silver method, in The Neurobiology of the Amygdala, edited by Eleftheriou BE. New York, Plenum, 1972, pp 145–204
147. Schwaber JS, Kapp BS, Higgins GA, et al: Amygdaloid and basal forebrain direct connections with the nucleus of the solitary tract and the dorsal motor nucleus. J Neurosci 1982; 2:1424–1438
148. Tribollet E: Vasopressin and oxytocin receptors in the rat brain, in Handbook of Chemical Neuroanatomy, vol 11, Neuropeptide Receptors in the CNS, edited by Björklund A, Hökfelt T, Kuhar MJ. Amsterdam, Elsevier, 1992, pp 289–320
149. Veinante P, Freund-Mercier MJ: Distribution of oxytocin- and vasopressin-binding sites in the rat extended amygdala: an histoautoradiographic study. J Comp Neurol (in press)
150. Paxinos G, Watson C: The Rat Brain in Stereotaxic Coordinates, 2nd edition. Sydney, Academic Press, 1986
151. Grove EA: Efferent connections of the substantia innominata in the rat. J Comp Neurol 1988; 277:347–364
152. Alheid GF, Beltramino CA, Braun A, et al: Transition areas of the striatopallidal system with the extended amygdala in the rat and primate: observations from histochemistry and experiments with mono- and transsynaptic tracer, in The Basal Ganglia IV, vol 41: New Ideas and Data on Structure and Function, edited by Percheron G, McKenzie JS, Féger J. New York, Plenum, 1994, pp 95–107
153. Shammah-Lagnado SJ, Alheid GF, Beltramino CA, et al: Afferents to the interstitial nucleus of the posterior limb of the anterior commissure: evidence for a novel component of the extended amygdala in comparison with adjacent amygdalostriatal transition areas (abstract). Society for Neuroscience Abstracts 1996; 22:806.12
154. Alheid GF, Shammah-Lagnado SJ, Beltramino CA, et al: Efferent projections of the "interstitial nucleus of the posterior limb of the anterior commissure": PHA-L transport from a dopamine-rich lateral wing of the extended amygdala (abstract). Society for Neuroscience Abstracts 1996; 22:806.13

5

Neurobiology of Fear Responses
The Role of the Amygdala

Michael Davis, Ph.D.

Fear is a hypothetical construct that is used to explain the cluster of behavioral effects that are observed and experienced when an organism faces a life-threatening situation. If suddenly confronted by a stranger holding a gun to your face, you will realize instantly that you are in danger, that you could be beaten or even killed. Your hands will sweat, your heart will pound, and your mouth will feel very dry. You will begin to tremble and feel like you can't catch your breath. You may feel the hair standing out on the back of your neck, and your mind will race, trying to decide whether you should hold still, run, or try to take the gun out of the assailant's hand. Your senses of smell, sight, and hearing will heighten, and your pupils will dilate. Later, if you survive, you will remember this terrible incident over and over again, seeing your assailant's face or the gun in apparently vivid detail. Returning to the place where the incident happened will revive those awful memories, often to the point where you will want to avoid that place forever. Thus, fear is a complex set of behavioral reactions that includes both the *expression* and the *experience* of the emotional event. Sweaty palms, increased heart rate, altered respiration, hair standing on end, and dilated pupils are part of the expression of fear. The feelings of dread of potentially being killed, and of having to decide whether to hold still or run, as well as the feeling of your heart pounding or the hair standing upright on the back of your neck, are part of the experience of fear.

Very similar reactions can be seen in animals. If a cat confronts a vicious dog, the cat will assume the familiar "Halloween posture," with its back arched, hair standing on end, and teeth bared. These expressions of fear can be seen easily and measured objectively. We would also presume, based on our own experience, that the cat is experiencing a feeling of fear, of impending death and threat to survival. However, unlike humans, with whom it is possible to discuss the experience of fear and how it feels, such a conversation is not possible with the cat. Hence, we can only infer that the cat is feeling fearful from looking at the situation and the set of behaviors displayed by the cat.

The author thanks Dr. Changjun Shi for information concerning afferent connections of the amygdala and Shari Birnbaum for her critical reading of the manuscript. Research reported in this chapter was supported by National Institute of Mental Health Grants MH25642 and MH47840, Research Scientist Development Award MH00004, a grant from the Air Force Office of Scientific Research, and the State of Connecticut.

Although the concept is still controversial, it is generally believed that most fears are not innate but instead are learned through experience. For example, monkeys in the wild are terrified by snakes, whereas monkeys bred in captivity are indifferent to snakes. However, once laboratory-raised monkeys see the fear reaction to a snake by another monkey bred in the wild and brought into the laboratory, they rapidly learn the same fear reaction and display it to the snake thereafter.[1] Hence, primates and many other lower animals readily acquire conditioned fear reactions via associations between formerly neutral stimuli and aversive events. Indeed, much of our behavior is determined by the accumulation of a long list of conditioned fears acquired over a lifetime. Very young babies are not afraid of snakes or of strangers holding guns. Such a conditioning mechanism is highly adaptive because it allows us to avoid bad things that have happened to us in the past and not to expose ourselves to things which other people tell us, by their reactions, are potentially dangerous.

Because conditioned fear occurs across so many different species, it can readily be studied in the laboratory by using a variety of animals. For example, when a light, which initially has no behavioral effect, is paired with an aversive stimulus such as a footshock, the light alone can then elicit a constellation of behaviors that are typically used to define a state of fear in animals. To explain these findings, it is generally assumed[2] that during light–shock pairings (training session) the shock elicits a variety of behaviors that can be used to infer a central state of fear (unconditioned responses; Figure 5–1). After pairing, the light can now produce the same central fear state, and thus the same set of behaviors, formerly produced by the shock (testing session). Moreover, the behavioral effects that are produced in animals by this formerly neutral stimulus (now called a conditioned stimulus; CS) are similar in many respects to the constellation of behaviors that are used to diagnose generalized anxiety in humans (Table 5–1).

A variety of animal models have been used to infer a central state of fear or anxiety. In some models, fear is inferred when an animal freezes, thus interrupting some ongoing behavior such as pressing a bar or interacting socially with other animals. In other models, fear is measured by changes in autonomic activity such as heart rate, blood pressure, or respiration, or the production of hormones such as corticosteroids. Fear also can be measured by changes in simple reflexes, such as an elevated startle reflex or a change in facial expression. Thus, fear appears to produce a complex pattern of behaviors that are often associated with each other.

The purpose of this chapter is to summarize data supporting the idea that the amygdala, and its many

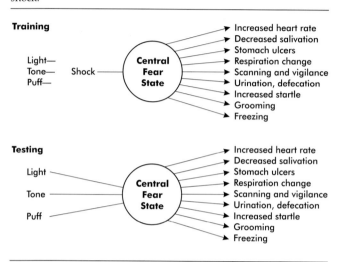

FIGURE 5–1. General scheme believed to occur during classical conditioning using an aversive conditioned stimulus. During training, the aversive stimulus (such as shock) activates a central fear system that produces a constellation of behaviors generally associated with aversive stimuli (unconditioned responses). After consistent pairings of some neutral stimulus such as a light or tone or puff of air with shock during the training phase, the neutral stimulus is now capable of producing a similar fear state and hence the same set of behaviors (conditioned responses) formerly only produced by the shock.

TABLE 5–1. Correspondence between measures of fear in animals and criteria used to diagnose generalized anxiety in humans

Measures of Fear in Animals	DSM Criteria: Generalized Anxiety
Increased heart rate	Heart pounding
Decreased salivation	Dry mouth
Stomach ulcers	Upset stomach
Respiration change	Increased respiration
Scanning and vigilance	Scanning and vigilance
Increased startle	Jumpiness, easy startle
Urination	Frequent urination
Defecation	Diarrhea
Grooming	Fidgeting
Freezing	Apprehensive expectation

efferent projections, may represent a central fear system involved in both the expression and the acquisition of conditioned and unconditioned fear. The chapter will review how lesions or electrical stimulation of the amygdala or local infusion of drugs into the amygdala alter several components of fear and/or anxiety. These components include various autonomic and hormonal measures (heart rate, blood pressure, respiration, colonic motility, gastric ulcers, adrenocorticotropin [ACTH] and corticosteroid release). Measures of attention and vigilance, as well as various motor behaviors (freezing, reflex facilitation, elevated plus maze, social interaction, bar pressing or licking in the conflict test, high-frequency vocalization) and hypoalgesia will also be covered. This chapter will thus update and extend previous reviews on this topic.[3]

Because most of the data have been gathered in rodents, much of the review will focus on rodents, although relevant research in primates will also be included. However, it should be emphasized that it is highly probable that the brain systems that have evolved to produce the autonomic and motor effects indicative of fear, so necessary for survival, have been highly conserved across evolution.

In the interest of space, except where noted, the extensive literature on the role of the amygdala in inhibitory avoidance (see reviews[4,5]) and the emerging literature on the role of the amygdala in conditioned taste aversion[6] and opiate withdrawal[7–10] will not be reviewed. Also, the prominent role of other brain areas, such as the central gray (see reviews[11,12]), will not be reviewed specifically except where noted. Finally, again only in the interest of space, this review will not include data on recording cellular activity in the amygdala.

ANATOMICAL CONNECTIONS BETWEEN THE CENTRAL NUCLEUS OF THE AMYGDALA AND BRAIN AREAS INVOLVED IN FEAR AND ANXIETY

The amygdala consists of several separate cell groups (nuclei), which receive input from many different brain areas (Table 5–2). Highly processed sensory information from various cortical areas reaches the amygdala through its lateral and basolateral nuclei.[13,14] In turn, these nuclei project to the central nucleus of the amygdala,[13,15–17] which then projects to hypothalamic and brainstem target areas that directly mediate specific signs of fear and anxiety. A great deal of evidence now indicates that the amygdala and its many efferent projections may represent a central fear system involved in both the expression and the acquisition of conditioned fear.[18–23]

Figure 5–2 summarizes many reports indicating that the central nucleus of the amygdala has direct projections to various anatomical areas that are likely to be involved in specific symptoms of fear or anxiety.

Autonomic and Hormonal Measures

Direct projections from the central nucleus of the amygdala to the lateral hypothalamus[24,25] appear to be involved in activation of the sympathetic autonomic nervous system seen during fear and anxiety.[26] Direct projections to the dorsal motor nucleus of the vagus, nucleus of the solitary tract, and ventrolateral medulla[25,27–30] may be involved in amygdala modulation of heart rate and blood pressure, which are known to be regulated by these brainstem nuclei. Projections of the central nucleus of the amygdala to the parabrachial nucleus[24,25,31,32] may be involved in respiratory changes during fear; electrical stimulation or lesions of this nucleus are known to alter various measures of respiration.

Direct projections of the central nucleus of the amygdala to the paraventricular nucleus of the hypothalamus,[33] or indirect projections by way of the bed nucleus of the stria terminalis and preoptic area (which receive input from the amygdala[24,34] and project to the paraventricular nucleus of the hypothalamus),[35] may mediate the prominent neuroendocrine responses to fearful or stressful stimuli.

TABLE 5–2. Inputs to the amygdala

Source of Inputs	Inputs To		
	Lateral Nucleus	Basolateral Nucleus	Central Nucleus
Cortex			
Temporal cortex	+		
Perirhinal cortex	+	+	
Entorhinal cortex	+	+	
Hippocampus		+	
Piriform cortex	+	+	
Insular cortex	+	+	+
Medial prefrontal cortex	+	+	+
Peri-amygdaloid cortex	+	+	+
Basal forebrain			
Lateral amygdaloid nucleus	+	+	
Basolateral amygdaloid nucleus	+	+	+
Basomedial amygdaloid nucleus	+	+	+
Ventromedial hypothalamic nucleus	+		+
Lateral hypothalamus			+
Bed nucleus of stria terminalis			+
Thalamus			
Lateral posterior nucleus	+		
Medial geniculate complex	+		
Gustatory thalamic nucleus		+	+
Posterior thalamic nucleus	+	+	+
Midline thalamic nucleus	+	+	+
Brainstem			
Ventral tegmental area	+	+	+
Locus coeruleus	+	+	+
Raphe	+	+	+
Parabrachial nucleus		+	+
Central gray			+
Nucleus of solitary tract			+

Attention and Vigilance

Projections from the amygdala to the ventral tegmental area[36] may mediate stress-induced increases in dopamine metabolites in the prefrontal cortex.[37] Direct amygdaloid projections to the dendritic field of the locus coeruleus,[36,38] or indirect projections via the paragigantocellularis nucleus,[39] may mediate the response of cells in the locus coeruleus to conditioned fear stimuli[40] and may also be involved in other actions of the locus coeruleus linked to fear and anxiety.[41] Direct projections of the amygdala to the lateral dorsal tegmental nucleus[31] and parabrachial nuclei, which have cholinergic neurons that project to the thalamus,[42] may mediate increases in synaptic transmission in thalamic sensory relay neurons[42] during states of fear. This cholinergic activation, along with increases in thalamic transmission accompanying activation of the locus coeruleus,[43] may thus lead to increased vigilance and superior signal detection in a state of fear or anxiety.

As emphasized by Kapp et al.,[44] in addition to its direct connections to the hypothalamus and brainstem, the central nucleus of the amygdala also has the potential for indirect widespread effects on the cortex via its projections to cholinergic neurons located within the sublenticular substantia innominata, which in turn project to the cortex. In fact, the rapid development of conditioned bradycardia during aversive conditioning, critically dependent on the amygdala, may not simply be a marker of an emotional state of fear, but instead may be a more general process reflecting an increase in attention. In the rabbit, low-voltage fast EEG activity, generally considered a state of cortical readiness for processing sensory information, is acquired during aversive conditioning at the same rate as conditioned bradycardia.[45]

Motor Behavior

Release of norepinephrine onto motor neurons, either via amygdala activation of the locus coeruleus or amygdaloid projections to serotonin-containing raphe neurons,[46] could lead to enhanced motor performance during a state of fear, since both norepinephrine and serotonin facilitate excitation of motor neurons.[47,48]

Direct projections of the central nucleus of the amygdala to the nucleus reticularis pontis caudalis,[49,50] as well as indirect projections to this nucleus via the central gray, probably are involved in fear potentiation of the startle reflex.[51,52] Direct projections to the lateral tegmental field, including parts of the trigeminal and facial motor nuclei,[53,54] may mediate some of the facial expressions of fear as well as potentiation of the eyeblink reflex.[54,55]

FIGURE 5–2. Schematic diagram showing direct connections between the central nucleus of the amygdala and a variety of target areas that may be involved in different animal tests of fear and anxiety. N. = nucleus; Hypothal. = hypothalamus; CER = conditioned emotional response.

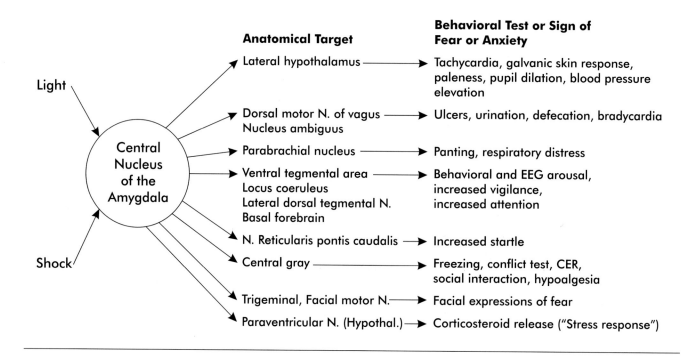

The amygdala also projects to regions of the central gray[56] that appear to be a critical part of a general defense system[57–60] and that have been implicated in conditioned fear in a number of behavioral tests involving freezing,[61] sonic and ultrasonic vocalization,[62] and stress-induced hypoalgesia.[61,63,64]

ELICITATION OF FEAR BY ELECTRICAL OR CHEMICAL STIMULATION OF THE AMYGDALA

Electrical stimulation or abnormal electrical activation of the amygdala (via temporal lobe seizures) can produce a complex pattern of behavioral and autonomic changes that, taken together, highly resemble a state of fear.

Autonomic and Hormonal Measures

As outlined by Gloor,[65]

> The most common affect produced by temporal lobe epileptic discharge is fear.... It arises "out of the blue." Ictal fear may range from mild anxiety to intense terror. It is frequently, but not invariably, associated with a rising epigastric sensation, palpitation, mydriasis, and pallor and may be associated with a fearful hallucination, a frightful memory flashback, or both. (p. 513)

In humans, electrical stimulation of the amygdala elicits feelings of fear or anxiety as well as autonomic reactions indicative of fear.[66,67] Although other emotional reactions occasionally are produced, the major reaction is one of fear or apprehension.

In unanesthetized animals, increases in blood pressure have been found following local infusion of either L-glutamate,[68] the cholinergic muscarinic agonist carbachol,[69] or the $GABA_A$ antagonist bicuculline methiodide.[70] Repeated infusion of initially subthreshold doses of bicuculline into the anterior basolateral nucleus led to a "priming" effect in which increases in heart rate and blood pressure were observed after 3 to 5 infusions.[71] This change in threshold lasted at least 6 weeks and could not be ascribed to mechanical damage or generalized seizure activity based on EEG measurements. It is possible, therefore, that long-term stress could lead to similar priming effects, which would then make the amygdala more reactive to subsequent stressors, thereby leading to certain types of psychiatric disorders.

Amygdala stimulation can also produce gastric ulceration,[72] which can be associated with chronic fear or anxiety. Electrical stimulation of the amygdala also alters respiration,[73] a prominent symptom of fear, especially in panic disorder. Electrical stimulation of the amygdala has been shown to increase plasma levels of corticosterone, indicating an excitatory effect of the amygdala on the hypothalamic-pituitary-adrenal axis.[74,75]

Attention and Vigilance

Electrical stimulation of sites in the central nucleus that produce bradycardia[76] also produce low-voltage fast EEG activity in both rabbits[77] and rats,[78] which can be blocked by systemic administration of cholinergic antagonists.[77,78] In the cat, electrical stimulation of the dorsal amygdala, including some sites in the central nucleus, elicited EEG desynchronization; this was not blocked by complete midbrain transection,[79] suggesting that it involved rostral projections from the amygdala to the basal forebrain. In fact, EEG desynchronization produced by amygdala stimulation can be blocked by local infusion of lidocaine into the substantia innominata–ventral pallidum of the basal forebrain ipsilateral, but not contralateral, to the site of stimulation.[78] In addition, electrical stimulation of the central nucleus elicits pupillary dilation and pinna orientation,[80,81] both of which would be associated with an increase in sensory processing; indeed, an attention or orienting reflex was the most common response elicited by electrical stimulation of the amygdala in cats described in references.[80,81] These and other observations have led Kapp et al.[44] to hypothesize that the central nucleus and its associated structures "function, at least in part, in the acquisition of an increased state of nonspecific attention or arousal manifested in a variety of CRs [conditioned responses] which function to enhance sensory processing" (p. 241).

Motor Behavior

Electrical or chemical stimulation of the central nucleus of the amygdala produces a cessation of ongoing behavior,[20,69,80–84] a critical component in several animal models such as freezing, the operant conflict test, the conditioned emotional response, and the social interaction test. Electrical stimulation of the amygdala also elicits jaw movements[20,80,85,86] and activation of facial motoneurons,[87] which probably mediate some of the facial expressions seen during the fear reaction. These motor effects may be indicative of a more general effect of amygdala stimulation, namely that of modulating brainstem reflexes such as the massenteric,[88,89] baroreceptor,[90–92] nictitating membrane,[54] eyeblink,[55] and startle reflexes.[93,94] The startle reflex is also increased by local infusion of N-methyl-D-aspartate (NMDA)[95] as well as the metabotropic glutamate receptor agonist trans-(±)-1-aminocyclopentane-1,3, dicarboxylate (trans-ACPD)[96] into the central nucleus of the amygdala.

Viewed in this way, the pattern of behaviors seen during fear may result from activation of a single area of the brain (the amygdala), which then projects to a variety of target areas that are themselves critical for each of the specific symptoms of fear (the expression of fear) as well as the experience of fear. Moreover, it must be assumed that all of these connections are already formed in an adult organism, because electrical stimulation produces these effects in the absence of prior explicit fear conditioning.

Thus, much of the complex behavioral pattern seen during a state of "conditioned fear" has already been "hard wired" during evolution. In order for a formerly neutral stimulus to produce the constellation of behavioral effects used to define a state of fear or anxiety, it is only necessary for that stimulus to now activate the amygdala, which in turn will produce the complex pattern of behavioral changes by virtue of its innate connections to different brain target sites.

Viewed in this way, plasticity during fear conditioning probably results from a change in synaptic inputs to or in the amygdala, rather than from a change in its efferent target areas. The ability to produce long-term potentiation in the amygdala[97-100] and the finding that local infusion of NMDA antagonists into the amygdala blocks the acquisition[101-106] and extinction[107] of fear conditioning are both consistent with this hypothesis.

EFFECTS OF AMYGDALA LESIONS ON CONDITIONED FEAR

The Klüver-Bucy Syndrome

In 1939, following earlier work,[108] Klüver and Bucy[109] described the now classic behavioral syndrome of monkeys with bilateral removal of the temporal lobes, including the amygdala, hippocampus, and surrounding cortical areas. Following such lesions, the monkeys developed "psychic blindness": they would approach animate and inanimate objects without hesitation and examine these objects by mouth rather than by hand, whether the object was a piece of food, feces, a snake, or a light bulb. They also had a strong tendency, almost a compulsion, to attend to and examine every visual stimulus that came into their field of view and showed a marked change in emotional behavior. These monkeys had a striking absence of emotional motor and vocal reactions normally associated with stimuli or situations eliciting fear and anger. As described by Klüver and Bucy,

The typical reaction of a "wild" monkey when suddenly turned loose in a room consists in getting away from the experimenter as rapidly as possible. It will try to find a secure place near the ceiling or hide in an inaccessible corner where it cannot be seen. If seen, it will either crouch and, without uttering a sound, remain in a state of almost complete immobility or suddenly dash away to an apparently safer place. This behavior is frequently accompanied by other signs of strong emotional excitement. In general, all such reactions are absent in the bilateral temporal monkey. Instead of trying to escape, it will contact and examine one object after another or other parts of the objects, including the experimenter, stranger, or other animals.... Expressions of emotions, such as vocal behavior, "chattering," and different facial expressions, are generally lost for several months. In some cases, the loss of fear and anger is complete. (p. 991)

In addition, many monkeys showed striking increases in heterosexual and homosexual behavior never previously observed in this monkey colony.

Lesions of the temporal lobe also were reported to cause profound changes in the social behavior of monkeys both in the laboratory and in the wild. Following temporal lobe lesions, monkeys rapidly fell in rank within dominance hierarchies established in monkey colonies (see review[110]). Lesioned monkeys now tried to fight with more dominant, larger monkeys, leading to frequent and often severe wounds. In the wild, these inappropriate interactions with other monkeys led to repeated attacks, social isolation, and eventual death.[111,112]

Subsequent studies have shown that all of the emotional components of the Klüver-Bucy syndrome can be reproduced by damage to the amygdala and the surrounding cortical tissue found in the perirhinal and entorhinal cortex.[113-118] The tameness and excessive orality can be reproduced by lesions restricted to only the amygdala.[119] In an extensive series of experiments evaluating both the emotional and the memory effects of lesions of the amygdala versus the hippocampus versus surrounding cortical areas, Zola-Morgan et al.[120] found that lesions of the amygdala disrupted emotional behavior to a set of novel objects, whereas lesions of the hippocampus or surrounding cortical areas did not. Conversely, damage to the hippocampus and the anatomically related perirhinal and parahippocampal cortex impaired memory but not emotional behavior. Moreover, combined damage to the amygdala and hippocampus had no greater effect on memory or emotion than did damage to either structure alone.

Although humans only rarely show the full-blown Klüver-Bucy syndrome following lesions restricted to the amygdala, they consistently show a blunting of emotional reactivity. This finding, along with the frequent change in emotional behaviors seen in Alzheimer's disease and other neurological diseases associated with amygdala pathology, is further evidence for the role of the amygdala in human emotion.[121,122] In fact, recent data using magnetic resonance imaging have shown bilateral activation of the amygdala when subjects view slides with high negative emotional content.[123]

It is not surprising, therefore, that several authors have seen a connection between the social inappropriateness following temporal lobe damage in monkeys and some of the negative or deficit symptoms in schizophrenia, such as inappropriate mood, flat affect, social isolation, poverty of speech, and difficulty in identifying the emotional status of other people.[121,124]

Facial Recognition and Classical Fear Conditioning in Humans

In nonhuman primates[125-127] and humans,[128,129] cells have been found that respond selectively to faces or direction of gaze.[130] In humans, removal of the amygdala has been associated with an impairment of memory for faces[131-134] and deficits in recognition of emotion in people's faces and interpretation of gaze angle.[134] In a very rare case involving bilateral calcification confined to the amygdala (Urbach-Wiethe disease), Patient S. M. could not identify the emotion of fear in pictures of human faces and could not draw a fearful face, even though other emotions such as happiness, sadness, anger, and disgust were identified and drawn within the normal range. Furthermore, she had no difficulty in identifying the names of familiar faces.[135,136] The deficit in recognizing facial expressions of fear only seemed to occur after bilateral amygdala damage, whereas several patients with unilateral lesions had difficulty in naming familiar faces.[136] On the basis of this double dissociation, Adolphs et al.[136] propose that "the amygdala is required to link visual representation of facial expression, on the one hand, with representations that constitute the concept of fear, on the other" (p. 5879). Two other patients with Urbach-Wiethe disease did not show the normal enhancement in memory for emotional material[137,138] which is known from animal work to be dependent on activation of beta-noradrenergic receptors in the amygdala.[139]

Autonomic and Hormonal Measures

Patients with unilateral[140] or bilateral[141] lesions of the amygdala also have been reported to have deficits in classical fear conditioning in studies using the galvanic skin response as a measure. In monkeys, removal of the amygdala decreases reactivity to sensory stimuli measured with the galvanic skin response.[142,143] In both adult[144-149] and infant mammals,[150] lesions of the central nucleus block conditioned changes in heart rate. In birds, lesions of the archistriatum, believed to be homologous with the mammalian amygdala, block heart rate acceleration in response to a cue paired with a shock.[151] Ibotenic acid lesions of the central nucleus of the amygdala,[152] localized cooling of this nucleus,[153] or lesions of the lateral amygdala nucleus[154,155] also block conditioned changes in blood pressure. Ablation of the central nucleus can reduce the secretion of ACTH,[156] corticosteroids,[157,158] and prolactin.[157] Neurotoxic lesions of the central and basolateral nuclei also block conditioned increases in corticosteroid release.[159] Lesions of the amygdala reduce stress-induced increases in dopamine release in the frontal cortex following mild footshock or exposure to a novel environment[160] or to a cue previously paired with footshock.[159]

Attention and Vigilance

Gallagher, Holland, and co-workers have found results consistent with an attentional role of the central nucleus of the amygdala.[161-163] In these studies, a CS such as a light or a tone is paired with receipt of food. Initially, rats rear when the light goes on or show small startle responses when the tone goes on, both of which habituate with stimulus repetition. When these stimuli are then paired with food, these initial orienting responses return (CS-generated CRs), along with approach behavior to the food cup (unconditioned stimulus [US]–generated responses). Neurotoxic lesions of the central nucleus of the amygdala severely impair CS-generated responses without having any effect on unconditioned orienting responses or US-generated responses. On the basis of these data, the authors concluded that the central nucleus of the amygdala modulates attention to a stimulus that signals a change in reinforcement. Further work seemed to confirm this hypothesis. For example, rats with lesions of the central nucleus fail to benefit from procedures that normally facilitate attention to conditioned stimuli.[162,163]

Differential roles of the central and basolateral nuclei have been found in a phenomenon known as taste-potentiated odor aversion learning. In this test, which requires processing information in two sensory modalities, rats develop aversions to a novel odor paired with illness only when the odor is presented in compound with a distinctive gustatory stimulus. Electrolytic[164] or chemical lesions[165] of the basolateral but not the central nucleus of the amygdala blocked taste-potentiated odor aversion learning even though they had no effect on taste aversion learning itself. Depletion of dopamine and norepinephrine in the amygdala via local infusion of 6-hydroxydopamine (4 μg/0.5 μl) also blocked odor aversion but not taste aversion.[166] Local infusion of NMDA antagonists into the basolateral nucleus also blocked the acquisition, but not the expression, of taste-potentiated odor aversion but had no effect on taste aversion learning itself.[102] More recently, in a very important study, neurotoxic lesions of the basolateral nucleus, but not the central nucleus, of the amygdala were reported to interfere with second-order conditioning and reinforcer devaluation.[167]

On the basis of these and other data, Hatfield et. al.[167] suggest that the central nucleus of the amygdala regulates attentional processing of cues during associative conditioning, whereas the basolateral nucleus of the amygdala is critically involved in "associative learning processes that give conditioned stimuli access to the motivation value of their associated unconditioned stimuli" (p. 5264).

Somewhat similar conclusions have been reached by Halgren,[168] based on recording of stimulus-evoked electrical activity in the amygdala in epileptic patients. In these studies, subjects are presented with a series of visual or auditory stimuli and are instructed to ignore some of them and attend to others. Averaged evoked responses show a prominent negative-positive component occurring roughly 200–300 ms after stimulus onset (N200/P300). These components, especially N200, are prominent within the amygdala and are much larger when elicited by a stimulus to which the subject is asked to attend. Halgren summarizes the cognitive conditions that evoke the N200/P300 as being stimuli that are novel or that are signals for behavioral tasks and hence require to be attended to and processed. Moreover, these components, along with other autonomic measures of the orienting reflex, seem to form an overall reaction of humans to stimuli that demand their evaluation.

Motor Behavior

Lesions of the amygdala eliminate or attenuate conditioned freezing normally seen in response to a stimulus formerly paired with shock,[26,147–149,159,169–179] a novel environment,[180] or a dominant male rat,[181,182] or presented during a continuous passive avoidance test.[183] Inactivation of the amygdala by direct infusion of lidocaine[171] or muscimol[184] prior to testing reduced conditioned freezing. However, the same doses infused prior to training did not fully block acquisition of contextual fear conditioning.[171,184]

Lesions of the amygdala counteract the normal reduction of bar pressing or licking in the operant conflict test[185–187] and the conditioned emotional response paradigm.[188,189] In birds, lesions of the archistriatum also block the development of a conditioned emotional response.[190]

Lesions of the central nucleus or of the lateral and basolateral nuclei of the amygdala block high-frequency vocalizations,[159] as well as reflex facilitation such as fear-potentiated startle[150,172,191–195] or tone-enhanced excitability of the nictitating membrane response.[196] Lesions of the amygdala also produce a dramatic decrease in shock-probe avoidance[197] but do not affect more active kinds of anxiogenic behaviors such as open arm time in the plus maze or burying a noxious shock probe, which are affected by lesions of the septum[197] as well as anxiolytic drugs. Furthermore, the magnitude of the anxiolytic effects after combined lesions of both structures was comparable to their magnitude after individual lesions, suggesting that the septum and amygdala independently control different fear-related behaviors.

Hypoalgesia

Lesions of the central nucleus of the amygdala block conditioned analgesia produced by reexposure to cues associated with noxious stimulation.[170,198,199] This effect does not seem to be due to a blockade of learning because the lesions can be made after training and still block the expression of conditioned analgesia.[200] NMDA lesions of the central, but not the basolateral or medial, nucleus of the amygdala blocked antinociception produced by a low dose of morphine in the formalin[201] or heat-evoked tail-flick test.[202] Direct infusion of lidocaine into the central nucleus had the same effect in the tail-flick test.[202]

These findings, along with a large literature implicating the amygdala in many other measures of fear, such as active and passive avoidance[5,23,82,139,203,204] and evaluation and memory of emotionally significant sensory stimuli,[139,205–217] provide strong evidence for a crucial role of the amygdala in fear.

EFFECTS OF AMYGDALA LESIONS ON UNCONDITIONED FEAR

Autonomic and Hormonal Measures

Lesions of the amygdaloid complex inhibit adrenocortical responses following olfactory or sciatic nerve stimulation[218] or exposure to visual or auditory stimuli.[219] In the study of exposure to stimuli,[219] lesions of the medial or central nuclei blocked these effects on the hypothalamic-pituitary axis, whereas lesions of the basal nucleus did not. Lesions of the amygdaloid complex also attenuate the compensatory hypersecretion of ACTH that normally occurs following adrenalectomy.[220] Lesions of the central nucleus have been found to significantly attenuate ulceration produced by restraint[221,222] or shock stress[223] or elevated levels of plasma corticosterone produced by restraint stress.[156,224] Lesions of the medially projecting component of the ventroamygdalofugal pathway, which carries the fibers connecting the central nucleus of the amygdala to the hypothalamus, attenuate the increase in ACTH secretion following adrenalectomy, whereas lesions of the stria terminalis do not.[220] Finally, lesions of the amygdala have been reported to block the capability of high levels of noise, which may be an unconditioned fear stimulus,[225] to produce hypertension[226] or activation of tryptophan hydroxylase.[227]

Motor Behavior

Lesions of the amygdala are known to block several measures of innate fear in different species.[169,204] Lesions of the cortical amygdaloid nucleus and perhaps the central nucleus markedly reduce emotionality in wild rats, measured in terms of flight and defensive behaviors.[228,229] Large amygdala lesions, or those which have damaged the cortical, medial, and in several cases the central nucleus, dramatically increase the number of contacts a rat will make with a sedated cat.[169] In fact, some of these lesioned animals crawl all over the cat and even nibble its ear, a behavior never shown by the nonlesioned animals. Following lesions of the archistriatum, birds become docile and show little tendency to escape from humans,[230,231] consistent with a general taming effect of amygdala lesions reported in many species.[232]

Lesions of the central nucleus of the amygdala block the increase in acoustic startle amplitude often observed after a series of footshocks.[233] The increase may represent an unconditioned effect of shock on the startle reflex,[234] although it might also represent a very rapid conditioned increase in startle due to contextual conditioning.[235]

Hypoalgesia

Lesions of the central nucleus block unconditioned analgesia to cat exposure,[198] loud noise,[236] or footshock,[198] but see Watkins et al.[199] Lesions of the central nucleus, but not the basolateral amygdala, tend to blunt analgesic effects of systemic administration of flumazenil as measured with the tail-flick test.[237]

Other data indicate that the amygdala appears to be involved in some types of aversive conditioning, but the extent of its involvement may depend on the exact unconditioned aversive stimulus that is used. For example, electrolytic lesions of the basolateral nucleus[238] or fiber-sparing chemical lesions of most of the amygdaloid complex[239] attenuate thirsty rats' avoidance of an electrified water spout through which they previously were accustomed to receive water. Importantly, however, these same lesioned animals did not differ from controls in the rate at which they found the water spout over successive test days or their avoidance of the water spout when quinine was added to the water.[239] This result led Cahill and McGaugh[239] to suggest that "the degree of arousal produced by the unconditioned stimulus, and not the aversive nature per se, determined the level of amygdala involvement" (p. 541). Perhaps this formulation may explain some of the apparently contradictory results concerning the effects of amygdala lesions on conditioned taste aversion (see reviews[6,240,241]).

EFFECTS OF LOCAL INFUSION OF DRUGS INTO THE AMYGDALA ON MEASURES OF FEAR AND ANXIETY

Clinically, fear is considered to be more stimulus-specific than anxiety, despite very similar symptoms. Figure 5–2 (p. 74) suggests that spontaneous activation of the central nucleus of the amygdala would produce a state resembling fear in the absence of any obvious eliciting stimulus. In fact, as mentioned earlier, fear and anxiety often precede temporal lobe epileptic seizures,[65,67] which are usually associated with abnormal electrical activity of the amygdala.[242]

An important implication of this distinction is that treatments that block conditioned fear might not necessarily block anxiety. For example, if a drug decreased transmission along a sensory pathway required for a conditioned stimulus to activate the amygdala, then that drug might be especially effective in blocking conditioned fear. However, if anxiety resulted from activation of the amygdala *not* involving that sensory pathway, then that drug might not be especially effective in reducing anxiety. On the other hand, drugs that act specifically in the amygdala should affect both conditioned fear and anxiety. Moreover, drugs that act at various target areas might be expected to provide selective actions affecting some, but not all, of the somatic symptoms associated with anxiety.

It is also probable that certain neurotransmitters within the amygdala especially may be involved in fear and anxiety. For example, the amygdala has a high density of corticotropin-releasing hormone (CRH) receptors[243] and CRH nerve endings,[244] and several recent papers indicate that stress, as well as conditioned fear, can induce a release of CRH in the amygdala that results in various anxiogenic effects. For example, 20 minutes of restraint stress led to an increase of extracellular CRH-like immunoreactivity levels in the amygdala as measured by microdialysis.[245] In ethanol-dependent rats there was an increase in CRH-like immunoreactivity in microdialysis samples 6 to 8 hours after ethanol removal. The CRH release reached a peak 10 to 12 hours after ethanol removal, at the same time that anxiogenic-like behaviors were observed in the elevated plus maze. CRH-like immunoreactivity also was reported to have decreased by 58% in the amygdala 48 hours after withdrawal from chronic cocaine use, suggesting increased release and degradation of CRH following drug withdrawal.[246]

Individual differences in general levels of fear or anxiety may be related to differences in the amount of CRH found in the amygdala. For example, Fawn-hooded rats, a strain derived from Wistar, Long-Evans, and brown rats, show more freezing in response to stress, have an increased preference for alcohol, develop adult-onset hypertension, and have elevated levels of urinary catecholamines. Compared with Wistar rats, this strain also has higher levels of CRH mRNA in the central nucleus of the amygdala.[247] In addition, rats that experienced prenatal stress, which apparently developed high levels of anxiety, had higher CRH levels in amygdala tissue compared with nonstressed controls, and the higher CRH levels were associated with a higher depolarization (KCl)–induced CRH release from amygdala minces.[248]

If the amygdala is critically involved in fear and anxiety, then drugs that reduce fear or anxiety clinically may well act within the amygdala. A variety of measures suggest that the anxiolytic effects of both opiates and benzodiazepines may result from binding to receptors in the amygdala.

Autonomic and Hormonal Measures

CRH (30 ng) infused into the central nucleus of the amygdala increased heart rate compared with heart rates in vehicle-infused animals.[249] Pretreatment with alpha-helical CRH [9-41], a CRH antagonist, dose-dependently reduced the CRH-induced tachycardia at doses that had no effects on heart rate by themselves. On the other hand, there were no effects of local infusion of CRH or alpha-helical CRH [9-41] on plasma levels of epinephrine, norepinephrine, or corticosterone. Boadle-Biber et al.[250] found that CRH infused into the central nucleus increased tryptophan hydroxylase activity measured in the cortex and that this effect of CRH could be blocked by prior administration of alpha-helical CRH [9-41] into the central nucleus. Moreover, like amygdala lesions,[227] infusion of alpha-helical CRH [9-41] into the amygdala blocked the increase in tryptophan hydroxylase in response to loud noise.[250]

Motor Behaviors

Benzodiazepines, GABA Agonists and Antagonists: Many studies have shown that local infusion of benzodiazepines into the amygdala has anxiolytic effects in the operant conflict test,[251–258] freezing,[259,260] the light-dark box measure in mice,[261] shock probe avoidance,[262] and the elevated plus maze,[263,264] and that it antagonizes the discriminative stimulus properties of pentylenetetrazol.[265] Infusion of diazepam into the amygdala also accelerated the rate of between-session habituation of the startle response.[260] This finding is consistent with the idea that loud startle stimuli produce contextual fear conditioning that competes with the expression of long-term habituation.[266] A reduction of contextual fear conditioning via diazepam infusion into the amygdala should thus increase long-term habituation, as it does after systemic administration.[260]

The anticonflict effect[251,254,257] or the decrease in shock probe avoidance[262] can be reversed by systemic administration of the benzodiazepine antagonist flumazenil or coadministration into the amygdala of the γ-aminobutyric acid (GABA) antagonist bicuculline,[255] and these effects can be mimicked by local infusion into the amygdala of GABA[251] or the GABA agonist muscimol.[255] In general, anxiolytic effects of benzodiazepines occur after local infusion into the lateral and basolateral nuclei[253–255,258,263] (the nuclei of the amygdala that have high densities of benzodiazepine receptors) and not after local infusion into the central nucleus,[253,255,263] although effects have been reported after infusion into the central nucleus.[187,256,267] More recently, consistent with earlier work,[263] Pesold and Treit[264] reported that local infusion of midazolam into the basolateral nucleus had an anxiolytic effect in the plus maze but did not impair shock-probe avoidance, whereas infusion into the central nucleus impaired shock-probe avoidance but did not affect plus maze performance. Both of the site-specific effects of midazolam could be blocked by systemic administration of flumazenil.

Sanders and Shekhar[71] found that infusion of the $GABA_A$ antagonists bicuculline or picrotoxin into the anterior basolateral nucleus had anxiogenic effects in the social interaction test. These same doses had no effect when infused into the central nucleus. Conversely, infusion of the GABA$_A$ agonist muscimol into the central nucleus had an anxiolytic effect, whereas it had no effect when infused into the basolateral nucleus. These data suggest a tonic, and perhaps maximal, level of GABA inhibition in the basolateral, but not the central, nucleus of the amygdala. Repeated infusion of initially subthreshold doses of bicuculline into the anterior basolateral nucleus led to a "priming" effect in which increases in heart rate and blood pressure were observed after 3 to 5 infusions.[71] The increases were accompanied by anxiogenic effects in the conflict and social interaction tests.

Taken together, these results suggest that drug actions in the amygdala may be sufficient to explain both fear-reducing and anxiety-reducing effects of various drugs given systemically. In fact, local infusion into the amygdala of the benzodiazepine antagonist flumazenil significantly attenuated the anticonflict effect of the benzodiazepine agonist chlordiazepoxide given systemically.[251] Similarly, Sanders and Shekhar[268] found that local infusion of the benzodiazepine antagonist flumazenil or the GABA antagonist bicuculline into the anterior part of the basolateral nucleus, at doses that had no anxiogenic effects by themselves, blocked the anxiolytic effect of systemically administered chlordiazepoxide on the social interaction test. These are very powerful experimental designs and strongly implicate the amygdala in mediating the anxiolytic effects of benzodiazepines.

Nonetheless, it should be emphasized that benzodiazepines can still have anxiolytic effects in animals with lesions of the amygdala.[185,269–271] Although these important results could be interpreted to indicate that the amygdala is not necessary for mediating anxiolytic effects of benzodiazepines, such a conclusion is difficult to reconcile with the many studies outlined above. Hence, it may be that other brain structures take over for the amygdala after it is lesioned (see Kim and Davis[194]) and that benzodiazepine binding in these other structures accounts for the anxiolytic effects after amygdala lesions.

Corticotropin-Releasing Hormone: Several studies now suggest an important role for CRH in the amygdala in mediating various anxiogenic effects in the plus maze as well as other tests. Local infusion of alpha-helical CRH [9-41] (250 ng) in the central nucleus attenuated the anxiogenic effect of social defeat[272] or ethanol withdrawal in ethanol-dependent rats[273] in the plus maze. Higher doses were not effective, perhaps due to partial agonist effects of this compound. Doses effective in the plus maze had no effect on plasma ACTH or corticosterone release, although these values returned to baseline earlier than in controls after antagonist infusion. The antagonist had no effect on overall activity or percentage of time in open arms of the maze in rats not dependent on ethanol. Liebsch et al.[274] found that local infusion into the central nucleus of the CRH receptor mRNA antisense oligodeoxynucleotide had an anxiolytic effect in the plus maze in rats that had previously experienced defeat stress. Infusion of the scrambled sequence oligodeoxynucleotide had no effect.

Using freezing as a measure, Swiergiel et al.[275] found that local infusion of low doses (50 and 100 ng) of alpha-helical CRH [9-41] into the central nucleus reduced the duration of freezing to an initial shock treatment. A higher dose (200 ng) was not effective. Infusion of the antagonist into the central nucleus immediately prior to reexposure to the shock box 24 hours later also attenuated freezing duration, indicating that the reduction in freezing by alpha-helical CRH [9-41] was not due to an alteration in sensitivity to the footshock.

In all of the above examples, infusion of the CRH antagonists produced behavioral effects in animals that had undergone prior stress, a condition that may be necessary to detect effects following local infusion into the amygdala. For example, although CRH infused into the central nucleus could produce increased grooming and exploration in animals tested under stress-free conditions (that is, in the home cage of the rat), local infusion of alpha-helical CRH [9-41] (0.1 and 1.0 µg/cannula) had no effect on activity under these same stress-free conditions.[276]

On the other hand, enhancement of the startle reflex, either by infusion of CRH intraventricularly (CRH-enhanced startle) or by conditioned fear, does not seem to depend on activation of CRH receptors in the amygdala, at least not in its central nucleus. Although large electrolytic lesions of the amygdala were found to block CRH-enhanced startle,[277] local infusion of CRH into the amygdala failed to increase startle in this study involving a large number of animals and several placements within the amygdala. Moreover, recent experiments using fiber-sparing lesions of the central and/or basolateral nuclei of the amygdala failed to block CRH-enhanced startle.[278] In contrast, neurotoxic lesions of the bed nucleus of the stria terminalis completely block CRH-enhanced startle, and direct infusion of CRH into this nucleus increases acoustic startle.[278] In addition, local infusion of alpha-helical CRH [9-41] into the central nucleus of the amygdala did not block fear-potentiated startle.[278]

Other Compounds: On the basis of a series of observations, Deakin and Graeff[279] hypothesized that serotonin

(5-hydroxytryptamine; 5-HT) enhances fear or anxiety in the amygdala, whereas it has the opposite effect in the dorsal central gray. A great deal of evidence supports the anti-anxiety effects of 5-HT in the dorsal central gray (see review[12]), although fewer direct data are available concerning the role of 5-HT in the amygdala. Local infusion of 5-HT or the 5-HT$_{1A}$ agonist 8-hydroxydipropylaminotetralin (8-OH-DPAT) into the amygdala have been reported to produce anxiogenic effects in the conflict test,[251,267] whereas infusion of the 5-HT$_2$ antagonist ketanserin has an anxiolytic effect.[251] On the other hand, infusion into the amygdala of the 5-HT$_{1A}$ agonists 8-OH-DPAT, buspirone, or ipsapirone reduced shock-induced vocalization.[280] In this case, however, the 5-HT$_{1A}$ agonists were infused into the posteromedial cortical amygdaloid nucleus rather than the basolateral nucleus[251] or the central nucleus.[267] Interestingly, the corticomedial nucleus, rather than the central or basolateral nuclei, was also the most effective site for morphine analgesia in the shock-induced jump response tested in freely moving rats[281,282] (see the next section).

Compounds acting as 5-HT$_3$ receptor subtype antagonists have been reported to produce anxiolytic effects after local infusion into the amygdala.[283,284] Such infusions also can block some of the signs of withdrawal following subchronic administration of diazepam, ethanol, nicotine, or cocaine[285] or increases in levels of dopamine or the serotonin metabolite 5-hydroxyindoleacetic acid (5-HIAA) in the amygdala after activation of dopamine neurons in the ventral tegmental area.[286] In addition, local infusion of a 5-HT$_{1A}$ antagonist into the central nucleus has been reported to have an anticonflict effect.[267]

Activation of neuropeptide Y$_1$ receptors in the central nucleus has been reported to produce selective anxiolytic effects in the conflict test,[287] and this effect could be blocked by prior intraventricular administration of antisense inhibition of Y$_1$ receptor expression,[288] which itself produced an anxiogenic effect.[288,289] Local infusions of opiate agonists in the central nucleus were reported to have anxiolytic effects in the social interaction test.[290]

Hypoalgesia

Infusion of morphine,[281,282,291] enkephalinase inhibitors,[292] or mu opioid agonists[293] into the amygdala has antinociceptive effects, although this has not always been found.[294] Morphine and mu agonists seem to be most effective in the basolateral nucleus when the tail-flick test is measured in anesthetized rats.[291,293] In waking rats, when jump threshold is used as a measure, the most sensitive placements seem to occur in the corticomedial nucleus and not in the basolateral or central nucleus.[281,282] Morphine infused into the corticomedial nucleus also reduced open field defecation but had no effect on tail-flick latency.[282] Antinociception also occurs after injection of neurotensin[295] into the central nucleus of the amygdala or carbachol into the basolateral or medial amygdala nuclei.[296] Unilateral local infusion of morphine into the central nucleus of the amygdala attenuated both the acquisition and expression of inhibitory avoidance and conditioned hypoalgesia measured with the formalin test,[297] which is sensitive to naloxone. These effects of morphine in the amygdala could be reversed by coadministration of naloxone into the amygdala, and the effects on acquisition could not be explained by state-dependent learning. Interestingly, morphine infusions did not block conditioned analgesia when rats were tested on the heated floor, a naloxone-insensitive form of hypoalgesia. Hence, the amygdala did not seem critical for all types of conditioned analgesia. Although Good and Westbrook[297] did not find effective sites for morphine in the basolateral amygdala, Harris and Westbrook[298] did see an attenuation of both hypoalgesia and inhibitory avoidance after infusion of midazolam in the basolateral nucleus.

THE ROLE OF CONNECTIONS BETWEEN THE AMYGDALA AND THE CORTEX IN FEAR AND ANXIETY

Thus far we have concentrated on connections between the central nucleus of the amygdala and brain areas related to the expression of fear and anxiety. It is also the case, however, that the amygdala, especially the basolateral amygdala, has extensive connections to many cortical areas such as the frontal cortex (see Amaral et al.[299] for review of primate data) that could be involved in the experience or perception of fear and anxiety. As outlined earlier, only in humans is it possible to directly measure the experience of fear because this can be understood only through verbal report. Moreover, to actually determine if changes in the experience or perception of emotion result from a disconnection of the amygdala from a given cortical structure, cases would have to be found that included an intact amygdala on one side of the brain combined with unilateral damage to a cortical region on the other side of the brain known to receive amygdala input. To my knowledge, such an analysis has not been systematically carried out in humans. However, exactly this approach is being used experimentally in monkeys to study the role of amygdala–frontal connections in an appetitive memory task.[300] It would thus be extremely interesting to know how such disconnections might affect fear-related behaviors in monkeys and other species.

Recently, connections between the prefrontal cortex and the amygdala have been implicated in extinction of fear, with freezing in rats used as a measure of conditioned fear. Lesions of the ventral medial prefrontal cortex slowed the rate of extinction,[301,302] whereas repeated presentation of a conditioned fear stimulus normally leads to a loss of the fear reaction to that stimulus. In these same animals, extinction of conditioned fear to contextual cues was not impaired,[302] suggesting that the role of the medial prefrontal cortex in fear inhibition must be highly specific. However, in an extensive series of experiments, we have found normal rates of extinction of conditioned fear (using both freezing and fear-potentiated startle) to both explicit and contextual cues after total removal of the ventral medial prefrontal cortex.[303] Moreover, lesions of the ventral medial prefrontal cortex did not interfere with conditioned inhibition, which is a more direct measure of fear inhibition than is extinction.

Because the lesions in the studies by Morgan et al.[301,302] were performed before fear conditioning, the apparent blockade of extinction following ventral medial prefrontal cortex lesions may have been produced by an increase in the strength of original fear conditioning rather than by interference with the process of extinction. Although the two groups did not differ significantly in their level of freezing before the initiation of the extinction sessions, freezing to explicit cues often becomes maximal after a very few training trials, so that ceiling effects might well have been operating. Because rate of extinction can be a more sensitive index of the strength of original conditioning than the terminal level of performance prior to the initiation of extinction (see, for example, Annau and Kamin[304]), the slower rate of extinction in the lesioned animals may have simply reflected a stronger degree of original learning. The fact that the lesions had no effect on the rate of extinction of context conditioning, which clearly was not at the ceiling of the freezing scale, is consistent with this interpretation. Moreover, when ventral prefrontal cortex lesions are made after fear conditioning, but before extinction, the lesions have no effect on the rate of extinction.[305] These results indicate that the ventral medial prefrontal cortex is not essential for the inhibition of fear under a variety of circumstances. Clearly, further work needs to be done regarding the role of the cortex in fear and inhibition of fear.

CONCLUSIONS

An impressive amount of evidence from many different laboratories using a variety of experimental techniques indicates that the amygdala plays a crucial role in conditioned fear and anxiety, as well as attention. Many of the amygdaloid projection areas are critically involved in specific signs that are used to measure fear and anxiety. Electrical stimulation of the amygdala elicits a pattern of behaviors that mimic natural or conditioned states of fear. Lesions of the amygdala block innate or conditioned fear, as well as various measures of attention, and local infusion of drugs into the amygdala has anxiolytic effects in several behavioral tests. It is possible that long-term potentiation in the amygdala may mediate the development of fear conditioning. An NMDA-dependent form of long-term potentiation has been observed in the amygdala, and local infusion of NMDA antagonists into the amygdala blocks the formation of conditioned fear memories, as measured with several different tests of fear. A better understanding of brain systems that inhibit the amygdala, as well as the role of its very high levels of peptides, may eventually lead to the development of more effective pharmacological strategies for treating clinical anxiety disorders, and perhaps memory disorders as well.

REFERENCES

1. Mineka S, Davidson M, Cook M, et al: Observational conditioning of snake fear in rhesus monkeys. J Abnorm Psychol 1984; 93:355–372
2. McAllister WR, McAllister DE: Behavioral measurement of conditioned fear, in Aversive Conditioning and Learning, edited by Brush FR. New York, Academic Press, 1971, pp 105–179
3. Davis M: The role of the amygdala in fear and anxiety. Annu Rev Neurosci 1992; 15:353–375
4. Izquierdo I, Medina JH: Correlation between the pharmacology of long-term potentiation and the pharmacology of memory. Neurobiol Learn Mem 1995; 63:19–32
5. McGaugh J, Cahill L, Parent MB, et al: Involvement of the amygdala in the regulation of memory storage, in Plasticity in the Central Nervous System, edited by McGaugh J, Bermudez-Rattoni F, Praco-Alcala RA. Hillsdale, NJ, Lawrence Erlbaum, 1995, pp 17–39
6. Yamamoto T, Shimura T, Sako N, et al: Neural substrates for conditioned taste aversion in the rat. Behav Brain Res 1994; 65:123–137
7. Calvino B, Lagowska J, Ben-Ari Y: Morphine withdrawal syndrome: differential participation of structures located within the amygdaloid complex and striatum of the rat. Brain Res 1979; 177:19–34
8. Heinrichs SC, Menzaghi F, Schulteis G, et al: Suppression of corticotropin-releasing factor in the amygdala attenuates aversive consequences of morphine withdrawal. Behav Pharmacol 1995; 6:74–80
9. Kelsey JE, Arnold SR: Lesions of the dorsomedial amygdala, but not the nucleus accumbens, reduce the aversiveness of morphine withdrawal in rats. Behav Neurosci 1994; 108:1119–1127
10. Lagowska J, Calvino B, Ben-Ari Y: Intra-amygdaloid applications of

naloxone elicits severe withdrawal signs in morphine dependent rats. Neurosci Lett 1978; 8:241–245

11. Bandler R, Shipley MT: Columnar organization in the midbrain periaqueductal gray: modules for emotional expression? Trends Neurosci 1994; 17:379–389

12. Graeff FG, Silveira MCL, Nogueira RL, et al: Role of the amygdala and periaqueductal gray in anxiety and panic. Behav Brain Res 1993; 58:123–131

13. Amaral D: Memory: anatomical organization of candidate brain regions, in Handbook of Physiology, vol. 5: Higher Functions of the Brain, edited by Plum F. Bethesda, MD, American Physiological Society, 1987, pp 211–294

14. Burwell RD, Witter MP, Amaral DG: Perirhinal and postrhinal cortices of the rat: a review of the neuroanatomical literature and comparison with findings from the monkey brain. Hippocampus 1995; 5:390–408

15. Aggleton JP: A description of intra-amygdaloid connections in the old world monkeys. Exp Brain Res 1985; 57:390–399

16. Pitkanen A, Stefanacci L, Farb CR, et al: Intrinsic connections of the rat amygdaloid complex: projections originating in the basolateral nucleus. J Comp Neurol 1995; 356:288–310

17. Savander V, Go C-G, LeDoux JE, et al: Intrinsic connections of the rat amygdaloid complex: projections originating in the basal nucleus. J Comp Neurol 1995; 361:345–368

18. Davis M: The role of the amygdala in conditioned fear, in The Amygdala: Neurobiological Aspects of Emotion, Memory and Mental Dysfunction, edited by Aggleton J. New York, Wiley, 1992, pp 255–305

19. Gray TS: Autonomic neuropeptide connections of the amygdala, in Neuropeptides and Stress, edited by Tache Y, Morley JE, Brown MR. New York, Springer-Verlag, 1989, pp 92–106

20. Gloor P: Amygdala, in Handbook of Physiology, edited by Field J. Washington, DC, American Physiological Society, 1960, pp 1395–1420

21. Kapp BS, Pascoe JP, Bixler MA: The amygdala: a neuroanatomical systems approach to its contribution to aversive conditioning, in The Neuropsychology of Memory, edited by Butters N, Squire LS. New York, Guilford, 1984, pp 473–488

22. LeDoux JE: Emotion, in Handbook of Physiology, vol 5: Higher Functions of the Brain, edited by Plum F. Bethesda, MD, American Physiological Society, 1987, pp 416–459

23. Sarter M, Markowitsch HJ: Involvement of the amygdala in learning and memory: a critical review, with emphasis on anatomical relations. Behav Neurosci 1985; 99:342–380

24. Krettek JE, Price JL: Amygdaloid projections to subcortical structures within the basal forebrain and brainstem in the rat and cat. J Comp Neurol 1978; 178:225–254

25. Price JL, Amaral DG: An autoradiographic study of the projections of the central nucleus of the monkey amygdala. J Neurosci 1981; 1:1242–1259

26. LeDoux JE, Iwata J, Cicchetti P, et al: Different projections of the central amygdaloid nucleus mediate autonomic and behavioral correlates of conditioned fear. J Neurosci 1988; 8:2517–2529

27. Gray TS, Magnusson DJ: Neuropeptide neuronal efferents from the bed nucleus of the stria terminalis and central amygdaloid nucleus to the dorsal vagal complex in the rat. J Comp Neurol 1987; 262:365–374

28. Schwaber JS, Kapp BS, Higgins GA, et al: Amygdaloid basal forebrain direct connections with the nucleus of the solitary tract and the dorsal motor nucleus. J Neurosci 1982; 2:1424–1438

29. Takeuchi Y, Matsushima S, Hopkins DA: Direct amygdaloid projections to the dorsal motor nucleus of the vagus nerve: a light and electron microscopic study in the rat. Brain Res 1983; 280:143–147

30. Veening JG, Swanson LW, Sawchenko PE: The organization of projections from the central nucleus of the amygdala to brain stem sites involved in central autonomic regulation: a combined retrograde transport-immunohistochemical study. Brain Res 1984; 303:337–357

31. Hopkins DA, Holstege G: Amygdaloid projections to the mesencephalon, pons and medulla oblongata in the cat. Exp Brain Res 1978; 32:529–547

32. Takeuchi Y, McLean JH, Hopkins DA: Reciprocal connections between the amygdala and parabrachial nuclei: ultrastructural demonstration by degeneration and axonal transport of horseradish peroxidase in the cat. Brain Res 1982; 239:538–588

33. Gray TS, Carney ME, Magnuson DJ: Direct projections from the central amygdaloid nucleus to the hypothalamic paraventricular nucleus: possible role in stress-induced adrenocorticotropin release. Neuroendocrinology 1989; 50:433–446

34. DeOlmos J, Alheid GF, Beltramino CA: Amygdala, in The Rat Nervous System, edited by Paxinos G. Orlando, FL, Academic Press, 1985, pp 223–334

35. Sawchenko PE, Swanson LW: The organization of forebrain afferents to the paraventricular and supraoptic nucleus of the rat. J Comp Neurol 1983; 218:121–144

36. Wallace DM, Magnuson DJ, Gray TS: Organization of amygdaloid projections to brainstem dopaminergic, noradrenergic, and adrenergic cell groups in the rat. Brain Res Bull 1992; 28:447–454

37. Goldstein LE, Rasmusson AM, Bunney BS, et al: Role of the amygdala in the coordination of behavioral, neuroendocrine, and prefrontal cortical monoamine responses to psychological stress in the rat. J Neurosci 1996; 16:4787–4798

38. VanBockstaele EJ, Chan J, Pickel VM: Input from central nucleus of the amygdala efferents to pericoerulear dendrites, some of which contain tyrosine hydroxylase immunoreactivity. J Neurosci Res 1996; 45:289–302

39. Aston-Jones G, Ennis M, Pieribone VA, et al: The brain nucleus locus coeruleus: restricted afferent control of a broad efferent network. Science 1986; 234:734–737

40. Rasmussen K, Jacobs BL: Single unit activity of locus coeruleus in the freely moving cat, II: conditioning and pharmacologic studies. Brain Res 1986; 371:335–344

41. Redmond DE Jr: Alteration in the function of the nucleus locus: a possible model for studies on anxiety, in Animal Models in Psychiatry and Neurology, edited by Hanin IE, Usdin E. Oxford, UK, Pergamon, 1977, pp 292–304

42. Pare D, Steriade M, Deschenes M, et al: Prolonged enhancement of anterior thalamic synaptic responsiveness by stimulation of a brainstem cholinergic group. J Neurosci 1990; 10:20–33

43. Rogawski MA, Aghajanian GK: Modulation of lateral geniculate neuron excitability by noradrenaline microintophoresis or locus coeruleus stimulation. Nature 1980; 287:731–734

44. Kapp BS, Whalen PJ, Supple WF, et al: Amygdaloid contributions to conditioned arousal and sensory information processing, in The Amygdala: Neurobiological Aspects of Emotion, Memory, and Mental Dysfunction, edited by Aggleton JP. New York, Wiley-Liss, 1992, pp 229–254

45. Yehle A, Dauth G, Schneiderman N: Correlates of heart rate classical conditioning in curarized rabbits. Journal of Comparative and Physiological Psychology 1967; 64:98–104

46. Magnuson DJ, Gray TS: Central nucleus of amygdala and bed nucleus of stria terminalis projections to serotonin or tyrosine hydroxylase immunoreactive cells in the dorsal and median raphe nucleus in the rat (abstract). Society for Neuroscience Abstracts 1990; 16:121

47. McCall RB, Aghajanian GK: Serotonergic facilitation of facial motoneuron excitation. Brain Res 1979; 169:11–27

48. White SR, Neuman RS: Facilitation of spinal motoneuron excitability by 5-hydroxytryptamine and noradrenaline. Brain Res 1980; 185:1–9

49. Koch M, Ebert U: Enhancement of the acoustic startle response by stimulation of an excitatory pathway from the central amygdala/basal nucleus of Meynert to the pontine reticular formation. Exp

Brain Res 1993; 93:231–241
50. Rosen JB, Hitchcock JM, Sananes CB, et al: A direct projection from the central nucleus of the amygdala to the acoustic startle pathway: anterograde and retrograde tracing studies. Behav Neurosci 1991; 105:817–825
51. Fendt M, Koch M, Schnitzler H-U: Lesions of the central gray block conditioned fear as measured with the potentiated startle paradigm. Behav Brain Res 1996; 74:127–134
52. Hitchcock JM, Davis M: The efferent pathway of the amygdala involved in conditioned fear as measured with the fear-potentiated startle paradigm. Behav Neurosci 1991; 105:826–842
53. Holstege G, Kuypers HGJM, Dekker JJ: The organization of the bulbar fibre connections to the trigeminal, facial and hypoglossal motor nuclei, II: an autoradiographic tracing study in cat. Brain 1977; 100:265–286
54. Whalen PJ, Kapp BS: Contributions of the amygdaloid central nucleus to the modulation of the nictitating membrane reflex in the rabbit. Behav Neurosci 1991; 105:141–153
55. Canli T, Brown TH: Amygdala stimulation enhances the rat eyeblink reflex through a short-latency mechanism. Behav Neurosci 1996; 110:51–59
56. Beitz AJ: The organization of afferent projections to the midbrain periaqueductal gray of the rat. Neuroscience 1982; 7:133–159
57. Bandler R, Carrive P: Integrated defence reaction elicited by excitatory amino acid microinjection in the midbrain periaqueductal grey region of the unrestrained cat. Brain Res 1988; 439:95–106
58. Blanchard DC, Williams G, Lee EMC, et al: Taming of wild *Rattus norvegicus* by lesions of the mesencephalic central gray. Physiology and Psychology 1981; 9:157–163
59. Fanselow MS: The midbrain periaqueductal gray as a coordinator of action in response to fear and anxiety, in The Midbrain Periaqueductal Gray Matter: Functional, Anatomical and Neurochemical Organization, edited by Depaulis A, Bandler R. New York, Plenum, 1991, pp 151–173
60. Graeff FG: Animal models of aversion, in Selected Models of Anxiety, Depression and Psychosis, edited by Simon P, Soubrie P, Wildlocher D. Basel, Switzerland, Karger, 1988, pp 115–141
61. Liebman JM, Mayer DJ, Liebeskind JC: Mesencephalic central gray lesions and fear-motivated behavior in rats. Brain Res 1970; 23:353–370
62. Borszcz GS, Johnson CP, Thorp MV: The differential contribution of spinopetal projections to increases in vocalization and motor reflex thresholds generated by the microinjection of morphine into the periaqueductal gray. Behav Neurosci 1996; 110:368–388
63. Fanselow MS, Helmstetter FJ: Conditioned analgesia, defensive freezing, and benzodiazepines. Behav Neurosci 1988; 102:233–243
64. Watkins LR, Mayer DJ: Organization of endogenous opiate and nonopiate pain control systems. Science 1982; 216:1185–1192
65. Gloor P: Role of the amygdala in temporal lobe epilepsy, in The Amygdala: Neurobiological Aspects of Emotion, Memory and Mental Dysfunction, edited by Aggleton J. New York, Wiley-Liss, 1992, pp 505–538
66. Chapman WP, Schroeder HR, Guyer G, et al: Physiological evidence concerning the importance of the amygdaloid nuclear region in the integration of circulating function and emotion in man. Science 1954; 129:949–950
67. Gloor P, Olivier A, Quesney LF: The role of the amygdala in the expression of psychic phenomena in temporal lobe seizures, in The Amygdaloid Complex, edited by Ben-Ari Y. New York, Elsevier North-Holland, 1981, pp 489–507
68. Iwata J, Chida K, LeDoux JE: Cardiovascular responses elicited by stimulation of neurons in the central amygdaloid nucleus in awake but not anesthetized rats resemble conditioned emotional response. Brain Res 1987; 418:183–188
69. Ohta H, Watanabe S, Ueki S: Cardiovascular changes induced by chemical stimulation of the amygdala in rats. Brain Res Bull 1991; 36:575–581
70. Sanders SK, Shekhar A: Blockade of $GABA_A$ receptors in the region of the anterior basolateral amygdala of rats elicits increases in heart rate and blood pressure. Brain Res 1991; 576:101–110
71. Sanders SK, Shekhar A: Regulation of anxiety by $GABA_A$ receptors in the rat amygdala. Pharmacol Biochem Behav 1995; 52:701–706
72. Henke PG: The telencephalic limbic system and experimental gastric pathology: a review. Neurosci Biobehav Rev 1982; 6:381–390
73. Harper RM, Frysinger RC, Trelease RB, et al: State-dependent alteration of respiratory cycle timing by stimulation of the central nucleus of the amygdala. Brain Res 1984; 306:1–8
74. Dunn JD, Whitener J: Plasma corticosterone responses to electrical stimulation of the amygdaloid complex: cytoarchitectural specificity. Neuroendocrinology 1986; 42:211–217
75. Mason JW: Plasma 17-hydroxycorticosteroid levels during electrical stimulation of the amygdaloid complex in conscious monkeys. Am J Physiol 1959; 196:44–48
76. Kapp BS, Wilson A, Pascoe JP, et al: A neuroanatomical systems analysis of conditioned bradycardia in the rabbit, in Neurocomputation and Learning: Foundations of Adaptive Networks, edited by Gabriel M, Moore J. New York, Bradford, 1990, pp 55–90
77. Kapp BS, Supple WF, Whalen PJ: Effects of electrical stimulation of the amygdaloid central nucleus on neocortical arousal in the rabbit. Behav Neurosci 1994; 108:81–93
78. Dringenberg HC, Vanderwolf CH: Cholinergic activation of the electrocorticogram: an amygdaloid activating system. Exp Brain Res 1996; 108:285–296
79. Kreindler A, Steriade M: EEG patterns of arousal and sleep induced by stimulating various amygdaloid levels in the cat. Arch Ital Biol 1964; 102:576–586
80. Applegate CD, Kapp BS, Underwood MD, et al: Autonomic and somatomotor effects of amygdala central n. stimulation in awake rabbits. Physiol Behav 1983; 31:353–360
81. Ursin H, Kaada BR: Functional localization within the amygdaloid complex in the cat. Electroencephalogr Clin Neurophysiol 1960; 12:109–122
82. Kaada BR: Stimulation and regional ablation of the amygdaloid complex with reference to functional representations, in The Neurobiology of the Amygdala, edited by Eleftheriou BE. New York, Plenum, 1972, pp 205–281
83. Roozendaal B, Wiersma A, Driscoll P, et al: Vasopressinergic modulation of stress responses in the central amygdala of the Roman high-avoidance and low-avoidance rat. Brain Res 1992; 596:35–40
84. Willcox BJ, Poulin P, Veale WL, et al: Vasopressin-induced motor effects: localization of a sensitive site in the amygdala. Brain Res 1992; 596:58–64
85. Kaku T: Functional differentiation of hypoglossal motoneurons during the amygdaloid or cortically induced rhythmical jaw and tongue movements in the rat. Brain Res Bull 1984; 13:147–154
86. Ohta M: Amygdaloid and cortical facilitation or inhibition of trigeminal motoneurons in the rat. Brain Res 1984; 291:39–48
87. Fanardjian VV, Manvelyan LR: Mechanisms regulating the activity of facial nucleus motoneurons, III: synaptic influences from the cerebral cortex and subcortical structures. Neuroscience 1987; 20:835–843
88. GaryBobo E, Bonvallet M: Amygdala and masseteric reflex, I: facilitation, inhibition and diphasic modifications of the reflex induced by localized amygdaloid stimulation. Electroencephalogr Clin Neurophysiol 1975; 39:329–339
89. Bonvallet M, GaryBobo E: Amygdala and masseteric reflex, II: mechanism of the diphasic modifications of the reflex elicited from the "defence reaction area": role of the spinal trigeminal nucleus (pars oralis). Electroencephalogr Clin Neurophysiol 1975; 39:341–352
90. Lewis SJ, Verberne AJM, Robinson TG, et al: Excitotoxin-induced lesions of the central but not basolateral nucleus of the amygdala

modulate the baroreceptor heart rate reflex in conscious rats. Brain Res 1989; 494:232–240
91. Schlor KH, Stumpf H, Stock G: Baroreceptor reflex during arousal induced by electrical stimulation of the amygdala or by natural stimuli. J Auton Nerv Syst 1984; 10:157–165
92. Pascoe JP, Bradley DJ, Spyer KM: Interactive responses to stimulation of the amygdaloid central nucleus and baroreceptor afferents in the rabbit. J Auton Nerv Syst 1989; 26:157–167
93. Rosen JB, Davis M: Enhancement of acoustic startle by electrical stimulation of the amygdala. Behav Neurosci 1988; 102:195–202
94. Rosen JB, Davis M: Temporal characterizations of enhancement of startle by stimulation of the amygdala. Physiol Behav 1988; 44:117–123
95. Koch M, Kungel M, Herbert H: Cholinergic neurons in the pedunculopontine tegmental nucleus are involved in the mediation of prepulse inhibition of the acoustic startle response in the rat. Exp Brain Res 1993; 97:71–82
96. Koch M: Microinjections of the metabotropic glutamate receptor agonist, trans-($\pm$)-1-amino-cyclopentane-1,3, dicarboxylate (trans-ACPD) into the amygdala increase the acoustic startle response of rats. Brain Res 1993; 629:176–179
97. Clugnet MC, LeDoux JE: Synaptic plasticity in fear conditioning circuits: induction of LTP in the lateral nucleus of the amygdala by stimulation of the medial geniculate body. J Neurosci 1990; 10:2818–2824
98. Chapman PF, Kairiss EW, Keenan CL, et al: Long-term synaptic potentiation in the amygdala. Synapse 1990; 6:271–278
99. Chapman PF, Bellavance LL: Induction of long-term potentiation in the basolateral amygdala does not depend on NMDA receptor activation. Synapse 1992; 11:310–318
100. Gean PW, Chang FC, Huang CC, et al: Long-term enhancement of EPSP and NMDA receptor-mediated synaptic transmission in the amygdala. Brain Res Bull 1993; 31:7–11
101. Fanselow MS, Kim JJ: Acquisition of contextual Pavlovian fear conditioning is blocked by application of an NMDA receptor antagonist D,L-2-amino-5-phosphonovaleric acid to the basolateral amygdala. Behav Neurosci 1994; 108:210–212
102. Hatfield T, Gallagher M: Taste-potentiated odor conditioning: impairment produced by infusion of an N-methyl-D-aspartate antagonist into basolateral amygdala. Behav Neurosci 1995; 109:663–668
103. Izquierdo I, Da Cunha C, Rosat R, et al: Neurotransmitter receptors involved in post-training memory processing by the amygdala, medial septum, and hippocampus of the rat. Behavioral and Neural Biology 1992; 58:16–26
104. Kim M, McGaugh JL: Effects of intra-amygdala injections of NMDA receptor antagonists on acquisition and retention of inhibitory avoidance. Brain Res 1992; 585:35–48
105. Liang KC, Hon W, Davis M: Pre- and post-training intra-amygdala infusions of N-methyl-D-aspartate receptor antagonists impair memory in an inhibitory avoidance task. Behav Neurosci 1994; 108:241–253
106. Miserendino MJD, Sananes CB, Melia KR, et al: Blocking of acquisition but not expression of conditioned fear-potentiated startle by NMDA antagonists in the amygdala. Nature 1990; 345:716–718
107. Falls WA, Miserendino MJD, Davis M: Excitatory amino acid antagonists infused into the amygdala block extinction of fear-potentiated startle (abstract). Society for Neuroscience Abstracts 1990; 16:767
108. Brown S, Schafer EA: An investigation into the functions of the occipital and temporal lobes of the monkey's brain. Philos Trans R Soc Lond B Biol Sci 1988; 179:303–327
109. Klüver H, Bucy PC: Preliminary analysis of functions of the temporal lobes in monkeys. Archives of Neurology and Psychiatry 1939; 42:979–1000
110. Kling AS, Brothers LA: The amygdala and social behavior, in The Amygdala: Neurobiological Aspects of Emotion, Memory and Mental Dysfunction, edited by Aggleton J. New York, Wiley, 1992, pp 353–377
111. Dicks D, Meyers RE, Kling A: Uncus and amygdala lesions: effects on social behavior in the free-ranging rhesus monkey. Science 1969; 165:69–71
112. Kling A, Lancaster J, Benitone J: Amygdalectomy in the free-ranging vervet. J Psychiatr Res 1970; 7:191–199
113. Horel JA, Keating EG, Misantone LJ: Partial Klüver-Bucy syndrome produced by destroying temporal neocortex and amygdala. Brain Res 1975; 94:347–359
114. Meyers RE, Swett C: Social behaviour deficits of free-ranging monkeys after anterior temporal cortex removals: a preliminary report. Brain Res 1970; 19:39
115. Mishkin M, Pribram KH: Visual discrimination performance following partial ablations of the temporal lobe, I: ventral vs. lateral. Journal of Comparative and Physiological Psychology 1954; 47:14–20
116. Pribram KH, Bagshaw M: Further analysis of the temporal lobe syndrome utilizing fronto-temporal ablations. J Comp Neurol 1953; 99:347–374
117. Schwartzbaum JS: Discrimination behavior after amygdalectomy in monkeys: learning set and discrimination reversals. Journal of Comparative and Physiological Psychology 1965; 60:314–319
118. Weiskrantz L: Behavioral changes associated with ablation of the amygdaloid complex in monkeys. Journal of Comparative and Physiological Psychology 1956; 49:381–391
119. Aggleton JP, Passingham RE: Syndrome produce by lesions of the amygdala in monkeys (Macaca mulatta). Journal of Comparative and Physiological Psychology 1981; 95:961–977
120. Zola-Morgan S, Squire LR, Alvarez-Royo P, et al: Independence of memory functions and emotional behavior: separate contributions of the hippocampal formation and the amygdala. Hippocampus 1991; 1:207–220
121. Aggleton J: The contribution of the amygdala to normal and abnormal emotional states. Trends Neurosci 1993; 8:328–333
122. Kromer Vogt LJ, Hyman BT, Van Hoesen GW, et al: Pathological alterations in the amygdala in Alzheimer's disease. Neuroscience 1990; 37:377–385
123. Irwin W, Davidson RJ, Lowe MJ, et al: Human amygdala activation detected with echo-planar functional magnetic resonance imaging. Neuroreport 1996; 7:1765–1769
124. Kirkpatrick B, Buchanan RW: The neural basis of the deficit syndrome of schizophrenia. J Nerv Ment Dis 1990; 178:545–555
125. Leonard CM, Rolls ET, Wilson FAW, et al: Neurons in the amygdala of the monkey with responses selective for faces. Behav Brain Res 1985; 15:159–176
126. Nakamura K, Mikami A, Kubota K: Activity of single neurons in the monkey amygdala during performance of a visual discrimination task. J Neurophysiol 1992; 67:1447–1463
127. Rolls ET: Neurons in the cortex of the temporal lobe and in the amygdala of the monkey with responses selective for faces. Human Neurobiology 1984; 3:209–222
128. Allison T, McCarthy G, Nobre A, et al: Human extrastriate visual cortex and the perception of faces, words, numbers, and colors. Cereb Cortex 1994; 4:544–554
129. Heit G, Smith ME, Halgren E: Neural encoding of individual words and faces by the human hippocampus and amygdala. Nature 1988; 333:773–775
130. Brothers L, Ring B: Mesial temporal neurons in the macaque monkey with responses selective for aspects of social stimuli. Behav Brain Res 1993; 57:53–61
131. Aggleton JP: The functional effects of amygdala lesions in humans: a comparison with findings from monkeys, in The Amygdala: Neurobiological Aspects of Emotion, Memory and Mental Dysfunction, edited by Aggleton JP. New York, Wiley-Liss, 1992, pp 485–503
132. Jacobson R: Disorders of facial recognition, social behaviour and

affect after combined bilateral amygdalotomy and subcaudate tractotomy—a clinical and experimental study. Psychol Med 1986; 16:439–450
133. Tranel D, Hyman BT: Neuropsychological correlates of bilateral amygdala damage. Arch Neurol 1990; 47:349–355
134. Young AW, Aggleton JP, Hellawell DJ, et al: Face processing impairments after amygdalotomy. Brain 1995; 118:15–24
135. Adolphs R, Tranel D, Damasio H, et al: Impaired recognition of emotion in facial expressions following bilateral damage to the human amygdala. Nature 1994; 372:669–672
136. Adolphs R, Tranel D, Damasio H, et al: Fear and the human amygdala. J Neurosci 1995; 15:5879–5891
137. Cahill L, Babinsky R, Markowitsch HJ, et al: The amygdala and emotional memory. Nature 1995; 377:295–296
138. Markowitsch HJ, Calabrese P, Wurker M, et al: The amygdala's contribution to memory: a study on two patients with Urbach-Wiethe disease. Neuroreport 1994; 5:1349–1352
139. McGaugh JL, Introini-Collison IB, Nagahara AH, et al: Involvement of the amygdaloid complex in neuromodulatory influences on memory storage. Neurosci Biobehav Rev 1990; 14:425–432
140. LeBar KS, LeDoux JE, Spencer DD, et al: Impaired fear conditioning following unilateral temporal lobectomy in humans. J Neurosci 1995; 15:6846–6855
141. Bechara A, Tranel D, Damasio H, et al: Double dissociation of conditioning and declarative knowledge relative to the amygdala and hippocampus in humans. Science 1995; 269:1115–1118
142. Bagshaw MH, Kimble DP, Pribram KH: The GSR of monkeys during orienting and habituation and after ablation of the amygdala, hippocampus and inferotemporal cortex. Neuropsychologia 1965; 3:111–119
143. Bagshaw MH, Benzies S: Multiple measures of the orienting reaction and their dissociation after amygdalectomy in monkeys. Exp Neurol 1968; 20:175–187
144. Gentile CG, Jarrel TW, Teich A, et al: The role of amygdaloid central nucleus in the retention of differential Pavlovian conditioning of bradycardia in rabbits. Behav Brain Res 1986; 20:263–273
145. McCabe PM, Gentile CG, Markgraf CG, et al: Ibotenic acid lesions in the amygdaloid central nucleus but not in the lateral subthalamic area prevent the acquisition of differential Pavlovian conditioning of bradycardia in rabbits. Brain Res 1992; 580:155–163
146. Kapp BS, Frysinger RC, Gallagher M, et al: Amygdala central nucleus lesions: effect on heart rate conditioning in the rabbit. Physiol Behav 1979; 23:1109–1117
147. Roozendaal B, Koolhaas JM, Bohus B: Differential effect of lesioning of the central amygdala on the bradycardiac and behavioral response of the rat in relation to conditioned social and solitary stress. Behav Brain Res 1990; 41:39–48
148. Roozendaal B, Koolhaas JM, Bohus B: Attenuated cardiovascular, neuroendocrine, and behavioral response after a single footshock in central amygdaloid lesioned male rats. Physiol Behav 1991; 50:771–775
149. Roozendaal B, Koolhaas JM, Bohus B: Central amygdala lesions affect behavioral and autonomic balance during stress in rats. Physiol Behav 1991; 50:777–781
150. Sananes CB, Campbell BA: Role of the central nucleus of the amygdala in olfactory heart rate conditioning. Behav Neurosci 1989; 103:519–525
151. Cohen DH: Involvement of the avian amygdala homologue (archistriatum posterior and mediale) in defensively conditioned heart rate change. J Comp Neurol 1975; 160:13–36
152. Iwata J, LeDoux JE, Meeley MP, et al: Intrinsic neurons in the amygdala field projected to by the medial geniculate body mediate emotional responses conditioned to acoustic stimuli. Brain Res 1986; 383:195–214
153. Zhang JX, Harper RM, Ni H: Cryogenic blockade of the central nucleus of the amygdala attenuates aversively conditioned blood pressure and respiratory responses. Brain Res 1986; 386:136–145
154. LeDoux JE: Information flow from sensation to emotion: plasticity in the neural computation of stimulus value, in Learning and Computational Neuroscience, edited by Gabriel M, Moore J. Cambridge, MA, Bradford Books/MIT Press, 1990, pp 3–51
155. Romanski LM, Clugnet MC, Bordi F, et al: Somatosensory and auditory convergence in the lateral nucleus of the amygdala. Behav Neurosci 1993; 107:444–450
156. Beaulieu S, DiPaolo T, Cote J, et al: Participation of the central amygdaloid nucleus in the response of adrenocorticotropin secretion to immobilization stress: opposing roles of the noradrenergic and dopaminergic systems. Neuroendocrinology 1987; 45:37–46
157. Roozendaal B, Koolhaas JM, Bohus B: Central amygdaloid involvement in neuroendocrine correlates of conditioned stress responses. J Neuroendocrinol 1992; 4:483–489
158. Van de Kar LD, Piechowski RA, Rittenhouse PA, et al: Amygdaloid lesions: differential effect on conditioned stress and immobilization-induced increases in corticosterone and renin secretion. Neuroendocrinology 1991; 15:89–95
159. Goldstein LE, Rasmusson AM, Bunney BS, et al: Role of the amygdala in the coordination of behavioral, neuroendocrine, and prefrontal cortical monoamine responses to psychological stress in the rat. J Neurosci 1996; 16:4787–4798
160. Davis M, Hitchcock JM, Bowers MB, et al: Stress-induced activation of prefrontal cortex dopamine turnover: blockade by lesions of the amygdala. Brain Res 1994; 664:207–210
161. Gallagher M, Graham PW, Holland PC: The amygdala central nucleus and appetitive Pavlovian conditioning: lesions impair one class of conditioned behavior. J Neurosci 1990; 10:1906–1911
162. Holland PC, Gallagher M: The effects of amygdala central nucleus lesions on blocking and unblocking. Behav Neurosci 1993; 107:235–245
163. Holland PC, Gallagher M: Amygdala central nucleus lesions disrupt increments, but not decrement, in conditioned stimulus processing. Behav Neurosci 1993; 107:246–253
164. Bermudez-Rattoni F, Grijalva CV, Kiefer SW, et al: Flavor-illness aversion: the role of the amygdala in acquisition of taste-potentiated odor aversions. Physiol Behav 1986; 38:503–508
165. Hatfield T, Graham PW, Gallagher M: Taste-potentiated odor aversion: role of the amygdaloid basolateral complex and central nucleus. Behav Neurosci 1992; 106:286–293
166. Fernandez-Ruiz J, Miranda MI, Bermudez-Rattoni F, et al: Effects of catecholaminergic depletion of the amygdala and insular cortex on the potentiation of odor by taste aversions. Behavioral and Neural Biology 1993; 60:189–191
167. Hatfield T, Han J-S, Conley M, et al: Neurotoxic lesions of basolateral, but not central, amygdala interfere with Pavlovian second-order conditioning and reinforcer devaluation effects. J Neurosci 1996; 16:5256–5265
168. Halgren E: Emotional neurophysiology of the amygdala within the context of human cognition, in The Amygdala: Neurobiological Aspects of Emotion, Memory, and Mental Dysfunction, edited by Aggleton JP. New York, Wiley-Liss, 1992, pp 191–228
169. Blanchard DC, Blanchard RJ: Innate and conditioned reactions to threat in rats with amygdaloid lesions. Journal of Comparative and Physiological Psychology 1972; 81:281–290
170. Helmstetter FJ: The amygdala is essential for the expression of conditioned hypoalgesia. Behav Neurosci 1992; 106:518–528
171. Helmstetter FJ: Contribution of the amygdala to learning and performance of conditional fear. Physiol Behav 1992; 51:1271–1276
172. Kiernan M, Cranney J: Excitotoxic lesions of the central nucleus of the amygdala but not of the periaqueductal gray block integrated fear responding as indexed by both freezing responses and augmentation of startle (abstract). Society for Neuroscience Abstracts 1992; 18:1566
173. Kim JJ, Fanselow MS: Modality-specific retrograde amnesia of fear.

Science 1992; 256:675–677
174. Kim JJ, Rison RA, Fanselow MS: Effects of amygdala, hippocampus, and periaqueductal gray lesions on short- and long-term contextual fear. Behav Neurosci 1993; 107:1093–1098
175. LeDoux JE: Emotion and the amygdala, in The Amygdala: Neurobiological Aspects of Emotion, Memory, and Mental Dysfunction, edited by Aggleton JP. New York, Wiley-Liss, 1992, pp 339–352
176. Lorenzini CA, Bucherelli C, Giachetti LM, et al: Effects of nucleus basolateralis amygdalae neurotoxic lesions on aversive conditioning in the rat. Physiol Behav 1991; 49:765–770
177. Maren S, Aharonov G, Fanselow MS: Retrograde abolition of conditioned fear after excitotoxic lesions in the basolateral amygdala of rats: absence of a temporal gradient. Behav Neurosci 1996; 110:718–726
178. Phillips RG, LeDoux JE: Differential contribution of amygdala and hippocampus to cued and contextual fear conditioning. Behav Neurosci 1992; 106:274–285
179. Romanski LM, LeDoux JE: Equipotentiality of thalamo-amygdala and thalamo-cortico amygdala circuits in auditory fear conditioning. J Neurosci 1992; 12:4501–4509
180. Burns LH, Annett L, Kelley AE, et al: Effects of lesions to amygdala, ventral subiculum, medial prefrontal cortex, and nucleus accumbens on the reaction to novelty: implications for limbic-striatal interactions. Behav Neurosci 1996; 110:60–73
181. Bolhuis JJ, Fitzgerald RE, Dijk DJ, et al: The corticomedial amygdala and learning in an agonistic situation. Physiol Behav 1984; 32:575–579
182. Luiten PGM, Koolhaas JM, deBoer S, et al: The cortico-medial amygdala in the central nervous system organization of agonistic behavior. Brain Res 1985; 332:282–297
183. Slotnick BM: Fear behavior and passive avoidance deficits in mice with amygdala lesions. Physiol Behav 1973; 11:717–720
184. Helmstetter FJ, Bellgowan PS: Effects of muscimol applied to the basolateral amygdala on acquisition and expression of contextual fear conditioning in rats. Behav Neurosci 1994; 108:1005–1009
185. Kopchia KL, Altman HJ, Commissaris RL: Effects of lesions of the central nucleus of the amygdala on anxiety-like behaviors in the rat. Pharmacol Biochem Behav 1992; 43:453–461
186. Selden NRW, Everitt BJ, Jarrard LE, et al: Complementary roles for the amygdala and hippocampus in aversive conditioning to explicit and contextual cues. Neuroscience 1991; 42:335–350
187. Shibata K, Kataoka Y, Yamashita K, et al: An important role of the central amygdaloid nucleus and mammillary body in the mediation of conflict behavior in rats. Brain Res 1986; 372:159–162
188. Kellicut MH, Schwartzbaum JS: Formation of a conditioned emotional response (CER) following lesions of the amygdaloid complex in rats. Psychol Rev 1963; 12:351–358
189. Spevack AA, Campbell CT, Drake L: Effect of amygdalectomy on habituation and CER in rats. Physiol Behav 1975; 15:199–207
190. Dafters RI: Effect of medial archistriatal lesions on the conditioned emotional response and on auditory discrimination performance of the pigeon. Physiol Behav 1976; 17:659–665
191. Campeau S, Davis M: Involvement of the central nucleus and basolateral complex of the amygdala in fear conditioning measured with fear-potentiated startle in rats trained concurrently with auditory and visual conditioned stimuli. J Neurosci 1995; 15:2301–2311
192. Hitchcock JM, Davis M: Lesions of the amygdala, but not of the cerebellum or red nucleus, block conditioned fear as measured with the potentiated startle paradigm. Behav Neurosci 1986; 100:11–22
193. Hitchcock JM, Davis M: Fear-potentiated startle using an auditory conditioned stimulus: effect of lesions of the amygdala. Physiol Behav 1987; 39:403–408
194. Kim M, Davis M: Electrolytic lesions of the amygdala block acquisition and expression of fear-potentiated startle even with extensive training, but do not prevent re-acquisition. Behav Neurosci 1993; 107:580–595
195. Kim M, Davis M: Lack of a temporal gradient of retrograde amnesia in rats with amygdala lesions assessed with the fear-potentiated startle paradigm. Behav Neurosci 1993; 107:1088–1092
196. Weisz DJ, Harden DG, Xiang Z: Effects of amygdala lesions on reflex facilitation and conditioned response acquisition during nictitating membrane response conditioning in rabbit. Behav Neurosci 1992; 106:262–273
197. Treit D, Pesold C, Rotzinger S: Dissociating the anti-fear effects of septal and amygdaloid lesions using two pharmacologically validated models of rat anxiety. Behav Neurosci 1993; 107:770–785
198. Fox RJ, Sorenson CA: Bilateral lesions of the amygdala attenuate analgesia induced by diverse environmental challenges. Brain Res 1994; 648:215–221
199. Watkins LR, Weirtelak EP, Maier SF: The amygdala is necessary for the expression of conditioned but not unconditioned analgesia. Behav Neurosci 1993; 107:402–405
200. Helmstetter FJ, Bellgowan PS: Lesions of the amygdala block conditioned hypoalgesia on the tail flick test. Brain Res 1993; 612:253–257
201. Manning BH, Mayer DJ: The central nucleus of the amygdala contributes to the production of morphine antinociception in the formalin test. Pain 1995; 63:141–152
202. Manning BH, Mayer DJ: The central nucleus of the amygdala contributes to the production of morphine antinociception in the rat tail-flick test. J Neurosci 1995; 15:8199–8213
203. McGaugh JL, Introini-Collison IB, Cahill L, et al: Neuromodulatory systems and memory storage: role of the amygdala. Behav Brain Res 1993; 58:81–90
204. Ursin H, Jellestad F, Cabrera IG: The amygdala, exploration and fear, in The Amygdaloid Complex, edited by Ben-Ari Y. Amsterdam, Elsevier North-Holland, 1981, pp 317–329
205. Bennett C, Liang KC, McGaugh JL: Depletion of adrenal catecholamines alters the amnestic effect of amygdala stimulation. Behav Brain Res 1985; 15:83–91
206. Bresnahan E, Routtenberg A: Memory disruption by unilateral low level, sub-seizure stimulation of the medial amygdaloid nucleus. Physiol Behav 1972; 9:513–525
207. Ellis ME, Kesner RP: The noradrenergic system of the amygdala and aversive information processing. Behav Neurosci 1983; 97:399–415
208. Gallagher M, Kapp BS, Frysinger RC, et al: β-Adrenergic manipulation in amygdala central n. alters rabbit heart rate conditioning. Pharmacol Biochem Behav 1980; 12:419–426
209. Gallagher M, Kapp BS: Effect of phentolamine administration into the amygdala complex of rats on time-dependent memory processes. Behavioral and Neural Biology 1981; 31:90–95
210. Gallagher M, Kapp BS, McNall CL, et al: Opiate effects in the amygdala central nucleus on heart rate conditioning in rabbits. Pharmacol Biochem Behav 1981; 14:497–505
211. Gallagher M, Kapp BS: Manipulation of opiate activity in the amygdala alters memory processes. Life Sci 1978; 23:1973–1978
212. Gold PE, Hankins L, Edwards RM, et al: Memory inference and facilitation with post-trial amygdala stimulation: effect varies with footshock level. Brain Res 1975; 86:509–513
213. Handwerker MJ, Gold PE, McGaugh JL: Impairment of active avoidance learning with posttraining amygdala stimulation. Brain Res 1974; 75:324–327
214. Kesner RP: Brain stimulation: effects on memory. Behavioral and Neural Biology 1982; 36:315–367
215. Liang KC, Bennett C, McGaugh JL: Peripheral epinephrine modulates the effects of post-training amygdala stimulation on memory. Behav Brain Res 1985; 15:93–100
216. Liang KC, Juler RG, McGaugh JL: Modulating effects of post-training epinephrine on memory: involvement of the amygdala noradrenergic systems. Brain Res 1986; 368:125–133
217. Mishkin M, Aggleton J: Multiple function contributions of the

amygdala in the monkey, in The Amygdaloid Complex, edited by Ben-Ari Y. New York, Elsevier North-Holland, 1981, pp 409–420
218. Feldman S, Conforti N: Amygdalectomy inhibits adrenocortical responses to somatosensory and olfactory stimulation. Neuroendocrinology 1981; 32:330–334
219. Feldman S, Conforti N, Itzik A, et al: Differential effect of amygdaloid lesions on CRF-41, ACTH and corticosterone responses following neural stimuli. Brain Res 1994; 658:21–26
220. Allen JP, Allen CF: Role of the amygdaloid complexes in the stress-induced release of ACTH in the rat. Neuroendocrinology 1974; 15:220–230
221. Henke PG: The amygdala and restraint ulcers in rats. Journal of Comparative and Physiological Psychology 1980; 94:313–323
222. Henke PG: Facilitation and inhibition of gastric pathology after lesions in the amygdala in rats. Physiol Behav 1980; 25:575–579
223. Henke PG: Attenuation of shock-induced ulcers after lesions in the medial amygdala. Physiol Behav 1981; 27:143–146
224. Beaulieu S, DiPaolo T, Barden N: Control of ACTH secretion by central nucleus of the amygdala: implication of the serotonergic system and its relevance to the glucocorticoid delayed negative feed-back mechanism. Neuroendocrinology 1986; 44:247–254
225. Leaton RN, Cranney J: Potentiation of the acoustic startle response by a conditioned stimulus paired with acoustic startle stimulus in rats. J Exp Psychol Anim Behav Process 1990; 16:279–287
226. Galeno TM, VanHoesen GW, Brody MJ: Central amygdaloid nucleus lesion attenuates exaggerated hemodynamic responses to noise stress in the spontaneously hypertensive rat. Brain Res 1984; 291:249–259
227. Singh VB, Onaivi ES, Phan TH, et al: The increases in rat cortical and midbrain tryptophan hydroxylase activity in response to acute or repeated sound stress are blocked by bilateral lesions to the central nucleus of the amygdala. Brain Res 1990; 530:49–53
228. Kemble ED, Blanchard DC, Blanchard RJ, et al: Taming in wild rats following medial amygdaloid lesions. Physiol Behav 1984; 32:131–134
229. Kemble ED, Blanchard DC, Blanchard RJ: Effects of regional amygdaloid lesions on flight and defensive behaviors of wild black rats (Rattus rattus). Physiol Behav 1990; 48:1–5
230. Phillips RE: Wildness in the Mallard duck: effects of brain lesions and stimulation on "escape behavior" and reproduction. J Comp Neurol 1964; 122:139–156
231. Phillips RE: Approach-withdrawal behavior of peach-faced lovebirds, Agapornis roseicolis, and its modification by brain lesions. Behavior 1968; 31:163–184
232. Goddard GV: Functions of the amygdala. Psychol Bull 1964; 62:89–109
233. Hitchcock JM, Sananes CB, Davis M: Sensitization of the startle reflex by footshock: blockade by lesions of the central nucleus of the amygdala or its efferent pathway to the brainstem. Behav Neurosci 1989; 103:509–518
234. Davis M: Sensitization of the acoustic startle reflex by footshock. Behav Neurosci 1989; 103:495–503
235. Kiernan MJ, Westbrook RF, Cranney J: Immediate shock, passive avoidance, and potentiated startle: implications for the unconditioned response to shock. Animal Learning and Behavior 1995; 23:22–30
236. Bellgowan PSF, Helmstetter FJ: Neural systems for the expression of hypoalgesia during nonassociative fear. Behav Neurosci 1996; 110:727–736
237. Grijalva CV, Levin ED, Morgan M, et al: Contrasting effects of centromedial and basolateral amygdaloid lesions on stress-related responses in the rat. Physiol Behav 1990; 48:495–500
238. Pellegrino L: Amygdaloid lesions and behavioral inhibition in the rat. Journal of Comparative and Physiological Psychology 1968; 65:483–491
239. Cahill L, McGaugh JL: Amygdaloid complex lesions differentially affect retention of tasks using appetitive and aversive reinforcement. Behav Neurosci 1990; 104:532–543
240. Dunn LT, Everitt BJ: Double dissociations of the effects of amygdala and insular cortex lesions on conditioned taste aversion, passive avoidance, and neophobia in the rat using the excitotoxin ibotenic acid. Behav Neurosci 1988; 102:3–23
241. Lamprecht R, Dudai Y: Transient expression of c-fos in rat amygdala during training is required for encoding conditioned taste aversion memory. Learning and Memory 1996; 3:31–41
242. Crandall PH, Walter RD, Dymond A: The ictal electroencephalographic signal identifying limbic system seizure foci. Proceedings of the American Association of Neurological Surgery 1971; 1:1
243. DeSouza EB, Insel TR, Perrin MH, et al: Corticotropin-releasing factor receptors are widely distributed within the rat central nervous system: an autoradiographic study. J Neurosci 1985; 5:3189–3203
244. Uryu K, Okumura T, Shibasaki T, et al: Fine structure and possible origins of nerve fibers with corticotropin-releasing factor-like immunoreactivity in the rat central amygdaloid nucleus. Brain Res 1992; 577:175–179
245. Pich EM, Lorang M, Yeganeh M, et al: Increase of extracellular corticotropin-releasing factor-like immunoreactivity levels in the amygdala of awake rats during restraint stress and ethanol withdrawal as measured by microdialysis. J Neurosci 1995; 15:5439–5447
246. Sarnyai Z, Biro E, Gardi J, et al: Brain corticotropin-releasing factor mediates "anxiety-like" behavior induced by cocaine withdrawal in rats. Brain Res 1995; 675:89–97
247. Altemus M, Smith MA, Diep V, et al: Increased mRNA for corticotrophin releasing hormone in the amygdala of fawn-hooded rats: a potential animal model of anxiety. Anxiety 1995; 1:251–257
248. Cratty MS, Ward HE, Johnson EA, et al: Prenatal stress increases corticotropin-releasing factor (CRF) content and release in rat amygdala minces. Brain Res 1995; 675:297–302
249. Wiersma A, Bohus B, Koolhaas JM: Corticotropin-releasing hormone microinfusion in the central amygdala diminishes a cardiac parasympathetic outflow under stress-free conditions. Brain Res 1993; 625:219–227
250. Boadle-Biber M, Singh VB, Corley KC, et al: Evidence that corticotropin-releasing factor within the extended amygdala mediates the activation of tryptophan hydroxylase produced by sound stress in the rat. Brain Res 1993; 628:105–114
251. Hodges H, Green S, Glenn B: Evidence that the amygdala is involved in benzodiazepine and serotonergic effects on punished responding but not on discrimination. Psychopharmacology 1987; 92:491–504
252. Nagy J, Zambo K, Decsi L: Anti-anxiety action of diazepam after intra-amygdaloid application in the rat. Neuropharmacology 1979; 18:573–576
253. Petersen EN, Scheel-Kruger J: The GABAergic anticonflict effect of intraamydaloid benzodiazepines demonstrated by a new water lick conflict paradigm, in Behavioral Models and the Analysis of Drug Action, edited by Spiegelstein MY, Levy A. Amsterdam, Elsevier Scientific, 1982, pp 467–473
254. Petersen EN, Braestrup C, Scheel-Kruger J: Evidence that the anticonflict effect of midazolam in amygdala is mediated by the specific benzodiazepine receptor. Neurosci Lett 1985; 53:285–288
255. Scheel-Kruger J, Petersen EN: Anticonflict effect of the benzodiazepines mediated by a GABAergic mechanism in the amygdala. Eur J Pharmacol 1982; 82:115–116
256. Shibata K, Kataoka Y, Gomita Y, et al: Localization of the site of the anticonflict action of benzodiazepines in the amygdaloid nucleus of rats. Brain Res 1982; 234:442–446
257. Shibata S, Yamashita K, Yamamoto E, et al: Effect of benzodiazepine and GABA antagonists on anticonflict effects of antianxiety drugs injected into the rat amygdala in a water-lick suppression test. Psychopharmacology 1989; 98:38–44

258. Thomas SR, Lewis ME, Iversen SD: Correlation of [^{3}H]diazepam binding density with anxiolytic locus in the amygdaloid complex of the rat. Brain Res 1985; 342:85–90
259. Helmstetter FJ: Stress-induced hypoalgesia and defensive freezing are attenuated by application of diazepam to the amygdala. Pharmacol Biochem Behav 1993; 44:433–438
260. Young BJ, Helmstetter FJ, Rabchenuk SA, et al: Effects of systemic and intra-amygdaloid diazepam on long-term habituation of acoustic startle in rats. Pharmacol Biochem Behav 1991; 39:903–909
261. Costall B, Jones BJ, Kelly ME, et al: Exploration of mice in a black and white test box: validation as a model of anxiety. Pharmacol Biochem Behav 1989; 32:777–785
262. Pesold C, Treit D: The septum and amygdala differentially mediate the anxiolytic effects of benzodiazepines. Brain Res 1994; 638:295–301
263. Green S, Vale AL: Role of amygdaloid nuclei in the anxiolytic effects of benzodiazepines in rats. Behav Pharmacol 1992; 3:261–264
264. Pesold C, Treit D: The central and basolateral amygdala differentially mediate the anxiolytic effects of benzodiazepines. Brain Res 1995; 671:213–221
265. Benjamin D, Emmett-Oglesby MW, Lah H: Modulation of the discriminative stimulus produced by pentylenetetrazol by centrally administered drugs. Neuropharmacology 1987; 26:1727–1731
266. Borszcz GS, Cranney J, Leaton RN: Influence of long-term sensitization on long-term habituation of the acoustic startle response in rats: central gray lesions, preexposure and extinction. J Exp Psychol Anim Behav Process 1989; 15:54–64
267. Takao K, Nagatani T, Kasahara K-I, et al: Role of the central serotonergic system in the anticonflict effect of d-AP159. Pharmacol Biochem Behav 1992; 43:503–508
268. Sanders SK, Shekhar A: Anxiolytic effects of chlordiazepoxide blocked by injection of GABA$_A$ and benzodiazepine receptor antagonists in the region of the anterior basolateral amygdala of rats. Biol Psychiatry 1995; 37:473–476
269. Davis M: The role of the amygdala in emotional learning, in International Review of Neurobiology 1994; 36:225–266
270. Treit D, Pesold C, Rotzinger S: Noninteractive effects of diazepam and amygdaloid lesions in two animal models of anxiety. Behav Neurosci 1993; 107:1099–1105
271. Yadin E, Thomas E, Strickland CE, et al: Anxiolytic effects of benzodiazepines in amygdala-lesioned rats. Psychopharmacology 1991; 103:473–479
272. Heinrichs SC, Pich EM, Miczek KA, et al: Corticotropin-releasing factor antagonist reduces emotionality in socially defeated rats via direct neurotropic action. Brain Res 1992; 581:190–197
273. Rassnick S, Heinrichs SC, Britton KT, et al: Microinjection of a corticotropin-releasing factor antagonist into the central nucleus of the amygdala reverses anxiogenic-like effects of ethanol withdrawal. Brain Res 1993; 605:25–32
274. Liebsch G, Landgraf R, Gerstberger R, et al: Chronic infusion of a CRH$_1$ receptor antisense oligodeoxynucleotide into the central nucleus of the amygdala reduced anxiety-related behavior in socially defeated rats. Regul Pept 1995; 59:229–239
275. Swiergiel AH, Takahashi LK, Kalin NH: Attenuation of stress-induced behavior by antagonism of corticotropin-releasing factor in the central amygdala of the rat. Brain Res 1993; 623:229–234
276. Wiersma A, Baauw AD, Bohus B, et al: Behavioural activation produced by CRH but not α-helical CRH (CRH-receptor antagonist) when microinfused into the central nucleus of the amygdala under stress-free conditions. Psychoneuroendocrinology 1995; 20:423–432
277. Liang KC, Melia KR, Campeau S, et al: Lesions of the central nucleus of the amygdala, but not of the paraventricular nucleus of the hypothalamus, block the excitatory effects of corticotropin releasing factor on the acoustic startle reflex. J Neurosci 1992; 12:2313–2320
278. Lee Y, Davis M: Role of the hippocampus, bed nucleus of the stria terminalis and amygdala in the excitatory effect of corticotropin releasing (CRH) hormone on the acoustic startle reflex. J Neurosci (in press)
279. Deakin JWF, Graeff FG: 5-HT and mechanisms of defence. J Psychopharmacol 1991; 5:305–315
280. Schreiber R, De Vry J: Neuronal circuits involved in the anxiolytic effects of the 5-HT$_{1A}$ receptor agonists 8-OH-DPAT, ipsapirone and buspirone in the rat. Eur J Pharmacol 1993; 249:341–351
281. Rodgers RJ: Elevation of aversive thresholds in rats by intra-amygdaloid injection of morphine sulfate. Pharmacol Biochem Behav 1977; 6:385–390
282. Rodgers RJ: Influence of intra-amygdaloid opiate injections on shock thresholds, tail-flick latencies and open field behaviour in rats. Brain Res 1978; 153:211–216
283. Costall B, Kelly ME, Naylor RJ, et al: Neuroanatomical sites of action of 5-HT$_3$ receptor agonist and antagonists for alteration of aversive behaviour in the mouse. Br J Pharmacol 1989; 96:325–332
284. Higgins GA, Jones BJ, Oakley NR, et al: Evidence that the amygdala is involved in the disinhibitory effects of 5-HT$_3$ receptor antagonists. Psychopharmacology 1991; 104:545–551
285. Costall B, Jones BJ, Kelly ME, et al: Sites of action of ondansetron to inhibit withdrawal from drugs of abuse. Pharmacol Biochem Behav 1990; 36:97–104
286. Hagan RM, Jones BJ, Jordan CC, et al: Effect of 5-HT-3 receptor antagonists on responses to selective activation of mesolimbic dopaminergic pathways in the rat. Br J Pharmacol 1990; 99:227–232
287. Heilig M, McLeod S, Brot M, et al: Anxiolytic-like action of neuropeptide Y: mediation by Y$_1$ receptors in amygdala, and dissociation from food intake effects. Neuropsychopharmacology 1993; 8:357–363
288. Heilig M: Antisense inhibition of neuropeptide Y (NPY)-Y$_1$ receptor expression blocks the anxiolytic-like action of NPY in amygdala and paradoxically increases feeding. Regul Pept 1995; 59:201–205
289. Wahlestedt C, Pich EM, Koob G, et al: Modulation of anxiety and neuropeptide Y-Y$_1$ receptors by antisense oligodeoxynucleotides. Science 1993; 259:528–531
290. File SE, Rodgers RJ: Partial anxiolytic actions of morphine sulphate following microinjection into the central nucleus of the amygdala in rats. Pharmacol Biochem Behav 1979; 11:313–318
291. Helmstetter FJ, Bellgowan PS, Tershner SA: Inhibition of the tail flick reflex following microinjection of morphine into the amygdala. Neuroreport 1993; 4:471–474
292. Al-Rodhan N, Chipkin R, Yaksh TL: The antinociceptive effects of SCH-32615, a neutral endopeptidase (enkephalinase) inhibitor, microinjected into the periaqueductal gray, ventral medulla and amygdala. Brain Res 1990; 520:123–130
293. Helmstetter FJ, Bellgowan PSF, Poore LH: Microinfusion of mu but not delta or kappa opioid agonists into the basolateral amygdala results in inhibition of the tail flick reflex in pentobarbital-anesthetized rats. J Pharmacol Exp Ther 1995; 275:381–388
294. Yaksh TL, Yeung JC, Rudy TA: Systematic examination in the rat of brain sites sensitive to the direct application of morphine: observation of differential effects within the periaqueductal gray. Brain Res 1976; 114:83–103
295. Kalivas PW, Gau BA, Nemeroff CB, et al: Antinociception after microinjection of neurotensin into the central amygdaloid nucleus of the rat. Brain Res 1982; 243:279–286
296. Klamt JG, Prado WA: Antinociception and behavioral changes induced by carbachol microinjected into identified sites of the rat brain. Brain Res 1991; 549:9–15
297. Good AJ, Westbrook RF: Effects of a microinjection of morphine into the amygdala on the acquisition and expression of conditioned fear and hypoalgesia in rats. Behav Neurosci 1995; 109:631–641
298. Harris JA, Westbrook RF: Effects of benzodiazepine microinjection into the amygdala or periaqueductal gray on the expression of conditioned fear and hypoalgesia in rats. Behav Neurosci 1995; 109:295–304

299. Amaral DG, Price JL, Pitkanen A, et al: Anatomical organization of the primate amygdaloid complex, in The Amygdala: Neurobiological Aspects of Emotion, Memory and Mental Dysfunction, edited by Aggleton JP. New York, Wiley-Liss, 1992, pp 1–66
300. Gaffan D, Murray EA, Fabre-Thorpe M: Interaction of the amygdala with the frontal lobe in reward memory. Eur J Neurosci 1993; 5:968–975
301. Morgan MA, Romanski LM, LeDoux JE: Extinction of emotional learning: contribution of medial prefrontal cortex. Neurosci Lett 1993; 163:109–113
302. Morgan MA, LeDoux JE: Differential contribution of dorsal and ventral medial prefrontal cortex to the acquisition and extinction of conditioned fear in rats. Behav Neurosci 1995; 109:681–688
303. Gewirtz JC, Falls WA, Davis M: Normal conditioned inhibition and extinction of freezing and fear potentiated startle following electrolytic lesions of medial prefrontal cortex. Behav Neurosci (in press)
304. Annau Z, Kamin LJ: The conditioned emotional response as a function of US intensity. Journal of Comparative and Physiological Psychology 1961; 54:428–432
305. Morgan MA, LeDoux JE: Medial prefrontal cortex (mPFC) and the extinction of fear: differential effects of pre- or post-training lesions (abstract). Society for Neuroscience Abstracts 1996; 22:1116

PART 2
Clinical Syndromes

6

Paroxysmal Limbic Disorders in Neuropsychiatry

Stephen Salloway, M.D., M.S., James White, M.D.

Dysfunction in medial temporal lobe/limbic circuits can give rise to a wide variety of transient clinical phenomena, such as disturbances of perception and memory, anxiety states, autonomic hyperarousal, and feelings of disconnection from the environment. Patients with medial temporal lobe dysfunction may complain of symptoms of *déjà vu*, depersonalization, derealization, auditory and visual hallucinations, fear, palpitations, numbness, loss of time, confusion, and "spacing out." The possibilities for differential diagnosis of these paroxysmal episodes include partial complex seizures, psychogenic seizures, panic disorder, transient global amnesia, migraine, and dissociative disorder, among others. The description of the events is often vague and confusing, and diagnosis and treatment of these disorders thus pose some of the most challenging problems in neuropsychiatry.

High-resolution structural MRI and video-EEG have aided in the diagnostic evaluation of these conditions. Depth electrode recordings in patients being evaluated for medial temporal lobe epilepsy have provided useful information about clinical-anatomic correlations in patients with the transient phenomena described above.

In this chapter we present a spectrum of paroxysmal disorders thought to involve limbic structures that are commonly seen in clinical practice. The cases chosen consist of short, discrete events that demonstrate an aspect of paroxysmal change in behavior thought to be due to dysfunction in limbic circuits. The clinical features that aid in diagnosis and treatment are highlighted, and clinical-anatomic correlations are provided whenever possible.

The cases are divided into paroxysmal disturbances of behavior caused by partial seizures, by psychiatric illness, by psychogenic seizures, by mixed epileptic and psychiatric syndromes, and by other causes of transient limbic impairment. The review is not meant to be comprehensive, but rather is intended to acquaint the reader with the diagnosis and treatment and localization of behavioral disturbance in paroxysmal disturbances commonly seen in a neurological or psychiatric outpatient setting.

Injury to medial temporal lobe structures can also give rise to a chronic disturbance of memory, perception, and behavior such as is seen in Alzheimer's disease, herpes encephalitis, medial temporal lobe stroke, and some cases of schizophrenia. These chronic disturbances of the limbic system are discussed elsewhere in this volume.

PAROXYSMAL BEHAVIORAL DISTURBANCE CAUSED BY EPILEPSY

Case 1: Episodes of fear, auditory hallucinations, automatic behavior, and loss of consciousness. A 28-year-old woman with refractory partial epilepsy was admitted for monitoring to an epilepsy evaluation service in preparation for possible epilepsy surgery. Depth electrodes were placed in her left medial temporal lobe, and 48 hours of videotaped recording were completed. While awake, she experienced typical seizures, which began with a feeling that a seizure was coming. She pressed the button for the nurse. Over the next 15 seconds she experienced a wave of fear, had a gnawing feeling in the pit of her stomach, and heard her mother calling her. She then lost contact with the environment, but did not pass out. Her right hand became dystonic, and she began picking at her dystonic right hand with her left hand in an automatic fashion and began smacking her lips. Her pupils became fixed and dilated, and she no longer followed commands, although she was awake. Her last memory of the event was the recollection of her mother calling her. Thirty seconds into the event her right arm began to shake, her eyelids began to flutter rhythmically, her eyes and neck deviated to the right, she called out, and then she had a brief generalized tonic-clonic seizure, with more shaking on the right side than the left. The entire seizure lasted 75 seconds. She experienced postictal confusion and fatigue and mild right arm paresis for 40 minutes after the event.

This patient had a brief, nonspecific warning that a seizure was coming, followed quickly by an epigastric sensation, fear, and an auditory hallucination before losing contact with the environment. Transient symptoms of autonomic, psychic, and perceptual disturbance are often seen in temporal lobe seizures, particularly those arising in the medial temporal lobe. Mullan and Penfield[1] were able to induce hallucinations by stimulating the temporal lobe in a subset of patients who were awake during their surgery for temporal lobe resection. In their study, auditory hallucinations were seen most commonly after stimulation in the auditory association cortex in the superior and upper middle temporal lobe gyri. Gloor[2] and Halgren and Chauvel[3] extended this line of research by recording spontaneous seizures from patients with electrodes placed in superficial and deep temporal structures. They found that autonomic and psychic phenomena were likely to occur in seizures beginning in the medial temporal lobe, particularly the amygdala.

Patient 1 recalled no more memories for the event after the aura. The impairment of memory with preservation of alertness suggests involvement of limbic memory circuits, particularly the hippocampus. She then developed dystonia of the right hand and automatic behaviors with her normal hand and lips. Automatisms are seen most commonly with seizure foci in the frontal and temporal lobes.[4] She then went on to develop a generalized tonic-clonic seizure; however, greater shaking on the right and head and eye deviation to the right suggested a left-sided focus. Depth electrodes revealed the origin of her seizure to be in the left amygdala. A recording from a similar case is shown in Figure 6–1A. MRI scan of the brain, utilizing an epilepsy protocol (fine T_1, T_2, and fluid attenuated inversion recovery [FLAIR] coronal cuts through the temporal lobes) showed an area of high signal in the left hippocampus and amygdala consistent with mesial temporal sclerosis (Figure 6–1B).

There are a number of features of her event that allow us to distinguish it from a psychogenic seizure. She had a brief, distinct psychic and autonomic aura, which was followed immediately by impairment of consciousness and minimal interaction with the environment when examined. She then had a visible autonomic change (pupillary dilatation), an ictal cry, rhythmic shaking, a postictal Todd's paresis, and confusion, with no recall for the event after the aura. Fear is one of the most common symptoms reported with temporal lobe seizures.[5] Ictal fear can at times be intense and is commonly associated with epigastric sensations, palpitations, mydriasis, and pallor. The central nucleus of the amygdala is likely to play a critical role in fear produced by an epileptic discharge. Its projections to the lateral hypothalamus appear to be involved in activation of the sympathetic autonomic nervous system seen during fear.[6-8] The central nucleus of the amygdala also projects to the dorsal motor nucleus of the vagus, nucleus of the solitary tract, and ventrolateral medulla, which may be involved with the tachycardia seen during ictal fear.[6]

Electrical or chemical stimulation of the central nucleus of the amygdala has been demonstrated to produce an apparent fear reaction in animal models. Electrical stimulation of the amygdala in cats resulted in a response that included hypervigilance, mydriasis, piloerection, hissing, ear retraction, and unsheathing claws. This response was accompanied by vasodilation in skeletal muscle, increase in heart rate and blood pressure, and rapid, shallow breathing.[9] Thus, it appears that a temporal lobe seizure may stimulate the central nucleus of the amygdala, which in turn excites the hypothalamus and brainstem, resulting in the signs and symptoms of fear. The role of the amygdala in fear responses is discussed in detail by Davis, and the role of the extended amygdala in the processing of emotion is reviewed by Heimer and by Heilman, elsewhere in this volume.

Case 2: Episodes of palpitations and left-sided numbness. A 40-year-old man, in good health, developed the abrupt onset of left-sided numbness, then waves of numbness, and had a generalized tonic-clonic seizure. In the hospital he had a mild fever, amnesia and confusion, and transient psychosis. CT scan of the brain without contrast was normal. MRI showed areas of high signal bilaterally in the medial temporal lobes (Figure 6–2). His cerebrospinal fluid had a mild lymphocytic pleocytosis. CSF polymerase chain reaction was

FIGURE 6–1A. Intracranial recording during a partial complex seizure in a patient similar to Case 1 who had right medial temporal lobe onset. Note the spike and wave, spike, sharp wave, and slow wave discharges arising from the right amygdala and hippocampus. The left amygdala and hippocampus show diffuse slow wave activity. R Am = right amygdala; R Ant hc = right anterior hippocampus; R Med hc = right medial hippocampus; R para hc = right parahippocampal gyrus; R Orbfr = right orbitofrontal gyrus; L Am = left amygdala; L Ant hc = left anterior hippocampus; L Orbfr = left orbitofrontal gyrus. EEG courtesy of Dr. Itzhak Fried, UCLA Epilepsy Surgery Program.

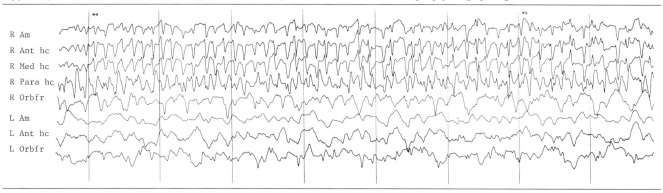

FIGURE 6–1B. T_2 coronal MRI through the medial temporal lobe in Case 1 demonstrates an area of high signal in the left hippocampus consistent with mesial temporal sclerosis.

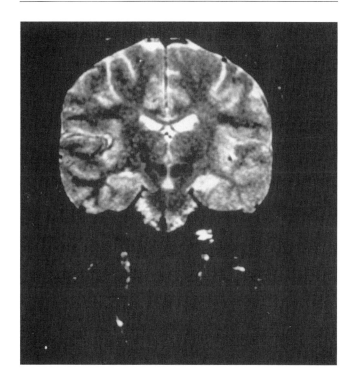

positive for herpes simplex. He was treated for 14 days with high-dose intravenous acyclovir, phenytoin, and carbamazepine, and he improved.

One year after the event he continued to have mild residual short-term memory loss and frequent episodes of numbness involving the left side of his body, which he called "chillies." These episodes began with palpitations and progressed to waves of left-sided numbness lasting approximately 10–120 seconds, and the events were often associated with a funny taste and smell and mild confusion. Mild sinus tachycardia at a rate of 110–120 beats per minute was documented during the episodes. Two to three times per year the episodes of numbness would evolve into a generalized tonic-clonic seizure with loss of consciousness.

His memory loss improved considerably, but simple partial seizures were hard to control despite treatment with three anticonvulsants. Follow-up MRI showed good resolution of high-signal abnormalities in the medial temporal lobes but development of generalized atrophy (Figure 6–3).

Herpes encephalitis most commonly affects the medial and anterior temporal lobes and frontal regions. Seizures and behavior change are often the presenting symptoms. Patients may develop extensive memory loss because of damage to limbic memory circuits, and refractory seizures may develop.

Overall this patient made a good recovery, but he was still troubled by persistent episodes of autonomic hyperarousal (palpitations), unilateral numbness, olfactory and gustatory sensations, and impaired consciousness. These symptoms are consistent with a simple partial seizure disorder, which can progress to partial complex seizures and may generalize. The seizure focus is in the right temporal lobe, most likely in the medial temporal area and insula.

Making the diagnosis of epilepsy is straightforward in this case because of the well-documented encephalitis, medial temporal lobe injury, right-sided seizure focus, lateralizing numbness to the right hemisphere, brief duration of events associated with other sensory auras and mild impairment of consciousness, and occasional generalized seizures. In the cases that follow, the diagnosis of epilepsy can be more difficult.

The central autonomic network is an internal regulation system through which the brain controls visceromotor, neuroendocrine, and behavioral responses important for survival.[10] Components of this network are involved in the autonomic hyperarousal associated with seizures.[11]

The insular cortex is an important area involved with autonomic hyperarousal in epilepsy. Human intraoperative stimulation studies have demonstrated that tachycardia and pressor responses occur more consistently with right anterior insular cortex stimulation, whereas bradycardia and depressor responses are usually seen with left anterior insular cortex stimulation.[12] Stimulation of the amygdala produces the autonomic changes described in Case 1. Stimulation of the medial prefrontal cortex can produce profound changes in blood pressure, heart rate, and gastrointestinal motility.[13]

Case 3: Psychic aura, perceptual distortion, and occasional impairment of consciousness. A 39-year-old man was hit on the right side of the head with a golf club at age 4 and had a motor vehicle accident requiring drainage of a right frontotemporal subdural hematoma at age 30. After the motor vehicle accident, he began to have episodes with electric-like sensations and the sense that what he was perceiving was distorted. He was not sure if his perceptions were accurate, but he was afraid they were not. Sometimes he was sure that objects in the room and his sense of his body were distorted, and this made him very nervous. He became very anxious that things were getting out of control, although he has never had a true panic attack. The events lasted approximately 10–20 seconds, and there was no postictal lethargy or confusion. These events were occurring 1–2 times per month before treatment with anticonvulsants. These events often occur during exercise, such as when he is playing tennis or on the step-climbing machine. Every 3 months he experienced a slightly longer event where he would stop at the net and stare for less than a minute. Once or twice a year, prior to anticonvulsant treatment, he had an event that included staring and wandering, with a clear impairment of consciousness followed by mild postictal lethargy and amnesia for the event.

This gentleman had significant right-sided frontotemporal head injury and experienced occasional brief, discrete events with electrical sensations and a sense that his view of the environment and his body were distorted. These symptoms were accompanied by fear and anxiety. Occasionally consciousness was impaired for a brief period. There was no other significant psychiatric or neurological history. His MRI revealed an old contusion in the lateral right frontotemporal region (Figure 6–4), and his EEG showed nonspecific slowing over right anterior brain areas. Taken together, these symptoms make the diagnosis of simple and complex partial seizures with a right frontal or temporal lobe focus likely. Perceptual disturbance, characterized by a sense that the environment has changed, is more commonly seen in the nondominant medial temporal lobe.

His seizures responded partially to phenytoin. Phenytoin was switched to carbamazepine, and with high therapeutic levels the partial complex seizures resolved and the auras decreased by 50%. His seizures are sensitive to changes in his medication regimen; missing one dose of carbamazepine is likely to lead to a simple or complex seizure. The addition of gabapentin to carbamazepine led to a further reduction in auras. However, the auras have not been completely eliminated. The phenomenon of persistence of auras despite good con-

FIGURE 6–2. Left: T_2 axial MRI at the level of the midbrain demonstrates bright signal in the hippocampus bilaterally, right > left, in Case 2 immediately after the onset of herpes encephalitis. Right: T_2 axial MRI at the same level 2 years after the onset of encephalitis shows resolution of high signal in the medial temporal lobe but atrophy of the hippocampus and dilatation of the temporal horns, right > left.

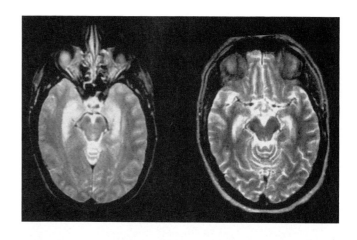

FIGURE 6–3. T_1 axial MRI at the level of the midbrain (left image) and T_1 coronal MRI at the level of the medial temporal lobes (right image) 2 years after the onset of encephalitis in Case 2 demonstrate atrophy of the hippocampus bilaterally.

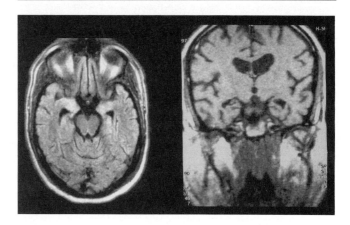

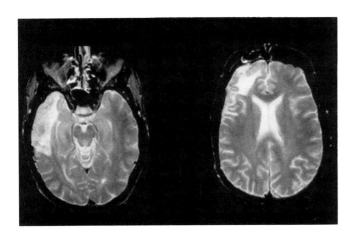

FIGURE 6–4. T_2 axial MRI at the level of the midbrain (left image) and at the level of the lateral ventricles (right image) in Case 3 demonstrates high signal in the right temporal lobe and lateral right frontal lobe consistent with remote head injury.

trol of more complex temporal lobe seizures is discussed by Fried elsewhere in this volume.

Case 4: Episodes of *déjà vu*. Now we are crossing the boundary between neurology and psychiatry. The next case is a 42-year-old woman with no past psychiatric or neurological history who has frequent, intense, and unpleasant episodes that she has been in the place she is in before. The events last 1 to 2 minutes, and afterwards she feels very tired. They may occur many times a day, once or twice a week, or every few months. Overall they are fairly frequent. On five or six occasions following an episode of familiarity she has become dizzy and lost consciousness without injuring herself. She has never had tonic-clonic activity.

She recalls the *déjà vu* phenomena all the way through the events, and she remembers getting dizzy, but she does not remember passing out. These events were verified by witnesses. There were no other psychiatric problems associated with her paroxysmal *déjà vu* experiences.

Her MRI utilizing an epilepsy protocol was normal. Routine EEG on one occasion was normal and on another occasion showed mild nonspecific slowing over the right temporal lobe. Six-hour video-EEG without a *déjà vu* experience was normal.

The patient suffers from frequent events of *déjà vu*. The events are brief and well circumscribed and, on a few occasions, have led to loss of consciousness. There are no other psychiatric symptoms. Are these seizures? Whatever the etiology of the events, can we identify the brain regions involved?

This patient was diagnosed with partial complex epilepsy of unclear etiology and was treated with carbamazepine with a good response. The partial complex seizures resolved, and the *déjà vu* phenomena decreased in frequency and intensity.

The *déjà vu* experience consists of disturbance of memory—thinking that you have been someplace before or that the current surroundings seem extremely familiar. It is a disturbance involving visual memory and visual perception: what the patient sees, associated with a prior experience, feels real. There is a disturbance of time and emotion. Sometimes *déjà vu* phenomena are associated with epigastric phenomena and fear. In the case of Patient 4, the phenomena were associated with dysphoria.

The anatomic localization of *déjà vu* is controversial. Several authors have examined spontaneous seizures and intraoperative stimulation studies that produce experiential phenomena such as *déjà vu*. Mullan and Penfield[1] argued that the temporal neocortex was the key structure in experiential phenomena such as *déjà vu*. In contrast, Gloor[2] found that stimulation of medial rather than lateral temporal regions evoked most experiential phenomena. Bancaud et al.[4] found that both medial and lateral temporal structures are involved in the production of experiential events. Although the evidence for lateralization of *déjà vu* is not clear, at least one study reported lateralization to the temporal lobe of the hemisphere nondominant for language,[3] and another study suggested lateralization to the temporal lobe of the hemisphere nondominant for handedness.[14]

Several investigators have proposed models to explain *déjà vu*.[4,2,15] Bancaud et al.[4] report that the lateral temporal neocortex receives major inputs from medial limbic structures as well as visual and auditory association cortices. A seizure in the medial or lateral cortex in the nondominant hemisphere may activate a memory that becomes associated with the patient's awareness of the environment, giving the current scene a sense of familiarity. Alternatively, feelings of familiarity may be stimulated without activation of a memory, giving the sensation that the current scene has been experienced before. See Fried elsewhere in this volume for more discussion of *déjà vu*.

PAROXYSMAL BEHAVIORAL DISTURBANCE CAUSED BY PSYCHIATRIC DISORDERS

Case 5: Episodes of anxiety, autonomic hyperarousal, and depersonalization. A 34-year-old woman has a 25-year history of frequent episodes of palpitations, intense anxiety associated with feelings of loss of control, depersonalization, and unreality. These episodes begin with palpitations and diaphoresis and a sense that she is about to lose control. She experiences extreme anxiety and discomfort that may last up to 10 minutes. There is no impairment of consciousness.

These events occur a few times per week despite treatment with alprazolam. Approximately one time per week these events progress to include marked symptoms of depersonalization and unreality. She feels as if she is not there or the room does not seem real. These events often occur while driving on the highway. During these episodes she may not be able to feel her foot pushing on the gas pedal. The sensations of depersonalization and unreality may last up to an hour without impairment of consciousness or respiratory difficulty. She has full recall for the events and has never had a generalized seizure. On a few occasions the light appeared to dim and shimmer during these episodes, and she was referred for evaluation of seizures. There is no history of migraine headache. MRI scan of the brain and routine EEGs were normal.

This patient has frequent paroxysmal disturbances characterized by autonomic dysfunction (palpitations and diaphoresis), psychic disturbance (extreme anxiety, fear and dysphoria) and at times perceptual disturbance related to her self-image and view of the environment (feeling unreal and disconnected from her environment). The events last for 10 minutes to 1 hour. There is no impairment of consciousness or amnesia and no history of motor seizures. The events would be long for a simple partial seizure, and the MRI and EEG are normal. The diagnosis is panic disorder. Her panic attacks greatly decreased in frequency and severity when the alprazolam was switched to moderate doses of clonazepam tid.

If there was concern about the diagnosis of epilepsy, she would be a good candidate for simultaneous video-EEG because of the frequency of her episodes. She could also benefit from 24-hour ambulatory EEG because she does not lose consciousness and could describe her events in detail in the log.

Panic attacks are a paroxysmal disorder that may have many similar features to the partial epileptic conditions described in the previous section, such as autonomic dysfunction, perceptual disturbance, fear, and anxiety. These symptoms make us think that the limbic system and its paralimbic cortical extensions, particularly the amygdala, insula, and adjacent temporal cortex, possibly in the nondominant hemisphere, are involved.

The neurobiology of panic disorder has been recently reviewed by Johnson and Lydiard.[16] Functional abnormalities in the medial temporal lobe structures have been implicated in subgroups of panic patients. PET and SPECT studies have demonstrated that some patients with lactate-sensitive panic attacks have abnormal activity in the hippocampal region, primarily in the right hemisphere, during the attacks, and this activity may play a role in triggering them. Increased blood flow in the temporal cortex may alter the interpretation of internally or externally perceived threats. Increased activity in the right prefrontal cortex between attacks has also been reported.[17] Stimulation and blockade of noradrenergic systems in the locus coeruleus can turn on or abolish fear-related behaviors.[18]

Case 6: Persistent sensation of depersonalization. A 38-year-old right-handed man has a 20-year history of the persistent sensation that nothing is behind him and that he is not here. He feels as if he is here, but the room is not as it is supposed to be, there is nothing behind him, and he does not physically fit into the world. These feelings occur every day, sometimes more intensely than others, and are always associated with dysphoria. These events began occurring at age 18 when he smoked marijuana one time, which he feels caused permanent brain damage.

At times he sees spots and feels disoriented, but these are minor problems. He has never had impairment of consciousness or a generalized seizure. A variety of anticonvulsant and psychotropic medications have been tried for this disturbance of his body image in space without relief. He also has chronic depression, which is well treated with sertraline. Despite these symptoms, he is able to work full-time. Cognitive testing was normal.

This patient's disturbance of the perception of his body image and space suggests the possibility of dysfunction of medial temporal lobe structures in the nondominant hemisphere. His symptoms are more persistent than paroxysmal, and he dates them back to the time he smoked marijuana. His symptoms have more of a delusional quality of depersonalization and are not suggestive of a paroxysmal epileptiform disorder. His EEG, including nasopharyngeal and sphenoid electrodes, was normal. MRI scan revealed enlarged CSF space in the right medial temporal lobe consistent with a choroidal fissure cyst (Figure 6–5), with a somewhat shrunken hippocampal/amygdaloid complex.

His findings are consistent with delusional disorder related to right medial temporal lobe dysfunction. The MRI finding is probably clinically significant. He has never had a documented seizure and his EEGs have been normal, but the sensitivity of detecting EEG abnormalities in small deep foci with surface EEG recordings may be low, even with sphenoidal and nasopharyngeal leads. Unfortunately, his delusional symptoms have been refractory to medication and psychological treatments.

Case 7: Dysthymia, emptiness, memory lapses, and occasional electrical shocks in the head. A 38-year-old woman is troubled by feelings of emptiness, a nagging sense of dysthymia, and mood lability. She has frequent "memory lapses," feelings of unreality and *déjà vu*, and occasional panic attacks. During the memory lapses she reports "spacing out" and losing track of what is said to her for minutes at a time, sometimes up to an hour. During these episodes she appears inattentive. They usually occur when she is upset. She has occasional auditory hallucinations, usually associated with worsening of her depressive symptoms. She has never had a generalized seizure. Two to three times per year she has nocturnal episodes of shaking, with electric-like shocks in her head lasting 2–5 minutes. Her EEG demonstrates a normal background with occasional bitemporal sharp wave discharges. Twenty-four-hour video-EEG recorded four events of "spacing out" without an accompanying EEG change. MRI scan of the brain was normal.

She has been diagnosed with borderline personality disorder and depression. Her current medications include sertraline for depression, clonazepam for anxiety, low-dose perphenezine for psychosis, carbamazepine for mood lability and seizure control, and trazodone for insomnia.

The patient occasionally has paroxysmal symptoms that are quite distinct from the ongoing, daily problems with her mood. Her EEG abnormalities and occasional electric shock episodes with shaking suggest a tendency toward partial epilepsy. Her memory lapses, *déjà vu* phenomena, and panic symptoms are consistent with temporal lobe dysfunction. The normal EEG during episodes of "spacing out" is not supportive of the diagnosis of epilepsy; however, a deep seizure focus may not be detected on surface EEG.

Her tendency toward seizures could be exacerbated at times by her complicated psychotropic regimen. Anticonvulsants control her partial seizures and help stabilize her mood somewhat, but they have not taken away her symptoms of unreality and *déjà vu*. Weekly psychotherapy sessions decrease the severity of her feelings of emptiness and depression.

This patient has a syndrome of chronic disturbance of mood and personality along with vague paroxysmal symptoms of memory loss and depersonalization, commonly seen by psychiatrists and psychologists. The syndrome is usually caused by temporal lobe dysfunction but not frank epilepsy. The same brain regions may be involved as in temporal lobe epilepsy, and the same medications may be used for both disorders. It is helpful to be accurate about the diagnosis so that patients and families do not focus on seeking treatments for epilepsy instead of psychiatric care.

PAROXYSMAL BEHAVIORAL DISTURBANCE CAUSED BY PSYCHOGENIC SEIZURES

Case 8: Akinetic episodes with fear. A 40-year-old right-handed woman experiences episodes of rigidity 2–3 times per day, 5 days per week. The episodes begin with generalized anxiety and fear but no palpitations, shaking, tachypnea, or diaphoresis. She feels as if she is withdrawing from the environment. Her body begins to stiffen and her back arches. All four limbs become stiff, and opisthotonic posturing may develop. No clonic movements or cyanosis have ever been witnessed. She does not speak during the episodes. Her pupils do not dilate, and there is no incontinence. She may resist examination by shutting her eyes or pulling away from the examiner during the episode. The events may last up to 20 minutes and end without postictal confusion or lethargy. She recalls only feeling far away and afraid during the events. She is not sure what brings them on, but the events are often preceded by emotional distress. Her routine EEG and extensive video-EEG monitoring during many events have been normal. MRI scan was also normal.

The diagnosis is psychogenic seizures. She experiences the akinetic rigid subtype of psychogenic seizures that Charcot induced in patients at the Salpêtrière in France around the turn of the century. The two main features of her episodes are a state of emotional discon-

FIGURE 6–5. T_2 coronal MRI at the level of the mid-anterior temporal lobes in Case 6 shows a choroidal fissure cyst in the right medial temporal area with decrease in size of the amygdala and hippocampus on the right. Note the normal left amygdala.

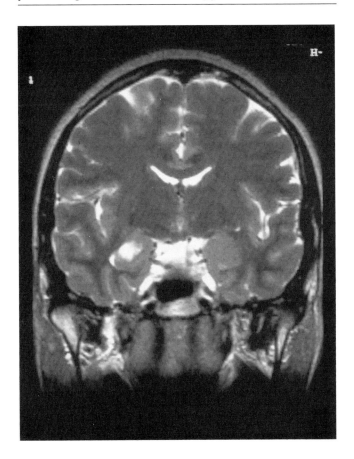

nection with fear, suggesting limbic disturbance, and an akinetic rigid state, suggesting involvement of the extrapyramidal motor system.

Psychogenic seizures are a common problem, comprising approximately 20% of evaluations in epilepsy monitoring centers.[19] Eighty percent of patients with psychogenic seizures are women, and at least 50% have a past history of sexual abuse, as was true in this case. Contrary to popular opinion, the majority of patients with psychogenic seizures do not have true epileptic seizures at other times. (Some demographic features of psychogenic seizures are listed in Table 6–1.)

Psychogenic seizures are considered a form of conversion disorder. The patient is usually not consciously aware of the psychological precipitants of the events, which makes psychotherapy in these cases difficult. The two main types of psychogenic seizures are the akinetic rigid type seen in this case and the thrashing type described in the next case. However, psychogenic seizures can take many forms (Table 6–2). The hardest cases to definitively diagnose involve psychic events such as *déjà vu*, memory lapses, and depersonalization.

The diagnosis of psychogenic seizures was facilitated in the case of Patient 8 by 1) the length of the events without cyanosis, incontinence, clonic activity, or postictal lethargy and 2) her resistance to examination during the event. The normal video-EEG during repeated rigid episodes was helpful in ruling out epilepsy.

It is extremely important to do a thorough evaluation to exclude the possibility of epilepsy. Once the diagnosis of psychogenic seizures has been made, it is important to inform the patient and family about the diagnosis. The patient and family should be encouraged to try to avoid coming to the emergency room for these events because this is likely to lead to unnecessary, potentially harmful treatments. We try to educate the patient and family to make them aware that these events can be controlled. Stress management and supportive counseling, particularly for the patient with limited insight, are often helpful. Psychotropic medication may help, depending on the symptoms. Benzodiazepines can help with anxiety, and antidepressants or mood-stabilizing agents may be indicated. The patient and family should be told the reason for each medicine. If anticonvulsants are used for mood stabilization, the patient should be informed that this is a treatment for stabilizing mood and not for treatment of epilepsy.

Case 9: Flashbacks of emotional trauma and wild shaking.
A 35-year-old right-handed woman experienced a motor vehicle accident with a brief loss of consciousness 10 years ago. She also has a history of incest and rape. Since the accident, she has experienced migraine-type headaches during inclem-

TABLE 6–1. Features of psychogenic seizures

- Approximately 20% of patients in epilepsy monitoring units have psychogenic seizures.
- Approximately 80% are female.
- Sexual or emotional abuse may have occurred in childhood.
- A mixed pattern of epileptic and psychogenic seizures occurs in a minority of patients.
- 40%–50% show resolution of psychogenic seizures.

TABLE 6–2. Types of psychogenic seizures

- Thrashing
- Akinesis, rigidity, opisthotonos
- Weakness and hypophonia
- Decreased responsiveness
- Drop attack
- Psychic: staring, loss of time, loss of focus, memory lapses, confusion, loss of control, depersonalization, flashbacks

ent weather. In addition, every 2 to 3 weeks she experiences dramatic episodes of passing out, followed by thrashing of all four extremities lasting 10 to 15 minutes. She makes loud, unintelligible sounds. During the episodes she appears very upset. She does not react meaningfully with family or examiners during the events. Gentle reassurance and comforting speed the resolution of the event. She recalls being very upset and afraid. Often she experiences flashbacks of prior sexual abuse or of the car accident. The events can occur during a heated argument, while riding as a passenger in a car, when her surroundings remind her of her past sexual abuse, or for no apparent reason. MRI scan and video-EEG recorded during these events have been normal.

The diagnosis is the thrashing type of psychogenic seizures related to posttraumatic stress disorder (PTSD) and dissociative disorder. She also has many symptoms of major depression and panic disorder. (See Figure 6–6 illustrating overlap of psychogenic nonepileptic events with various psychiatric diagnoses.) Her episodes were carefully evaluated and determined not to be epileptic seizures because of the length of the events, the clear emotional precipitants and dramatic emotional content of the episodes, the striking out-of-phase movements, and negative video-EEG. (See comparison in Table 6–3 of epileptic versus nonepileptic seizures.)

The patient was informed of the diagnosis of psychogenic seizures and placed on sodium valproate for mood stabilization and paroxetine for depression and anxiety. She participated in weekly psychotherapy aimed at reducing the distress related to her prior sexual abuse and encouraging practical stress management. PTSD symptoms continued to occur intermittently, but psychogenic seizures resolved.

The neurobiology of PTSD has been reviewed by Southwick et al.[20] As a group, patients with PTSD have been noted to have heightened sympathetic nervous system arousal. Multiple studies have demonstrated noradrenergic dysregulation in PTSD.[21,22] For example, 24-hour urine norepinephrine has been shown to be higher in veterans with PTSD compared with veterans with major depression or schizophrenia.[22] Studies also suggest an increase in glucocorticoid receptor responsivity, resulting in enhanced negative feedback and diminished basal cortisol excretion and heightened suppression of cortisol by dexamethasone.[20] These changes may cause PTSD patients to have a lower threshold for stress-induced triggering of traumatic memories and heightened autonomic responses.

FIGURE 6–6. Psychogenic nonepileptic events and DSM-III-R classifications. Reprinted, with permission, from James Rowan and John R. Gates (eds.), *Non-Epileptic Seizures* (Boston, Butterworth-Heinemann, 1993), p. 24.

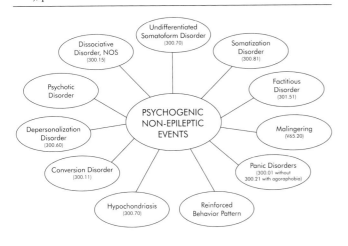

TABLE 6–3. Differentiating epileptic from psychogenic events

	Epileptic	Psychogenic
Length	Short	May be longer
Tongue biting, incontinence, self-injury	Common	May occur
During sleep	May occur	Rare
Neurological signs during seizure	May occur	No
Movements	In phase (GTS) Automatic (CPS)	Usually out of phase
Recall	Scant	May be detailed
Postictal confusion	Yes	Variable
Prolactin ↑	Usually occurs	No
Psychiatric diagnosis	May occur	Common

Note: GTS = generalized tonic-clonic seizures; CPS = complex partial seizures.

PAROXYSMAL BEHAVIORAL DISTURBANCE DUE TO A COMBINATION OF EPILEPSY AND PSYCHIATRIC ILLNESS

Case 10: Anxiety attacks and focal shaking episodes.

A 28-year-old right-handed woman experiences panic attacks with fear, intense anxiety, and palpitations lasting 10–30 minutes once a week. These are controlled with clonazepam. One to two times per month she has milder symptoms of anxiety followed by numbness and rhythmic shaking of her left arm, without loss of consciousness, lasting 1–2 minutes. She feels very upset about her arm shaking but cannot control it. Afterwards she feels tired but is not confused. Lack of sleep and emotional distress precipitate these events.

These events were treated with carbamazepine, which decreased the frequency but did not eliminate the events. Every 3 months she experiences a similar event with left arm shaking, but in these instances she feels as if she is going to pass out. She had two generalized seizures at age 13.

MRI scan of the brain is normal. The background of her EEG is normal, but she does have occasional asymmetrical theta slowing and sharp waves over the right temporal lobe, particularly during sleep. Video-EEG during one shaking episode was normal, but intermittent right temporal theta slowing and sharp waves appeared 30 minutes after the event.

The diagnosis in this case is a little confusing. The patient has discrete panic episodes, which are controlled with benzodiazepines. On other occasions she has stereotyped events beginning with anxiety, which are followed quickly by brief rhythmic shaking of the left arm. These sound like simple partial seizures with a right hemisphere focus. However, her EEG during one event was normal. She was given the diagnosis of psychogenic seizures by another neurologist, and her carbamazepine was tapered. When her carbamazepine level was low, she experienced a generalized tonic-clonic seizure with loss of consciousness.

Her remote history of two generalized seizures, a recurrence of a generalized seizure when her carbamazepine was lowered, the stereotyped unilateral quality to her event, with occasional feelings of near-syncope, and the intermittent mild paroxysmal discharges seen in the right temporal lobe suggest that she has an underlying partial seizure disorder as well as panic disorder. These events are brought on by poor sleep and emotional distress. Emotional distress is a nonspecific factor that can precipitate both epileptic and psychogenic seizures.

She remains on carbamazepine and clonazepam, with good control of her panic attacks and no further generalized seizures. Stress management has helped. She continues to have occasional, though less frequent, episodes of left arm numbness and shaking.

Case 11: Psychosis and agitation following seizures. A 39-year-old right-handed man has had frequent right-sided temporal lobe epilepsy since age 15. His seizures occur at night and often begin without a clear aura. He will stare, pick at his clothes, and wander around the house unable to communicate or follow commands. The seizure usually lasts approximately 3 minutes and ends with the automatic brushing of his teeth. Afterwards he feels very tired and goes to sleep. The seizures vary in frequency from many times per week to once every 2 months. He has never had a generalized seizure.

If the seizures occur nightly for a few nights in a row, his family notices a change in his behavior. On the morning of day 3 after the second night of seizures he becomes noticeably more withdrawn and less talkative. He seems to be in a bad mood and has a surly expression. On day 4 he becomes preoccupied and agitated. When questioned, he may report receiving special messages through his teeth and feeling the urge to move to another state. He tries to get in the car and drive out of state. If his father tries to prevent him from driving, he becomes extremely upset and irrational and may become aggressive. On a few occasions he has been admitted to the psychiatric hospital with florid psychosis and aggression on day 5 or 6 following his last seizure. His psychotic symptoms respond rapidly to neuroleptics. In between episodes, mild psychotic symptoms may be present, but he recognizes them as unreal perceptual events brought on by his seizures.

This case represents an example of delayed postictal psychosis following a flurry of partial complex seizures arising from the right temporal lobe. The psychotic events contain elements of affective disturbance and delusions. Much of the content of the delusions is based on real-life frustrations. One to 2 days after his last seizure the patient becomes keenly aware of the desire to have a normal life, and he feels the intense desire to move out of state away from his family to establish this new life. In this frame of mind he does not closely monitor his thinking or behavior and does not carefully scrutinize his plans. He also does not recognize the impact that his behavior has on others around him. He can become aggressive if people interfere with his plans. EEGs obtained during the interictal psychotic state show prominent bilateral moderate-voltage slow-wave activity consistent with a confusional state (Figure 6–7). His EEG returns to baseline after treatment with neuroleptics.

In this case the psychosis is a delayed postictal, not an ictal, event. The patient's psychotic symptoms are transient and are clearly related to the frequency of his seizures. His frequent seizures produce an encephalopathic state limited to mental processing, without impairment of alertness. The encephalopathy is demonstrated by the high-voltage slow-wave activity seen on EEG during these episodes of postictal psychosis.

For the past 36 months he has been seizure-free while maintained on high therapeutic levels of phenytoin, carbamazepine, and a low dose of clonazepam. During that time there have been no psychotic events, and he has been able to marry and return to work, providing further evidence that his psychosis is related to seizure frequency and is not an independent psychotic disorder.

The association of epilepsy and psychosis has been noted for more than a century. Postictal psychosis is described as a psychotic state that is temporally related to a preceding increase in epileptic activity, usually complex partial seizures with or without secondary generalization. Delusions, hallucinations, confusion, and aggression are common and are usually transient but tend to recur. There is usually a lucid interval between seizures and psychosis; thus, postictal psychosis is not simply due to a clouding of consciousness related to preceding seizures.[23] Interictal psychosis differs from postictal psychosis in that 1) it is not greatly influenced by seizure activity and 2) remission is unusual. Both postictal and interictal psychoses tend to develop after many years of seizures, often more than 20 years.[23] Some patients with postictal psychosis develop interictal psychosis.

Bilateral structural lesions in the temporal lobes have been reported at a significantly higher rate in patients with postictal and ictal psychosis than in other patients with temporal lobe epilepsy.[24] Bilateral independent temporal epileptogenic abnormalities also appear to be associated.[24] Wieser et al.[25] suggested that epileptic psychosis is due to biologically confined limbic status epilepticus. Wolf[26] argued that the generator of epileptic discharge is still active in such patients but that the usual pathways of propagation are blocked, resulting in a pattern of spread that results in psychotic symptoms. The psychotic behavior seen in Case 11 is most likely due to a prolonged postictal confusion involving limbic structures. This formulation is supported by the diffuse slow-wave activity seen on EEG during the psychosis, which resolved when the psychosis was treated.

OTHER CAUSES OF PAROXYSMAL BEHAVIORAL DISTURBANCE

Case 12: Headache, anxiety, confusion, and aphasia. A 52-year-old right-handed woman has a 15-year history of frequent migraine headaches lasting 1 to 2 days. Two times each year she experiences an episode that begins with intense anxiety and is followed by difficulty finding words and comprehending what is said to her. She has intermittent difficulty using her right hand. A severe left-sided throbbing headache follows, causing her to writhe in bed with pain, which is accompanied by nausea and occasional vomiting.

For the next 2 days she remains extremely anxious and has intermittent confusion, aphasia, and headache. She is depressed and fears that she is dying. During these episodes, head pain responds to opiate analgesics. Anxiety is partially relieved by benzodiazepines. After the episode she returns to normal with no residual deficits.

EEGs, including those obtained during the events, are normal except for occasional mild, nonspecific slow waves seen bilaterally. MRI scan of the brain is normal. The patient has never had a witnessed seizure. Her mother and maternal aunt suffered from migraines and depression.

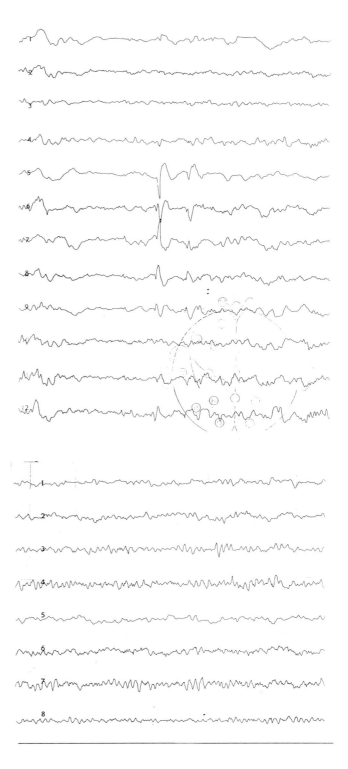

FIGURE 6–7. Top: Occasional spike and wave discharges from the right temporal lobe and intermittent background slowing in Case 11 during an episode of delayed postictal psychosis on day 6 after a flurry of partial complex seizures (longitudinal bipolar montage). Bottom: Return of the normal background EEG in Case 11 three days later, after successful treatment of psychosis with haloperidol (longitudinal bipolar montage, standard international 10-20 system).

This patient with a well-established migraine history presents with occasional episodes of intense anxiety, aphasia, confusion, right hand weakness, depression, headache, and nausea lasting 2 days. The most likely diagnosis is complicated migraine with prominent involvement of limbic elements, language cortex, and left lateral primary motor cortex. The vast majority of her headaches, though severe, are common migraines that do not involve the complicated features described above. The family history of migraines and depression in her mother and maternal aunt suggests an inherited predisposition to migraines and psychiatric disturbance.

The differential diagnosis of her complicated migraine episodes includes transient ischemic attacks (TIAs) involving the left middle cerebral artery territory, subarachnoid hemorrhage, and nonconvulsive seizures. The EEG was not consistent with epilepsy but did show some of the nonspecific changes often seen with migraine. TIAs are possible, but she has no vascular risk factors, and the diagnostic evaluation for stroke has been negative. The MRI did not show evidence of hemorrhage.

A full spectrum of prophylactic treatments for migraine has been ineffective at reducing the frequency and intensity of her headaches. Medication containing butalbital/caffeine/aspirin makes headache pain manageable if taken at the onset of the headache. Sumatriptan has been helpful in aborting the severe migraines but is relatively contraindicated in patients with complicated migraine because of the risk of stroke or myocardial infarction. Fluoxetine has been helpful in improving her mood and overall sense of well-being but has had no effect on her headaches. Stress management does help decrease the severity of her headaches somewhat.

The comorbidity of migraine, epilepsy, and affective disorders has recently been reviewed.[27] Migraine is associated with major depression, with anxiety disorders such as panic disorder, and, to a lesser degree, with bipolar disorder. The prevalence of each of the affective disorders is higher among subjects who have migraine without aura.[28] There is also an association between migraine and epilepsy. One study demonstrated that 20% of adult epileptics had migraine and 3% experienced seizures during or immediately following a migraine aura.[29] Certain medications, such as valproate, are effective in the treatment of migraine, affective disorders, and epilepsy, suggesting a common pathophysiology.

Case 13: Transient disturbance of memory. Two years ago, a 65-year-old right-handed woman had the abrupt onset of confusion lasting 4 hours. During the episode she was disoriented to time and place and kept asking where she was and what she and her husband were going to do that day. She was able to walk around without difficulty, although she seemed somewhat in a daze. There was no weakness or sensory disturbance. After the event she could not recall the 4-hour period of confusion and amnesia. Her short-term memory was noted to be impaired after the event. She could recall remote events but had trouble remembering names. She could register 3 items but could recall only 1 of 3 items after a 5-minute delay, even with prompting. Verbal memory was clearly more impaired than nonverbal memory. Her memory deficits have persisted.

MRI scan of the brain revealed an area of high signal on T_2-weighted images in the left anteromedial thalamus, consistent with infarction of the posterior cerebral perforators (Figure 6–8). The diagnosis was transient global amnesia (TGA) associated with left thalamic infarction causing memory loss. The patient was placed on daily aspirin for stroke prophylaxis.

One year later she developed a similar episode of transient amnesia and confusion lasting 4 hours. She was disoriented to time and place and kept telling her husband that she had to go to work, even though it was Sunday. She repeated the same phrases over and over and could not be redirected. There was no sensorimotor impairment. The episode resolved without any residual deficit. She had no memory for the event. There was no change in her short-term memory deficit on bedside testing. Repeat MRI scan and neuropsychological testing were unchanged from the prior studies. EEG showed mild nonspecific asymmetric slowing over the left hemisphere.

The diagnosis was recurrent TGA. Transient global amnesia is a mono-episodic event in two-thirds of cases. The etiology is usually thought to involve ischemia in medial temporal/limbic memory circuits, although medial temporal or limbic seizures can also produce TGA. Often a definitive cause cannot be determined.

In this case the first event was associated with a thalamic stroke causing persistent short-term memory loss. The anteromedial thalamus is an important part of the limbic memory circuit involving the hippocampus, fornix, mammillary bodies, thalamus, and cingulum. A lesion anywhere in the circuit can produce a disturbance of short-term memory. The second episode of TGA was felt to be caused most likely by a recurrent TIA in the same distribution as her first stroke, although the mechanism of the TIA is unclear. The differential diagnosis includes partial complex seizures or nonconvulsive seizures, although the EEG did not support this diagnosis and the event is long for a common partial complex seizure. Her aspirin was switched to ticlopidine, and there have been no further episodes of TGA, TIA, or stroke and no progression of her memory loss in 2 years.

The etiologies most commonly proposed for TGA include transient ischemia in the inferomedial parts of the temporal lobes, migraine, and hippocampal seizure activity. A case-controlled study by Kushner and Hauser[30] found an increase in vascular risk factors in patients with TGA. However, not all studies have found evidence of vascular risk factors in TGA patients.[31,32] Hodges and Warlow[31] compared 114 TGA patients with 109 normal community-based control subjects and 212 TIA control subjects and reported that TGA was not associated with vascular risk factors. Follow-up comparison of TGA and TIA groups revealed a significantly higher incidence of vascular events and mortality in the TIA group. Seizures could account for a small number of episodes of TGA.[32]

Cerebral blood flow studies of small numbers of TGA patients have shown hypoperfusion in the medial temporal lobes around the time of the attack. The mechanism of the hypoperfusion is unclear.

Oleson and Jorgenson[33] postulated that the phenomenon of spreading depression described by Leao,[34] which may be an important mechanism in migraine, may also play a role in TGA. Spreading depression can be elicited in the hippocampus, where it has been shown to cause amnesia, giving support to the migraine hypothesis of TGA.

Case 14: Prolonged confusional episode. A 65-year-old right-handed woman in good general health, with a history of well-controlled partial epilepsy, was admitted to the hospital for an episode of confusion. Her last seizure was more than 10 years ago, and she was compliant with phenytoin.

On the day of admission her family noted that she was staring and seemed confused and inattentive. A few hours later she became disoriented to time and place, and her speech, although clear, did not make sense. She denied weakness or sensory disturbance and was able to ambulate without difficulty, although she seemed to be wandering. No tonic-clonic movements were seen, and she did not lose consciousness.

She appeared confused and disoriented in the emergency room and would respond intermittently to the examiner, with long pauses in between responses. Many of the answers were repetitive, and she had little awareness that she was repeating herself.

She was afebrile and had normal vital signs. CT scan of the brain without contrast was normal. Screening laboratories for metabolic encephalopathy, including electrolytes, thyroid functions, BUN/creatinine, ammonia, liver function tests, B_{12}, folate, complete blood count, and toxicology screen, were normal. Her phenytoin level was 12, which was her usual level. She was on no other medications that could have contributed to this event, and the family was not aware of ingestion of any unprescribed medications.

There are many possible explanations for this patient's confusional episode. Toxic and metabolic causes are most likely in this age group, but no laboratory abnormalities were found. There was no evidence of underlying cognitive impairment or concurrent infection. A focal stroke such as the thalamic infarction seen in the last case is possible, and it may have been too early for the CT scan to detect a small infarction.

Given her prior history of epilepsy, there was concern that she was having postictal confusion or partial complex status epilepticus. Her EEG obtained during the episode revealed frequent spike and wave and slow-wave discharges in the right temporal lobe and an inability to maintain a normal background rhythm in the right hemisphere, consistent with nonconvulsive partial status epilepticus. The EEG normalized after 1 mg of lorazepam given intravenously (Figure 6–9).

Nonconvulsive status epilepticus (SE) usually presents as a confusional episode. The diagnosis is difficult to make without the EEG. Two types of nonconvulsive status epilepticus are commonly seen: absence status, showing irregular 3–4 Hz spike and wave discharges bilaterally, and partial complex status, with frequent epileptifom discharges arising out of the temporal or frontal regions. Both forms usually show prompt resolution following the administration of intravenous benzodiazepines.

Nonconvulsive status epilepticus has recently been reviewed by Cascino.[35] Absence SE is the most common form of nonconvulsive status. Most patients who develop absence SE have a primary or secondary generalized seizure disorder. The events may last minutes to days. It may be seen after head injury or other focal brain lesion or during withdrawal from alcohol or psychotropic or anticonvulsant medication.

It is unclear what caused this patient to develop the abrupt onset of nonconvulsive status epilepticus. It is important to consider this diagnosis, since it is a treatable cause of delirium. This patient's anticonvulsants were switched from phenytoin to carbamazepine, with levels maintained in the high therapeutic range. She has had no recurrence of seizures or nonconvulsive SE for the past 18 months.

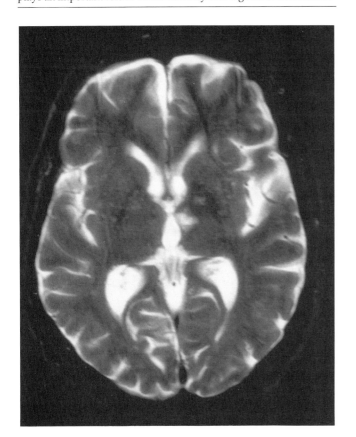

FIGURE 6–8. T_2 axial MRI at the level of the third ventricle in Case 13 demonstrating an area of high signal in the left anteromedial thalamus consistent with small vessel infarction. This area of the thalamus plays an important role in verbal memory and regulation of behavior.

Case 15: Violent attacks during sleep. A 67-year-old right-handed man had developed at age 65 a depression characterized by a decrease in spontaneity and a loss of interest in his usual activities. The depressive symptoms responded well to moderate doses of nortriptyline.

At age 66 he began having attacks during sleep that were characterized by loud cursing and thrashing, and he struck his wife during these attacks in bed on a number of occasions. At times he was so active during these events that he threw himself out of bed and landed on the floor, causing minor head injuries. He devised a belt to keep himself strapped in bed. These events occurred a few nights per week, sometimes more than once per night. The events usually lasted 2–10 minutes. There was no personal or family history of sleep disturbance, head injury, or seizures.

His wife noted that he seemed to be acting out a struggle in a violent dream during the episodes. She also noted that his eyes seemed to be moving under his lids. Though not a good describer, he recalled some dream-like fragments during some of the events.

Neurological and cognitive examinations were normal. Mood was euthymic. CT scan of the brain and routine EEG were normal. No events were recorded during an overnight sleep study, but this test documented persistence of tonic muscle activity on electromyography during epochs of REM sleep.

The frequent violent episodes during sleep beginning abruptly in a 66-year-old man raise the possibility of REM sleep behavior disorder (RBD). His wife's report of witnessing violent dreams associated with rapid eye movements and the persistence of tonic EMG activity during REM strongly support the diagnosis of REM sleep behavior disorder. Other diagnostic possibilities include partial complex seizures arising in sleep, slow-wave sleep parasomnias, or psychiatric disturbance such as PTSD or psychotic depression. His sleep attacks almost completely resolved on clonazepam 1 mg at bedtime, and he was able to remove the restraining belt from his bed.

The REM onset cells are located in the posterior aspect of the pons near the locus coeruleus. The atony center is located just anterior to this region. Normally these two systems act in unison, producing REM sleep with muscle atony. REM sleep behavior disorder is caused by the dissociation of REM sleep from the atony normally seen during REM. Acetylcholine can induce both REM and atony, and experimental use of anticholinergic agents can cause this dissociation of REM and atony in animals. Glutamate also plays a role in atony in the pons.[36]

The symptoms of REM sleep behavior disorder are intermittent. On some occasions patients are capable of acting out their dreams. This frequently leads to strange, often violent behavior resulting in injury to themselves or their bed partners. Men are more affected than women, and onset of symptoms usually begins in the patient's 50s and 60s.[37] There have been recent reports of older men with RBD developing parkinsonism, and the possibility has been raised that RBD represents an early form of dopaminergic dysfunction.[37]

Although the mechanism of disease is primarily in the brainstem, the manifestations of the disease often involve limbic elements of heightened emotion, fear, and violent behavior. The dream fragments often have strong emotional content. The mechanism of limbic activation in RBD has not been determined. The nearby locus coeruleus has rich noradrenergic connections to

FIGURE 6–9. Left: EEG in Case 14 showing persistent polyspike and wave discharges with phase reversal at the T_6-O_2 electrodes in the right posterior-temporal/occipital region during a prolonged confusional episode of nonconvulsive status epilepticus (longitudinal bipolar montage). Right: Resolution of epileptiform activity in Case 14 after treatment with phenytoin and phenobarbital. Mentation returned to normal after paroxysmal discharges stopped. Low-voltage beta activity is a barbiturate effect (longitudinal bipolar montage, standard international 10-20 system).

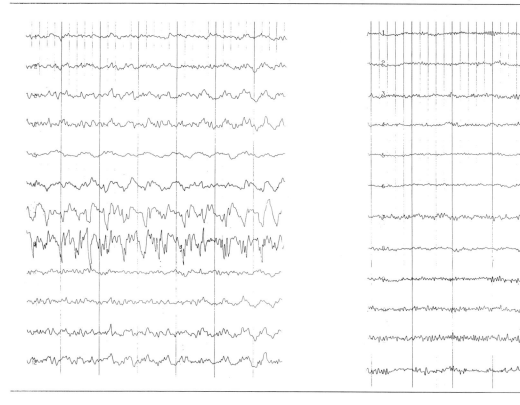

TABLE 6–4. Distinguishing features of paroxysmal limbic episodes

Category	Timing of Episode	Distinguishing Features
Temporal lobe seizures	Seconds to usually < 2 minutes	Discrete events Aura (autonomic hyperarousal, fear, olfactory or gustatory symptoms, etc.) Impaired level of consciousness and memory if complex May progress to include tonic-clonic activity Postictal lethargy Ictal EEG reveals seizure activity Postictal headache may occur
Panic disorder	10 minutes to 1 hour	Autonomic dysfunction (palpitations, diaphoresis) Psychic disturbance (anxiety, fear, dysphoria) No impairment in consciousness or memory Ictal EEG normal
Depersonalization	May be persistent	Disturbance in the perception of body image or the environment suggests dysfunction in the nondominant medial temporal lobe Can be caused by seizures, but often part of a psychiatric disorder
Psychogenic seizures	Often longer than epileptic seizures (> 2 minutes)	Movements may be out of phase, uncoordinated, or akinetic and rigid History of psychiatric disease Milder postictal confusion Ictal EEG normal
Postictal psychosis/ interictal psychosis	Days to weeks, may be persistent	Temporally related to increased seizures Delusions, hallucinations, confusion, and aggression are common
Complicated migraine	Aura lasts 10–30 minutes Headache may last several hours or longer	Headache usually present at some point during the episode Aura usually precedes headache, consists of confusion, anxiety, aphasia, or focal motor/sensory signs Headache is typically throbbing, associated with nausea, photophobia, and visual scotomas Personal or family history of migraine
Transient global amnesia	3–24 hours	Sudden onset of amnesia Personal identity retained, significant persons recognized, neurologic exam nonfocal Patients are confused and repetitive in their questions, like a "broken record"
Nonconvulsive status epilepticus	Hours, sometimes days	Confusion, which may wax and wane, is the most common symptom Ictal EEG reveals seizure activity
REM sleep behavior disorder	2–10 minutes	Frequent, often violent episodes during sleep associated with dream fragments and anxiety Age of onset is typically sixth or seventh decade, male predominance

medial temporal lobe limbic structures, and cholinergic systems in the brainstem have strong connections to the basal forebrain.

DISCUSSION

Although these cases of transient disturbance in behavior are caused by a variety of diagnostic entities, there are some important unifying features. It is important to make an accurate diagnosis of the condition in order to educate the patient and family and to develop an effective treatment plan. An accurate diagnosis rests on careful reconstruction of the history by examination of the patient and by interviewing people who witnessed the events, as was seen in the wife's report of rapid eye movements during the events in Case 15 that helped diagnose REM sleep disorder. Many evaluations fail because of inadequate attention to finding informants who can describe the events.

There are some important questions to address. Is the event brief and discrete or long and diffuse? Do the symptoms fit a focal pattern of dysfunction, such as the stereotyped left-arm shaking in Case 10, or a known syndrome, such as the transient memory loss in the case of TGA described in Case 13? Do the events make physiological sense; that is, are jerking movements in phase or thrashing? Is there prolonged shaking of all four extremities and impaired consciousness without cyanosis or postictal lethargy, as in the psychogenic seizures described in Case 9? Does the patient recall the event? Is there a past history of brain injury or well-documented seizures? Is there a personal or family history of psychiatric illness? What is the age of the patient, and what is the most likely explanation for the transient

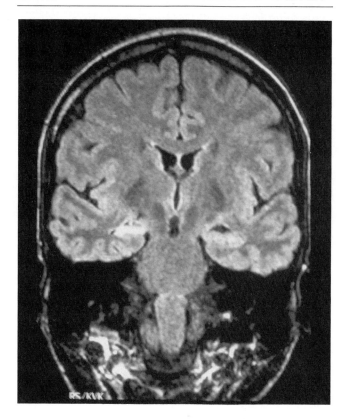

FIGURE 6–10. Fluid attenuated inversion recovery coronal sequence (FLAIR) at the level of the mid-pons shows a subtle area of high signal and a small right hippocampus in a patient with mesial temporal sclerosis. These findings are best seen by using FLAIR images in an epilepsy protocol with fine coronal cuts through the temporal lobes.

behavioral disturbance in this age group? For example, the average age of the 11 patients in the epilepsy, psychiatric, psychogenic seizures, and mixed psychiatric and epilepsy groups was 36, whereas the average age of the 4 patients in the other category of TGA—complicated migraine, nonconvulsive status, and REM sleep behavior disorder—was 62. (Distinguishing features of paroxysmal limbic episodes are described in Table 6–4.)

EEG, and particularly video-EEG, can be extremely helpful in making the diagnosis, especially in cases with frequent events. The clinician should be aware of the limitations of surface EEG in detecting small deep foci. Depth electrodes are helpful in diagnosis in specialized cases, and this technique has provided very valuable information for understanding the localization of transient behavioral phenomena. MRI with fine T_1 and T_2 coronal cuts through the temporal lobes provides a good view of the integrity of medial temporal lobe and other limbic structures. FLAIR sequences are especially sensitive at detecting subtle signal changes such as those seen in mesial temporal sclerosis (Figure 6–10).

Regardless of the etiology, the same limbic brain regions may be involved and the events may respond to similar treatment approaches. Stress reduction and common-sense approaches such as adequate sleep, regular meals, and careful medication compliance can help reduce the frequency and severity of both epileptic and nonepileptic events. If anticonvulsants are used, the reason for their use should be clearly communicated to the patient and family.

We have presented 15 cases of transient behavioral disturbance thought to involve limbic structures from five different categories of syndromes. Making an accurate diagnosis and treatment plan in these conditions is one of the most challenging problems in clinical neuropsychiatry. We have highlighted key features of each syndrome and described our approach to diagnosis and treatment. These disorders provide an important opportunity to localize the brain regions involved in producing the transient behavioral phenomena. We anticipate rapid growth in our understanding of limbic circuitry as our ability to carry out increasingly sophisticated in vivo morphometric and physiological monitoring techniques grows.

REFERENCES

1. Mullan S, Penfield W: Illusions of comparative interpretation and emotion: production by epileptic discharge and by electrical stimulation in the temporal cortex. Arch Neurol Psychiatry 1959; 81:269–284
2. Gloor P: Experiential phenomena of temporal lobe epilepsy. Brain 1990; 113:1673–1694
3. Halgren E, Chauvel P: Experiential phenomena evoked by human brain electrical stimulation. Adv Neurol 1992; 65:87–104
4. Bancaud J, Brunet-Bourgin F, Chauvel P, et al: Anatomical origin of déjà vu and vivid "memories" in human temporal lobe epilepsy. Brain 1994; 117:71–91
5. Gloor P: Role of the amygdala in temporal lobe epilepsy, in The Amygdala: Neurobiological Aspects of Emotion, Memory and Mental Dysfunction, edited by Aggleton JP. New York, Wiley-Liss, 1992, pp 505–538
6. Krettek JE, Price JL: Amygdaloid projections to subcortical structures within the basal forebrain and brainstem in the rat and cat. J Comp Neurol 1978; 178:225–254
7. Price JL, Amaral DG: An autoradiographic study of the projections of the central nucleus of the monkey amygdala. J Neurosci 1981; 1:1242–1259
8. Shiosaka S, Tokyama M, Takagi H, et al: Ascending and descending components of the medial forebrain bundle in the rat as demonstrated by the horseradish peroxidase-blue reaction. Exp Brain Res 1980; 29:377–388
9. Jordan D: Autonomic changes in affective behavior, in Central Regulation of Autonomic Functions, edited by Loewy AD, Spyer KM. New York, Oxford University Press, 1990, pp 349–366
10. Benarroch EE: The central autonomic network: functional organization, dysfunction, and perspective. Mayo Clin Proc 1993; 68:988–1001
11. Freeman R, Schachter SC: Autonomic epilepsy. Semin Neurol 1995; 15:158–166
12. Oppenheimer SM, Gelb A, Girvin JP, et al: Cardiovascular effects of

human insular cortex stimulation. Neurology 1992; 42:1727–1732
13. Hurley KM, Herbert H, Moga MM, et al: Efferent projections of the infralimbic cortex of the rat. J Comp Neurol 1991; 308:249–276
14. Weinand ME, Hermann B, Allen R, et al: Long-term subdural strip electrocorticographic monitoring of ictal *déjà vu*. Epilepsia 1994; 35:1054–1059
15. Sno HN, Linszen DH: The *déjà vu* experience: remembrance of things past? Am J Psychiatry 1990; 147:1587–1595
16. Johnson MR, Lydiard B: The neurobiology of anxiety disorders. Psychiatr Clin North Am 1995; 18:681–725
17. Nordahl TE, Semple WE, Gross M, et al: Cerebral glucose metabolic differences in patients with panic disorder. Neuropsychopharmacology 1990; 3:261–272
18. Reiman EM, Rachle ME, Robins E, et al: Neuroanatomical correlates of a lactate-induced anxiety attack. Arch Gen Psychiatry 1989; 46:493–500
19. Mirza WU, Penry JK, Riela AR, et al: Video EEG findings in 1,100 consecutive patients (abstract). Epilepsia 1990; 31(suppl 5):686
20. Southwick SM, Bremner D, Krystal JH, et al: Psychobiologic research in post-traumatic stress disorder. Psychiatr Clin North Am 1994; 17:251–263
21. Kosten TR, Mason JW, Giller EL, et al: Sustained urinary norepinephrine and epinephrine elevation in post-traumatic stress disorder. Psychoneuroendocrinology 1987; 12:13–20
22. Blanchard EB, Kolb LC, Prins A, et al: Changes in plasma norepinephrine to combat related stimuli among Vietnam veterans with post traumatic stress disorder. J Nerv Ment Dis 1991; 179:371–373
23. Kanemoto K, Kawasaki J, Kawai I: Postictal psychosis: a comparison with acute interictal and chronic psychoses. Epilepsia 1996; 37:551–556
24. Savard G, Andermann F, Olivier A, et al: Postictal psychosis after partial complex seizures: a multiple case study. Epilepsia 1991; 32:225–231
25. Wieser HG, Hailemariam S, Regard M, et al: Unilateral limbic epileptic status activity: stereo-EEG, behavioural, and cognitive data. Epilepsia 1985; 26:19–29
26. Wolf P: Acute behavioral symptomatology at disappearance of epileptiform EEG abnormality: paradoxical or "forced normalization." Adv Neurol 1991; 55:127–142
27. Post RM, Silberstein SD: Shared mechanisms in affective illness, epilepsy, and migraine. Neurology 1994; 44(suppl 7):S37–S47
28. Merikangas KR, Merikangas JR, Angst J: Headache syndromes and psychiatric disorders: association and familial transmission. J Psychiatr Res 1993; 27:197–210
29. Marks DA, Ehrenberg BL: Migraine-related seizures in adults with epilepsy, with EEG correlation. Neurology 1993; 43:2476–2483
30. Kushner MJ, Hauser WA: Transient global amnesia: a case-control study. Ann Neurol 1985; 18:684–691
31. Hodges JR, Warlow CP: The aetiology of transient global amnesia. Brain 1990; 113:639–657
32. Zorzon M, Antonutti L, Mase G, et al: Transient global amnesia and transient ischemic attack. Stroke 1995; 26:1536–1542
33. Olesen J, Jorgensen MB: Leao's spreading depression in the hippocampus explains transient global amnesia: a hypothesis. Acta Neurol Scand 1986; 73:219–220
34. Leao AAP: Spreading depression of activity in the cerebral cortex. J Neurophysiol 1944; 7:359–390
35. Cascino GD: Nonconvulsive status epilepticus in adults and children. Epilepsia 1993; 34(suppl 1):S21–S28
36. Siegel JM: Brainstem mechanisms generating REM sleep, in Principles and Practices of Sleep Medicine, 2nd edition, edited by Kinger M, Roth T, Dement W. Philadelphia, WB Saunders, 1994, pp 125–144
37. Mahowald MW, Schent CH: REM behavior disorder, in Principles and Practices of Sleep Medicine, 2nd edition, edited by Kinger M, Roth T, Dement W. Philadelphia, WB Saunders, 1994, pp 574–588

7

Auras and Experiential Responses Arising in the Temporal Lobe

Itzhak Fried, M.D., Ph.D.

The temporal lobe is a juncture for neuronal processes involved in the faculties of perception, memory, and affect, and it is thus critical to human experience. Epilepsy arising in the temporal lobe provides a unique window into mechanisms involved in the formation and maintenance of experience. Both complex partial seizures and the auras that often precede them are instances of alteration in the ongoing activity of the temporal lobe resulting in behavioral change. If we are to gain some insight from these phenomena into brain–behavior relationships, we need to localize the changes in electrical activity and at the same time characterize the behavioral changes.

COMPLEX PARTIAL SEIZURES

The most common seizures arising from the temporal lobe are complex partial seizures. The cardinal sign of these seizures is a temporary impairment of consciousness.[1] From a practical standpoint, the concept of "consciousness" presents a problem to the observer because it is difficult to define in objective terms. What is perhaps more important in complex partial seizures is the frequent impairment of memory consolidation that is associated with them. The patient has anterograde amnesia, which later is experienced as retrograde amnesia to the events of the seizure. The other feature of importance is the frequent abnormality in interaction with the environment. These impairments in memory for the seizure and in contact with the environment limit the access of an observer to the patient's experience during a seizure; that is, patients cannot report their experience while the seizure takes place, nor can they remember it later. The observer, then, is privy only to the behavioral manifestations of the seizure.

The behavioral manifestations of complex partial seizures of temporal lobe origin are quite variable. But observers must be careful not to confuse their interpretation of the behavior with an objective description of the behavior. The most common manifestations include the following:

1. *Unresponsiveness.* During complex partial seizures, patients often do not respond to external stimuli. As Gloor[2] points out, this should not be confused with "loss of consciousness." The lack of response may be due to a number of factors, such as behavioral arrest (described below), impairment of speech output or language comprehension (aphasia), or inattentiveness. Despite the unresponsiveness, it is important to present objects to the patient for future recall so that retrograde amnesia can be tested later.

2. *Automatism.* More than 50% of complex partial seizures of temporal lobe origin may be accompanied by specific motor manifestations. Oroalimentary automatism involving repetitive actions such as chewing, lipsmacking, or swallowing is frequent in seizures of mesial temporal lobe origin. Also commonly seen are repetitive motor acts using the upper extremities, such as picking at clothing, face or leg rubbing, pill rolling movements of the fingers, and manipulation of objects in the environment. Much more complex and highly stereotypical activities can take place, such as undressing or voicing complex but repetitive utterances consisting of several words or a short sentence. Automatic behavior such as aimless walking can be seen as well, often in the postictal phase, and in such instances attempts to restrain the patient may lead to violent behavior. Automatism can also be seen in seizures limited to the frontal lobe[3] or in parietal and occipital seizures with spread to the temporal lobe.[4,5]
3. *Behavioral arrest.* Although seizures often include active behavioral manifestations such as automatism, they may also include what appears to be arrest of behavior. Complex partial seizures of limbic origin often start with a motionless stare.[6] Arrest of neurological function is often apparent in language disturbances, including speech arrest and dysphasia, that are part of some seizures.

It is remarkable that during complex partial seizures patients are often able to execute quite elaborate behavior concurrently with the stereotypical manifestations of the seizure. Such behavior requires some contact with the environment and indicates that at least some cognitive functioning persists despite the ongoing ictus. A striking example is provided by a patient with temporal lobe epilepsy who described to us seizures that occurred while riding his bicycle to work. He stated that after setting out in the morning he would occasionally find himself riding back home. Clearly, in the space of a relatively short time he had turned around and ridden back home, all the while maintaining his ability to operate a bicycle with the concomitant demands of contact with the environment. Yet he was unable to recollect this period of time. Patients are occasionally able to conduct a telephone conversation while having a seizure,[2] although often the person at the other end of the line will notice a problem.

How does the occurrence of massive production of electrical activity result in complex behavior? Indeed, seizure activity can be looked upon as a disruption of neuronal function, but it can also be viewed as the production or perhaps overproduction of neural function. The behavioral manifestations that we see in complex partial seizures are often the result of disruption of higher cortical function such as speech, memory, and perceptual mechanisms. The result is unresponsiveness during some seizures and an inability to fully interact with the environment or consolidate memories for events. On the other hand, automatism may represent production of elaborate motor programs that have been "overlearned" and cannot be undone. These segments of behavior are then produced in obligatory fashion, in the same way that a memory is retrieved without conscious control. One cannot deliberately forget or "undo" a memory; nor can the programs involved in the behavioral manifestations of a seizure be undone.

THE LOCALIZING VALUE OF SEIZURE SEMIOLOGY

Complex partial seizures are behavioral sequences that are relatively well defined in their temporal boundaries. There have been several investigations of the correlation between the behavioral phenomenology of seizures and the presumed seizure origin, localized by electroencephalography (EEG) using extracranial or intracranial electrodes. Delgado-Escueta et al.[6] reported that complex partial seizures starting with a motionless stare or behavioral arrest were of mesial temporal lobe (limbic) origin. Maldonado et al.[7] showed that 39% of complex partial seizures of amygdalar and hippocampal origin started with motionless staring, 25% with nonfocal discrete movements, and 21% with oroalimentary automatism. Wieser[8] distinguished between several types of seizures of temporal and extratemporal origin. Of temporal lobe seizures, the unilateral temporobasal limbic type begins in the amygdala and hippocampal complex and is characterized by an aura followed by ipsilateral motor signs and, later, oroalimentary automatism or behavioral arrest. The temporal polar type involves the amygdala and lateral mesial anterior cortex and is characterized by autonomic changes indicating amygdalar involvement, early oroalimentary automatism, alteration in consciousness, and staring. The posterior temporal neocortical type involves association areas and is characterized by aphasia (in the dominant temporal lobe) and auditory, visual, and vestibular hallucinations.

A modified classification[9] of epileptic syndromes divides temporal lobe epilepsies into two groups: mesial and lateral temporal lobe epilepsies. The lateral type is characterized by aphasia, focal motor and sensory signs, and hallucination, implicating auditory and visual association areas. The definition of the syndrome of mesial temporal lobe epilepsy (or mesiolimbic epilepsy) has been refined over the last decade and is now well accepted.[10] The syndrome is characterized by history, clinical features of the seizures, EEG signature, and a number of neuroimaging findings. Seizure features include the presence of an aura, followed by staring and behavioral arrest. Oroalimentary automatism is common. The extracranial EEG often shows an initial or delayed 5–7 Hz rhythmic activity with maximal amplitude at the basal temporal derivation. The neuroimaging findings include unilateral temporal lobe hypometabolism on interictal positron emission tomography (PET) with fluorodeoxyglucose, as well as a characteristic pattern of hyper- and hypoperfusion on ictal single-photon emission computed tomography (SPECT); detailed magnetic resonance imaging (MRI) often shows hippocampal atrophy or loss of hippocampal architecture.[10] Neurocognitive evaluation usually shows material-specific memory deficit: verbal if the dominant temporal lobe is involved and visuospatial if the nondominant temporal lobe side is involved. The pathological substrate of the syndrome is sclerosis of the hippocampus, and surgical removal of the mesial temporal lobe structures on the involved side yields excellent results in terms of seizure control.[11]

The syndrome of temporal lobe epilepsy of lateral neocortical onset is not as well defined or understood. The semiology of the seizure is only one factor in distinguishing these two types, and by itself it is not very specific, so the separation of the two syndromes based on symptoms alone is difficult.[12] This difficulty underscores the general problem of identifying seizure onset based on its behavioral manifestations: seizures may spread quickly, so that a lateral neocortical onset may involve the mesial temporal structures quite early. Furthermore, seizures may start in a behaviorally "silent" area and only later spread to an area where activation yields observable behavioral changes.

AURAS

The term *aura* is used to describe the experience that immediately precedes some convulsive attacks. It has been defined as "that portion of the seizure which occurs before consciousness is lost and for which memory is retained afterwards".[1] The cardinal feature of the aura is the ability of the patient to consolidate it in memory and to retain memory for it once the seizure is over. This feature makes auras a singular avenue for investigating subjective experience and its anatomic substrate.

The auras are intriguing phenomena. Some of them are simple sensations that can be accurately described by the patients. Others are quite complex and have been described as highly elaborate mental states or "intellectual auras."[13] Similar "experiences" have occasionally been elicited in patients with epilepsy by using electrical stimulation. Some of the more elaborate auras have been attributed to the temporal lobe, either from observations in patients with temporal lobe epilepsy or from electrical stimulation of sites in the temporal lobe via depth electrodes.

Do auras represent a local phenomenon that can be attributed to a specific area in the brain? This question has clinical and theoretical implications. If the auras can be localized, perhaps the seizures can as well. From a practical standpoint, that may mean that the patient is a suitable candidate for focal surgical resection that can eradicate the seizures. From a theoretical standpoint, localization of the aura opens a window into the cortical networks participating in specific behavior.

Classification of Auras

The classification of auras is difficult because they vary not only by sensory modality but also in complexity; moreover, patients often find it difficult to express what they experience during auras. Auras can be categorized as follows: visual phenomena, auditory phenomena, olfactory and gustatory phenomena, viscerosensory phenomena, cephalic sensations, vertigo and dizziness, and diffuse warm and cold sensations.[14,15] *Visual* auras include simple phenomena, such as phosphenes, colors, and loss or "whitening" of vision, and complex phenomena such as a scene or a face. *Auditory* auras include simple auditory sensations, mainly ringing in the ears or humming, and more complex sensations such as voices or music. Somatosensory phenomena include tingling, numbness, or pain in a particular region or side of the body. Examples of *olfactory* or *gustatory* phenomena include "foul odor," "burning smell," "odd taste," "sour taste," and "bad taste." *Viscerosensory* auras include epigastric sensations (usually a feeling of "rising"), nausea or a "sickening feeling," and chest sensations. *Epigastric* sensations can be classified as a separate subgroup because of their frequent occurrence; they are limited to a specific sensation localized by the patient's report in the epigastric region (such as sensation in the "stomach" or "butterflies in my stomach"). *Cephalic* auras include sensations such as a "funny feeling in the head," "pressure in the head," and "something crawling in my head." Reports of *dizziness*, vertigo, or light-headedness prior to seizures can be classified as a separate group. Sometimes true vertigo is difficult to distinguish from dizziness.

The most intriguing auras have been those that patients reported as similar to real-life experience; hence the term *experiential* auras. The similarity is both in vividness and complexity of the experience, and it is accentuated by the individual nature of the auras. Jackson[13] regarded these auras as "the most elaborate psychical states" and used the term *intellectual aurae* or *dreamy states*. Experiential auras include experiences such as *déjà vu* (an intense feeling of visual familiarity), *jamais vu* (a feeling of visual strangeness), the auditory counterparts *déjà entendu* and *jamais entendu*, *sentiment de l'inconnu* (such as a "feeling of things being different"), forced thoughts, complex visual or auditory illusions or hallucinations, and emotional experiences, mostly fear.

Penfield and Jasper[16] further classified these auras into illusions, hallucinations, and emotions. They regarded illusions and emotions as "interpretative," in the sense that they involve interpretation of the present; hallucinations often involve reenactment of past events. Visual illusions may involve macroscopia or microscopia, as well as seeing objects nearer or farther, and auditory illusions may involve changes in the quality of sound, e.g., loud or soft, clear or faint. Visual hallucinations include images of objects, faces, or scenes, and auditory types include voices and music. In addition, auras sometimes include illusions involving self-image, such as depersonalization, remoteness, and "out-of-body" experiences. These auras are obviously difficult for the patient to describe.

Localization of Auras

As with other types of seizures, auras are often regarded as expressions of abnormal or excessive neural activity. Localization of this activity can then be the basis for identifying the neural structures involved in the seizures. However, it is often impossible to identify abnormalities on scalp EEG, or even intracranial EEG, that coincide with the auras. Sperling et al.[17] found that limbic EEG correlates of auras were lacking in patients who had good localization of seizure origin to limbic structures by intracranial electrodes. Several studies have examined the localizing value of auras by correlating the phenomenology of the aura with the presumed location of EEG onset of the seizure, established either by scalp EEG recordings[18,19] or by intracranial recordings.[14] Palmini and Gloor[14] found that the clinical features of auras were as useful in localizing the lobe of seizure onset as the EEG recording or other tools such as PET, computed tomography (CT), and MRI. However, auras could not be used in lateralizing the seizure onset. On the other hand, Gupta et al.[18] found that auras may have lateralizing value as well: right hemispheric EEG abnormalities were more likely in patients with psychic or autonomic auras.

We have taken a different approach to the localization of auras, attempting to correlate the auras with structural brain abnormalities.[15] Some patients with complex partial seizures have localized lesions or pathology that can be localized. Removal of the pathological region often results in elimination of the seizures. We have examined the relationship between the semiology of the auras in these patients and the location of the pathological substrate underlying the seizures. In our study, certain types of auras were associated with limbic system or temporal lobe structural pathology, although no significant correlation with the side of pathology was found.[15]

Epigastric auras were associated with hippocampal sclerosis.[15] These are perhaps the most common epileptic auras and have been described in detail by Jackson,[20] Gowers,[21] and others. The localization of these auras has been controversial. Similar sensations have been elicited with electrical stimulation of several regions, including the amygdala, pes hippocampus, uncus, area olfactoria, basal ganglia, internal capsule, and centromedian.[16,22,23] They have also been observed occasionally in extratemporal epilepsies.[24] However, we found no epigastric auras in patients with extratemporal structural pathology.

Gustatory and olfactory auras have been associated with temporal lobe pathology by several reports.[16,25] Occasionally these auras have been seen in patients with temporal lobe tumors.[16,25] Penfield and Jasper[16] regarded them as comparatively rare and ascribed localization of olfactory auras to the uncus and localization of gustatory auras to the deep perisylvian or insular region. In our study,[15] these auras were associated with hippocampal sclerosis. Auras of vertigo and dizziness were more frequent in patients with extratemporal lesions than in patients with hippocampal sclerosis or patients with temporal lobe masses.

Several authors have reported correlation of experiential auras with seizures of temporal lobe origin,[8,14,16,18,26,27] and experiential phenomena have been elicited with electrical stimulation in that area.[28,29] In our study,[15] 26% of patients with auras and temporal lobe pathology had experiential auras. These auras were present with medial as well as lateral temporal lobe pathology. Two patients (12%) with extratemporal pathology had experiential auras. Interestingly, the lesions involved occipital cortex in one patient and cingulate gyrus in the other.

The relationship of seizure localization to the localization of auras remains controversial. The aura bears a relatively constant temporal relationship to the seizure that it precedes. One might expect that where a seizure is associated with a discrete epileptogenic zone, the aura would be associated with that same brain region. Furthermore, it could be expected that surgical removal of the epileptogenic focus would eliminate the seizures, including the auras. Yet it is now well recognized that, following surgical removal of the presumed epileptogenic zone, as many as 35% of patients retain their auras in spite of being free of seizures.[15,17,30] This observation raises the possibility that the aura may be caused by a process separate from that causing the seizure or that it may arise in a different location. The hypothesis of pathophysiological dissociation of seizure and aura was first proposed in 1879 by Jackson,[20] who suggested that auras might arise connected to, but not at, the seizure focus. EEG correlates of auras are difficult to find even when there is good seizure localization.[17] In a study of 90 patients with complex partial seizures and auras who underwent epilepsy surgery, we found that nearly 20% of the patients with hippocampal sclerosis who were seizure-free following temporal lobe resection had persistent auras. In contrast, only 2.6% of patients with mass lesions who were rendered seizure-free following surgery continued to have auras.[15] These results suggest anatomic dissociation of seizures and auras that is unique to patients with hippocampal sclerosis.

RESPONSES TO ELECTRICAL STIMULATION

Seizures and auras are behavioral manifestations of electrical activity arising spontaneously. They may, therefore, be viewed as spontaneous electrical stimulation of cortical circuits. Artificial stimulation can be applied by means of electrodes placed on the surface of the cortex or in deep brain structures. These intracranial electrodes are used for chronic EEG monitoring and functional brain mapping in epilepsy patients who are being evaluated for surgical treatment of their medically refractory seizures when scalp EEG monitoring has failed to localize a resectable seizure focus. Intracranial monitoring is customarily performed on the medical ward for an average of 1 to 2 weeks. Occasionally, electrical stimulation is applied during the operation, with the patient awake, in order to obtain a functional map of the brain in the region of the planned resection.

Electrical stimulation can cause a seizure, but when applied at levels below local afterdischarges it can lead to discrete responses that are often similar to auras, although they rarely resemble the patient's habitual auras. Electrical stimulation can elicit responses that are characteristic of the function of the stimulated region. Thus, electrical stimulation of motor cortex yields discrete movements, and stimulation of sites in each of the primary sensory cortices yields the appropriate somatosensory, auditory, or visual responses. Stimulation in association cortex usually disrupts functions such as language and memory. This response is the basis for the mapping of language cortex in patients undergoing resections in that vicinity.

In a small number of patients, electrical stimulation of some sites in the temporal lobe evokes complex responses which as a group are similar to experiential auras or "dreamy states," although seldom are these responses similar to the patient's habitual auras. Complex hallucinations or illusions involving auditory, visual, and other sensory modalities can be evoked. As Gloor[31] has pointed out, these responses "create in the patient's mind experiences, usually from the past, that have compelling immediacy similar to those occurring in real life." In fact, sometimes the experience appears more vivid than real life. The patient may see a face, a scene, or a complex visual image, or hear a voice or a segment of music. This may be accompanied by a sense of familiarity and may be identified as a real-life experience. However, either the image or the feeling of familiarity can come in isolation. When the sense of familiarity comes in isolation, it is applied to the present and perceived as *déjà vu*. When both a sense of familiarity and a complex hallucination are present, the experience may be described as a memory or "flashback." Such experiences can also be accompanied by a specific feeling, often fear. Thus, a patient undergoing stimulation of a site in the left superior temporal gyrus while she was counting aloud reported that she heard her speech as an "echo" and at the same time felt "frightened." Perceptual illusions can be quite complex and are sometimes observed only when certain stimuli are provided. Upon electrical stimulation of a site in the left inferior temporal gyrus during a reading task, the above patient described a "grouping" illusion, where she felt that all the words were strongly "associated."

In general, then, experiential responses are percepts, memories, or emotion, or some or all of these in combination. The experiential responses are fragmentary in character, and, as Gloor[31] points out, they lack a continuous flow of time (with the exception of auditory responses). This stands in contrast to Penfield's[32] description of these phenomena as a stream of consciousness that faithfully reenacts past experience. Gloor[31] noted that these experiences, although vivid, often lack detail, and their main feature is the patient's feeling of "being there." At the same time, patients are always aware that these events are not actually occurring at the present time. This coexistence of being and not being "there" is what led Jackson to coin the term "doubling of consciousness" or "mental diplopia" to describe these states.[13] In the case of *déjà vu*, as pointed out by Bancaud et al.,[33] "the patients found it bizarre that they should feel the current situation had happened before even though they knew, at the same time, that it had not. Thus the experience was strange because of its very familiarity."

The best description of this unique experience can be found in the account[13] given by one of Jackson's patients, a highly educated medical man:

> The central feature has been mental, and has been a feeling of Recollection, i.e., of realizing that what is occupying the attention is what has occupied it before, and indeed has been familiar, but has been for a time forgotten, and now is recovered with a slight sense of satisfaction as if it had been sought for.... [Compared with the patient's normal memory,] the recollection is much more instantaneous, much more absorbing, more vivid, and for the moment more satisfactory, as filling up a void which I imagine at the time I had previously in vain sought to fill. At the same time... I am dimly aware that the recollection is fictitious. (p. 202)

The complexity of the experiential responses elicited by electrical stimulation is illustrated by the following reports by a patient who underwent electrical stimulation mapping of the left temporal cortex via a subdural electrode array placed to localize seizure onset. Upon stimulation at a site in the left inferior temporal gyrus, the patient reported hearing the theme music from the film *Star Wars*, while at the same time "seeing birds" and feeling like "going through a tunnel." On stimulation of a neighboring site 1 cm away, he first reported "something like a radio station"; then, when the stimulation was repeated, he reported hearing music, wedding reception bells, and "Here Comes the Bride." He then reported seeing "an angel on a Christmas tree." When the stimulation was repeated once more, he reported a scene of "throwing the wedding bouquet" and "some song." These vivid auditory and visual fragments were related to a wedding scene, yet they did not represent a particular recollection, and the patient was aware that they were "unreal." This experience was not related to the patient's habitual auras or seizures and was not accompanied by afterdischarges in the surrounding tissue.

The experience of hearing *Star Wars* music also illustrates the individual nature of these experiences. As the patient explained, that particular music was of great "personal meaning" to him, as it involved a composer whose work he greatly admired. Halgren et al.[28] concluded that the experiential phenomena elicited by electrical stimulation are often specific to the individual patient, and they minimized the importance of the anatomic location stimulated. The above example appears to support the importance of location as well as patient history and personality; the responses were elicited in temporal association areas involving both auditory and visual modalities, yet the experience was personal in meaning. Whereas electrical stimulation in primary cortices may lead to predictable sensory and motor responses, the relationship of the response to the anatomic location becomes more variable and individual when association areas are stimulated.

THE PATHOPHYSIOLOGY OF EXPERIENTIAL RESPONSES

As we have seen, the phenomena evoked by temporal lobe stimulation, whether spontaneous (as in complex partial seizures and auras) or extraneous (by application of small electrical currents), are segments of behavior that include motor, perceptual, mnemonic, and affective components. There are two contrasting views concerning the physiological substrate of these phenomena. According to one view, first suggested by Jackson,[20] these responses reflect release from inhibition, "a result of the removal of control of higher centers." Halgren et al.[28] proposed that these responses may reflect subconscious material, specific to the individual, that is allowed to surface as a result of removal of inhibition exerted by limbic cortex. The opposite hypothesis was put forward by Penfield,[29,32] who viewed these phenomena as the expression of active physiological processes represented in temporal cortex. As Gloor[31] points out, "temporal association and limbic structures are precisely those known to be substrates for the functions revealed by experiential phenomena." These phenomena include formation of visual-auditory percepts, encoding in memory, recall of past events, and the attachment of affective significance. These are all functions attributed to the normal temporal lobe.

In contrast to actual complex partial seizures, which are manifestations of robust and often propagated electrical activity, phenomena associated with auras and electrical stimulation are presumed to reflect more discretely localized activity. However, the anatomic localization of the experiential phenomena associated with auras and electrical stimulation is controversial. Several investigators have argued for lateral temporal lobe origin, while others favor medial temporal lobe localization.

On the basis of studies of intraoperative electrical stimulation, Penfield attributed experiential phenomena, with the possible exception of fear, to temporal neocortex.[16,27,29,32] Penfield's conclusions were at odds with Jackson's[34] earlier hypothesis that the origin of the "dreamy state" is in the uncus. Yet Jasper and Rasmussen[35] did not observe such responses with uncal stimulation. Several later studies reported experiential phenomena elicited at mesial temporal lobe sites.[28,36-40]

Contrary to Penfield's position, Gloor et al.[41] reported that most experiential phenomena were evoked by stimulation of medial rather than lateral temporal lobe sites. This was true not only for emotional responses, but also for complex visual and auditory hallucinations, activation of memory recall, and illusions of familiarity. Of the sites involved in experiential responses, the amygdala predominated, followed by the hippocampus and then the parahippocampal gyrus, and most were elicited in the absence of afterdischarges. In those cases where lateral temporal lobe stimulation evoked experiential phenomena, there was always spread of excitation to medial temporal lobe sites. Gloor et al.[41] concluded that "unless limbic structures are activated, either in the course of a spontaneous seizure or through artificial electrical stimulation, experiential phenomena do not occur." One explanation for the discrepancy between Penfield's observations and those reported by Gloor and colleagues is in the sampling of stimulated sites. Most of Penfield's stimulation was done in the lateral neocortex, whereas Gloor et al. had only limited sampling of the neocortex because most of the stimulation was done with depth electrodes introduced via the middle temporal gyrus.

Bancaud et al.[33] report that both medial and lateral temporal structures are involved in the dreamy state, with most of the responses located in the anterior hippocampus, amygdala, and superior temporal gyrus. These authors proposed a unified model that would 1) reconcile Jackson's and Penfield's positions concerning release versus direct activation of temporal lobe physiological mechanisms and 2) incorporate a role for both medial and lateral temporal lobe regions in the mediation of experiential phenomena. They drew on current models of memory that describe involvement of both medial and lateral temporal cortex in memory processes and that emphasize the importance of the anterior superior temporal gyrus. The latter area may be homologous to multisensory regions in the superior temporal area in primates that are heavily connected to the hippocampus. According to the model proposed by Bancaud et al., the permanent memory trace lies in neocortex, thus supporting Penfield's position: "Memories could be evoked by lateral temporal stimulation through activation of a medial structure that projects back to lateral temporal neocortex, and which contains the indices to specific declarative memories."[33] At the same time, "the mechanism of the dreamy state could involve both activation of the memory trace [and] alteration in cerebral state that would relatively privilege the entry of nonsensory data into awareness."[33]

Gloor[31] proposed a neurobiological substrate for experiential phenomena based on models of parallel distributed processing developed by McClelland[42] and others. In networks capable of parallel distributed processing, the entire network can be activated as a whole by any of its constituent units, and a whole pattern can be reactivated from a single fragment. Such networks tolerate a high level of noise as well as degradation. In the case of experiential phenomena in temporal lobe epilepsy, "a limited discharge in a group of neurons either in the amygdala or temporal isocortex may utilize previously strengthened connections underlying an experience to recreate a widely distributed matrix of excitation and inhibition in a disseminated neuronal network that is the substrate of that particular experience."[31]

Gloor's model provides a possible explanation for the persistence of auras following temporal lobe surgery in patients with complex partial seizures due to hippocampal sclerosis.[15] If the matrix that represents the substrate of the aura experience involves extrahippocampal structures (limbic or neocortical) that are not removed at surgery, then auras might persist even after removal of the pathological substrate for the seizures. In this case, spontaneous activity in the remaining tissue may be sufficient to activate the network to recreate the whole experience of the aura from its anatomic and cognitive fragments.

FRAGMENTS OF EXPERIENCE AND THEIR NEURAL REPRESENTATION

The dissection of human experience into components that are represented in different brain regions is a constant preoccupation of modern neuroscience. Localization of function has been investigated by observing the effects of lesions in patients with neurological injury, by cortical application of electrical stimulation during neurosurgical procedures, and by modern techniques of brain activation and imaging. The latter techniques often use linear methods to compare one experimental condition with another, in the hope of isolating individual components of behavior and correlating them with observable changes in the measured variable, be it regional blood flow or metabolism.

Auras and experiential responses to stimulation represent a unique way to isolate components of experience and correlate them with local changes in brain electrical activity. At the same time, they provide valuable insight into another central question in neuroscience, known as the "binding problem": namely, how are various components of behavior bound into a unified experience? This problem has traditionally been posed with respect to perception, but it is readily generalized to the entire domain of experience. The auras and "experiential responses" suggest that individual fragments of experience are readily "completed" or "filled in" by the brain, possibly by neuronal networks using parallel distributed processing,[31] resulting in vivid experiences. This process of filling in draws upon the present as well as on the history and personality of the patient, as is evident from the individual nature of these phenomena. At the same time, these neuronal processes in themselves do not lead to a unified experience with a normal time flow. The patient, then, is aware that something is amiss, leading to what Jackson termed "mental diplopia."

Auras and complex partial seizures are segments of behavior that persist, often in the face of pharmacological and surgical manipulation, in a remarkably reliable and consistent way. In that sense they are analogous to learning and memory phenomena. An experiential aura can be likened to a memory: it appears spontaneously, as the recollection of past experience might, and it cannot be eradicated at will, just as a memory cannot be intentionally forgotten. Furthermore, the explicit nature of the aura, involving conscious recollection that can be reported by the patient, implies the participation of brain systems that subserve declarative memory, primarily the medial temporal lobe. The failure to eliminate auras by local tissue removal reminds one of Lashley's[43] failure to localize the engram in a series of experiments where learning in rats could not be eradicated by local tissue removal. Lashley then resorted to the principle of mass action as a substitute for localization. In agreement with current models of parallel distributed processing, the aura, like memory, may be represented in a distributed network.[31]

According to current models, declarative memory appears to involve a complex network including both the hippocampus and its associated structures as well as neocortex.[44,45] After a certain period of time, yet undetermined, a memory may become hippocampal-independent, presumably stored in neocortex, and therefore resistant to medial temporal lobe resection. In the same manner, an aura may eventually achieve representation in the neocortex and therefore may be particularly resistant to mesial temporal lobe resection. The failure to eradicate auras by mesial temporal surgery thus may reflect not only the resilience of memory to fragmentation by virtue of the network capability of "filling in," but also the fact that representation of the aura has shifted from mesial temporal cortex to neocortex.

REFERENCES

1. Bancaud J, Henrikson O, Rubio-Donnadieu F, et al: Proposal for revised clinical and electroencephalographic classification of epileptic seizures. Epilepsia 1981; 22:489–501
2. Gloor P: Neurobiological substrates of ictal behavioral changes. Adv Neurol 1991; 55:1–34
3. Munari C, Stoffels C, Bossi L, et al: Automatic activities during frontal and temporal lobe seizures: are they the same? In Advances in Epileptology: XIIth Epilepsy International Symposium, edited by Dam M, Gram L, Penry JK. New York, Raven, 1981, pp 287–291
4. Geier S, Bancaud J, Talairach J, et al: Ictal postural changes and automatisms of the upper limb during epileptic parietal lobe discharges. Epilepsia 1977; 18:517–524
5. Williamson PD, Spencer SS, Spencer DD, et al: Complex partial seizures with occipital lobe onset. Epilepsia 1981; 22:247–248
6. Delgado-Escueta AV, Kunze U, Waddell G, et al: Lapse of consciousness and automatisms in temporal lobe epilepsy: a videotape analysis. Neurology 1977; 27:144–155
7. Maldonado HM, Delgado-Escueta AV, Walsh GO, et al: Complex partial seizures of hippocampal and amygdalar origin. Epilepsia 1988; 29:420–433
8. Wieser HG: Electroclinical Features of the Psychomotor Seizure. London, Butterworths, 1983
9. Dreifuss FE: Proposal for classification of epilepsies and epileptic syndromes. Epilepsia 1985; 26:268–278
10. Engel J Jr: Surgery for seizures. N Engl J Med 1996; 334:647–652
11. Engel J Jr, Van Ness PC, Rasmussen TB, et al: Outcome with respect to epileptic seizures, in Surgical Treatment of the Epilepsies, vol 2, edited by Engel J Jr. New York, Raven, 1993, pp 609–621
12. Kotagal P, Luders H, Williams G, et al: Psychomotor seizures of temporal lobe onset: analysis of symptom clusters and sequences. Epilepsy Res 1995; 20:49–67
13. Jackson JH: On a particular variety of epilepsy "intellectual aura": one case with symptoms of organic brain disease. Brain 1888; 11:179–207
14. Palmini A, Gloor P: The localizing value of auras in partial seizures: a prospective and retrospective study. Neurology 1992; 42:801–808
15. Fried I, Spencer DD, Spencer SS: The anatomy of epileptic auras: focal pathology and surgical outcome. J Neurosurg 1995; 83:60–66
16. Penfield W, Jasper H: Epilepsy and the Functional Anatomy of the Human Brain. Boston, Little, Brown, 1954
17. Sperling MR, Lieb JP, Engel J Jr, et al: Prognostic significance of independent auras in temporal lobe seizures. Epilepsia 1989; 30:322–331
18. Gupta AK, Jeavons PM, Hughes RC, et al: Aura in temporal lobe epilepsy: clinical and electroencephalographic correlation. J Neurol Neurosurg Psychiatry 1983; 46:1079–1083
19. Janati A, Nowack WJ, Dorsey S, et al: Correlative study of interictal electroencephalogram and aura in complex partial seizures. Epilepsia 1990; 31:41–46
20. Jackson JH: Lectures on the diagnosis of epilepsy, in Selected Writings of John Hughlings Jackson, edited by Taylor J. London, Hodder and Stoughton, 1931, pp 276–307
21. Gowers WR: Epilepsy and Other Chronic Convulsive Diseases: Their Causes, Symptoms and Treatment. London, Churchill, 1901
22. Feindel W, Penfield W: Localization of discharge in temporal lobe automatism. Arch Neurol 1954; 72:605–630
23. Van Buren JM: The abdominal aura: a study of abdominal sensations occurring in epilepsy and produced by depth stimulation. Elec-

troencephalogr Clin Neurophysiol 1963; 15:1–19
24. Rasmussen T: Surgical therapy of frontal lobe epilepsy. Epilepsia 1963; 4:181–198
25. Arseni C, Petrovici IN: Epilepsy in temporal lobe tumours. Eur Neurol 1971; 5:201–214
26. Ajmone-Marsan C: Commentary: clinical characteristics of partial seizures, in Surgical Treatment of the Epilepsies, vol 1, edited by Engel J Jr. New York, Raven, 1987, pp 121–127
27. Mullan S, Penfield W: Illusions of comparative interpretation and emotion: production by epileptic discharge and by electrical stimulation in the temporal cortex. Arch Neurol Psychiatry 1959; 81:269–284
28. Halgren E, Walter RD, Cherlow GD, et al: Mental phenomena evoked by electrical stimulation of the human hippocampal formation and amygdala. Brain 1978; 101:83–117
29. Penfield W, Perot P: The brain's record of auditory and visual experience: a final summary and discussion. Brain 1963; 86:595–696
30. Lund JS, Spencer SS: An examination of persistent auras in surgically treated epilepsy (abstract). Epilepsia 1992; 33(suppl):95
31. Gloor P: Experiential phenomena of temporal lobe epilepsy: facts and hypotheses. Brain 1990; 113:1673–1694
32. Penfield W: The permanent record of the stream of consciousness. Proceedings of the 14th International Congress on Psychology, Montreal. Acta Physiologica 1954; 11:47–69
33. Bancaud, J, Grunet-Bourgin F, Chauvel P, et al: Anatomical origin of *déjà vu* and vivid "memories" in human temporal lobe epilepsy. Brain 1994; 117:71–90
34. Jackson JH, Colman WS: Case of epilepsy with tasting movements and "dreamy state": very small patch of softening in the left uncinate gyrus. Brain 1898; 21:580–590
35. Jasper HH, Rasmussen T: Studies of clinical and electrical responses to deep temporal stimulation in man with some considerations of functional anatomy. Res Publ Assoc Res Nerv Ment Dis 1958; 36:316–334
36. Bickford RG, Mulder DW, Dodge HW, et al: Changes in memory function produced by electrical stimulation of the temporal lobe in man. Res Publ Assoc Res Nerv Ment Dis 1958; 36:227–243
37. Baldwin M: Electrical stimulation of the mesial temporal region, in The Amygdala: Neurobiological Aspects of Emotion, Memory, and Mental Dysfunction, edited by Aggleton JP. New York, Wiley-Liss, 1992, pp 1–66
38. Horowitz MJ, Adam JE, Rutkin BB. Visual imagery on brain stimulation. Arch Gen Psychiatry 1968; 19:469–486
39. Ferguson SM, Rayport M, Gardner R, et al: Similarities in mental content of psychotic states, spontaneous seizures, dreams, and responses to electrical brain stimulation in patients with temporal lobe epilepsy. Psychosom Med 1969; 31:478–498
40. Wieser HG: Ictal "psychical phenomena" and stereoelectroencephalographic findings, in EEG and Clinical Neurophysiology: Proceedings of the Second European Congress on EEG and Clinical Neurophysiology, edited by Lechner E, Aranibar A. Amsterdam, Excerpta Medica, 1979, pp 62–76
41. Gloor P, Olivier A, Quesney LF, et al: The role of the limbic system in experiential phenomena of temporal lobe epilepsy. Ann Neurol 1982; 12:129–144
42. Rumelhart DE, McClelland JL: Parallel Distributed Processing: Explorations in the Microstructure of Cognition, vol 1: Foundations. Cambridge, MA, MIT Press, 1986
43. Lashley K: In search of the engram. Society of Experimental Biology 1950; 4:454–482
44. Squire LR: Memory and hippocampus: a synthesis from findings with rats, monkeys and humans. Psychol Rev 1992; 99:195–231
45. Schacter DL, Tulving E (eds): Memory Systems. Cambridge, MA, MIT Press, 1994

8

Neuropsychiatric Symptoms From the Temporolimbic Lobes

Michael R. Trimble, M.D., F.R.C.P., F.R.C.Psych., Mario F. Mendez, M.D., Ph.D., Jeffrey L. Cummings, M.D.

Investigators have long recognized that neuropsychiatric symptoms can arise from temporal lobe lesions.[1] In the late 1930s, clinical observations of psychopathology from epileptic foci, brain tumors, strokes, and other lesions in the temporal lobes coincided with neuroanatomical studies of the temporolimbic system.[2,3] Behavioral studies in primates formed the link between observable behavioral disturbances and pathology that specifically involves the limbic system within the temporal lobes.[3,4] Despite this information, clinicians are incompletely aware of the concept of neuropsychiatric symptoms from the temporolimbic lobes.

A major reason for the lack of familiarity with these symptoms is that they are not linked to specific temporolimbic lesions in the classic focal or disconnection sense. Neurobehavioral disorders often result from disturbances in discrete centers or regions; for example, Wernicke's aphasia from lesions in Wernicke's area in the posterior superior left temporal gyrus or Gerstmann's syndrome from lesions in the left angular gyrus. Alternatively, many neurobehavioral disorders result from disconnection between these centers; for example, conduction aphasia from disconnection of Wernicke's from Broca's areas or the "split-brain" syndrome from lesions of the corpus callosum. Temporolimbic lesions, in contrast, lead to behavioral disorders by their effects on distributed networks.

In this chapter we aim to elucidate the neuropsychiatric symptoms of temporolimbic lesions. We first review the current behaviorally relevant aspects of temporolimbic neuroanatomy. We then describe positive, productive symptoms from the temporal lobes, such as the Klüver-Bucy syndrome, the Gastaut-Geschwind syndrome, emotional or mood disorders, delusions, anxiety and dissociative disorders, and neurovegetative symptoms. We also describe amnesia, the signature temporolimbic disturbance of cognition. Finally, we discuss pathologic entities associated with these temporolimbic symptoms.

NEUROANATOMICAL BASIS OF TEMPOROLIMBIC SYNDROMES

Neuroanatomical Networks and Circuits

The temporolimbic lobe is a classic example of a widely distributed circuit within the brain. Papez's limbic circuit, the neurological substrate for emotion, includes the amygdalae and hippocampi in the temporal lobes and

---

This work was supported by the U.S. Department of Veterans Affairs, an Alzheimer's Disease Center grant from the National Institute on Aging (AG10123), and the Sidell-Kagan Research Fund.

is functionally indivisible from other parts of the limbic circuit in the diencephalon and frontal lobes.[3] Other temporolimbic connections extend, via median forebrain bundle, to the frontal lobes, which serve as the emotional manager of the brain. Although the distributed aspect of these circuits decreases the neuroanatomical specificity of temporolimbic symptoms, these symptoms, nevertheless, occur with recognizable frequency as a result of temporolimbic lesions.

Recent advances in neuroanatomy have delineated the related concepts of parallel distributed processing (PDP) and cortical-subcortical circuits.[5] PDP involves different levels of neurons firing concurrently or in parallel with active feedback loops and servomechanisms. In PDP, the pattern of activation represents information. Behavioral information is stored in connections among processing units, each unit with a graded activity value and a specific probability of firing.

In addition to PDP, anatomically distinct cortical-subcortical connections represent specific functional units of behavior.[5] Investigators have placed much emphasis on frontal-striatal-thalamic circuits, but much less on temporolimbic connections. They have defined at least five discrete frontal-subcortical loops that link frontal cortex with the basal ganglia, then the thalamus, with a recoupling back to the frontal cortex.[5] Recently, neuroanatomical investigations have begun to elaborate similar circuitry involving the temporal lobes.[6-9] The limbic temporal–subcortical connectivity has direct influences not only back to temporal cortex via the Papez circuit, but also to the frontal lobes and to somatomotor and emotional motor systems.[6,10] Specific temporolimbic circuits include medial limbic circuits, lateral limbic circuits, and the "extended" amygdala.

Medial and Lateral Limbic Circuits

The original Papez limbic circuit was a loop comprising the hippocampus, the mammillary bodies of the hypothalamus, the anterior thalamic nuclei, and the cingulate gyrus, with a loop back to the hippocampus. This limbic system is composed of medial and lateral circuits.[11] The medial circuit is located toward the medial wall of the hemisphere and includes the hypothalamus, anterior thalamic nuclei, cingulate gyrus, hippocampus, and related tracts. The lateral (or basolateral) limbic circuit consists of the orbitofrontal-insular-temporal polar cortices, the amygdala, and the dorsomedial thalamic nuclei. The medial limbic circuit has rich connections with the brainstem reticular formation. The lateral limbic circuit has extensive connections with dorsolateral prefrontal cortex and posterior parietal association cortex.

As a consequence of their anatomic connections, the medial limbic circuits mediate information more closely related to internal states, whereas the lateral limbic circuit is more involved with information concerning the body surface, external world, and social-personal interactions. The medial circuit mediates important aspects of learning and memory; the lateral circuit mediates empathy and social cognition. Disorders of sexual-visceral and endocrine function occur with medial limbic dysfunction, whereas abnormalities of mood and civil behavior are more common with dysfunction of the lateral system.

Extended Amygdala and Related Concepts

Alheid and Heimer's[12] concept of the extended amygdala emphasizes the distributed circuitry associated with the temporolimbic lobe. They define the extended amygdala as an anatomical corridor that extends from the central and medial amygdaloid nuclei through subpallidal gray matter and bed nucleus of the stria terminalis to the caudomedial nucleus accumbens. Its afferents are largely cortical, especially from other amygdaloid nuclei, the hippocampus, and the temporal and prefrontal cortex. Amygdalar projections influence the ventral striatum, that part of the basal ganglia thought to be related to emotional motor behaviors. Other projections go to the ventral tegmental area and the substantia nigra and affect dopamine output to many other regions. Amygdalar projections also influence brain areas that coordinate autonomic and somatomotor behavior, such as the periaqueductal gray region and nuclei in the reticular formation.

TEMPOROLIMBIC SYNDROMES

There is not a single unified temporolimbic syndrome. As in the case of the frontal lobe syndromes, a number of clinicopathologic conditions exist (Table 8–1). This discussion emphasizes the Klüver-Bucy syndrome, the Gastaut-Geschwind syndrome, and the cognitive disorder of amnesia because of their putative specificity for the temporolimbic lobe. Paroxysmal epileptic conditions of temporolimbic origin are discussed elsewhere in this volume.

Klüver-Bucy Syndrome

The Klüver-Bucy syndrome is one of the most important temporolimbic syndromes because of its specificity for bilateral anterior temporal lesions. Klüver and Bucy[4] in 1939 made significant observations regarding brain–behavior relationships following placement of bilateral lesions in the temporal lobes of monkeys. The animals became tame, with loss of fear and aggression; they appeared to indiscriminately mount other animals; and they demonstrated excessive oral exploration of the environment (hypermetamorphosis). They also showed an associative visual agnosia. Because the lesions removed the uncus, amygdala, and part of the hippocampus, it was reasonable to assume that the intactness of these structures was a prerequisite for the organization and control of mood, sexual behavior, and visual perception. The main components of the syndrome are shown in Table 8–2.

Subsequent studies have demonstrated that the most important site of damage to produce these behaviors is the amygdalae in the temporolimbic lobes.[13] These data and others have led to the view that sensory stimuli are given affective and motivational significance by the amygdalae, the latter being part of a system for the development of stimulus-reward associations.[14]

Oral behaviors are especially prominent in the Klüver-Bucy syndrome. The monkeys with bilateral temporal lesions not only examined objects by mouth but also ate previously rejected foods.[15] Marlowe et al.[16] first described the Klüver-Bucy syndrome in a human patient with meningoencephalitis. Their patient examined objects, including his hands, by placing them in his mouth and sucking or chewing them. He had an insatiable appetite and ate almost everything within his grasp, including plastic wrappers, ink, cleaning materials, dog food, and feces. The extent of this hyperorality can be severe and life-threatening. Mendez and Foti[17] described two patients with the Klüver-Bucy syndrome; one asphyxiated after stuffing his mouth with surgical gauze, towels, toilet paper, and Styrofoam cups, and the other aspirated after overingestion of a large amount of food.

The complete Klüver-Bucy syndrome rarely occurs in people. The diagnosis requires the presence of at least 3 of the 6 symptoms listed in Table 8–2. Among humans, placidity, indiscriminate dietary behavior, and hyperorality appear to be the most common manifestations, whereas hypersexuality, beyond inappropriate sexual commentary, is rare. These patients often have accompanying aphasia, amnesia, echopraxia, dementia, and seizures. Etiologies of the human Klüver-Bucy syndrome include trauma, frontotemporal dementia, Alzheimer's disease, herpes simplex encephalitis, ischemia or anoxia, temporal lobectomies, progressive subcortical gliosis, adrenoleukodystrophy, Rett's syndrome, systemic lupus erythematosus, porphyria, limbic encephalitis, multicentric glioblastoma multiforme, and carbon monoxide intoxication.[18]

Gastaut-Geschwind Syndrome

The view that there is a specific interictal syndrome of temporal lobe epilepsy (TLE) received clinical support from the writings of Gastaut and Geschwind.[19,20] These authors called attention to disorders of sexual function, hyperreligiosity, hypergraphia, exaggerated philosophical concerns, and irritability, which form the kernel of this syndrome (Table 8–3).

TABLE 8–1. Neuropsychiatric symptoms from the temporolimbic lobes

Klüver-Bucy syndrome
  Complete
  Partial: combinations of placidity, hyperorality, sexual behavior changes, hypermetamorphosis, "psychic blindness"
Gastaut-Geschwind syndrome
  Complete
  Partial: combinations of hyposexuality, hyperreligiosity, hypergraphia, interpersonal "stickiness," circumstantiality
Amnestic disorders
  Verbal learning disturbances (left temporal)
  Nonverbal learning disturbances (right temporal)
Delusions and hallucinations
Emotional and mood disorders
  Rage and aggression
  Depression and mania
Anxiety and dissociative disorders
Neurovegetative disorders
  Oral aggression and other hyperoral behaviors
  Sexuality changes

TABLE 8–2. The Klüver-Bucy syndrome

Tameness: loss of fear/anxiety or diminished aggression
Dietary changes: indiscriminate dietary behavior
Altered sexuality: greatly increased autoerotic, homosexual, or heterosexual activity or inappropriate sexual object choice
Hypermetamorphosis: a tendency to attend to and react to every visual stimulus
Hyperorality: a tendency to examine all objects by mouth
Psychic blindness: visual agnosia

TABLE 8–3. The Gastaut-Geschwind syndrome

Hypergraphia
Hyposexuality
Hyperreligiosity
Exaggerated philosophical concern
Interpersonal "stickiness"
Circumstantiality
Mood changes (irritability, elation)

This group of traits, termed the Gastaut-Geschwind syndrome, occurs in a subset of patients with TLE. Bear[21] defined three subgroups of behaviors that contribute to this syndrome. The first relates to alteration of physiological drives, such as sexuality, aggression, and fear. Second, there are "nascent intellectual interests," with a preoccupation with religious, moral, and philosophical themes. Some epileptic patients with a temporolimbic focus develop a sense of the heightened significance of things. These patients are serious, humorless, and overinclusive, and have an intense interest in philosophic, moral, or religious issues. Occasionally, epileptic patients experience multiple religious conversions or experiences. Some have altered interpersonal dispositions, including increased (obsessive) preoccupation with detail, circumstantiality of speech, and a tendency to prolong interpersonal encounters ("viscosity"). Epileptic patients can spend a long time getting to the point, give detailed background information with multiple quotations, or write copiously about their thoughts and feelings (hypergraphia).

Viscosity may be the most important characteristic of the Gastaut-Geschwind syndrome. This refers to "stickiness" of thought processes or an adherence to an idea and to an interpersonal adhesiveness or increased social adhesion. Patients display circumstantiality, have difficulty terminating conversations, and tend to prolong interpersonal encounters (for example, interviews with physicians) beyond the time indicated by social cues. Investigators have reported viscosity more commonly in patients with left or bilateral seizure foci and have suggested that it represents a subtle interictal language disturbance.[22]

Studies of hypergraphia in patients with epilepsy emphasize a link to temporal lobe epilepsy[23,24] and possible associations with right-sided as opposed to left-sided EEG abnormalities.[25,26] In some patients it is a constant personality feature, while in others it is variable, sometimes precipitated or accelerated by a cluster of seizures. The hypergraphia of the Gastaut-Geschwind syndrome frequently leads to a written output that is meticulous, obsessional, and carried out with a compulsion. The content is often moral and religious. Repetition of words and sequences is common, and variants include excessive drawing, painting, or even, in one of Geschwind's cases, the hiring of a third party to write down information. The symptom is not confined to Western culture; hypergraphic outpourings from patients from non-Western cultures have been reported.[27]

Definitive proof is lacking that epileptic patients with temporal lobe foci are disproportionately prone to the Gastaut-Geschwind syndrome.[28,29] Most of the early studies used the Minnesota Multiphasic Personality Inventory (MMPI), a test that proved insensitive to most of the specific traits attributed to epilepsy. To overcome the shortcomings of the MMPI, Bear and Fedio[30] developed a rating scale of 18 behavioral features, drawn from a review of the literature, that were said to characterize patients with epilepsy. Studies with the Bear-Fedio Inventory found that epileptic patients with temporal lobe foci were sober and humorless, dependent, circumstantial, and strong on philosophical interests. In addition, those with a left-sided focus had a more reflective ideational style and maximized their problems, whereas those with a right-sided focus had emotional tendencies and minimized their problems. However, other applications of this inventory found the same characteristics in nonepileptic patients with psychiatric disorders or with comparable physical disabilities.[29] Nevertheless, these personality characteristics do occur in some epileptic patients with temporolimbic foci, and some authors have clearly distinguished the behavioral profile of TLE patients.[22,30,31]

Additional studies link these behaviors to the temporolimbic lobe. When assessed with the Bear-Fedio Inventory, TLE patients with a mediobasal electroencephalographic (EEG) focus appear to have significantly more hypergraphia, elation, guilt, and paranoia than those with a lateral focus.[31] In patients assessed with other instruments, the presence of TLE was related to the presence of viscosity and increased social cohesion.[22] Others have reported that psychiatric problems in patients with epilepsy are associated with limbic auras.[32] These data support the neuroanatomical observations reviewed above, implying the significance of cortico-subcortical connectivity in the expression of both normal and abnormal emotional behavior.

Amnestic Syndromes

Among temporal lobe syndromes, amnesia is one of the most well studied. There are several excellent general reviews of memory functions.[33,34] Amnesia is an acquired disturbance of memory in which new declarative information is not encoded and incorporated into long-term storage. Amnesia defined in this way is specific to medial temporolimbic dysfunction. Amnesia may occur with lesions of the hippocampus, fornix, mammillary bodies, or medial thalamus. Unilateral left temporal lobe lesions produce verbal learning disturbances. Unilateral right temporal lobe lesions cause predominantly nonverbal learning disturbances. Severe amnesia is observed in patients with bilateral lesions. Possible differences between diencephalic amnesia associated with hypothalamic or thalamic lesions and hippocampal amnesia are controversial, but it is certain that integrity of the medial limbic circuits is essential for laying down of new memory traces for later recall. The retrieval process requires frontal-subcortical circuits as well as the medial limbic structures.

Amnesia has been most extensively studied in a single patient, H.M., who was subjected to bilateral temporal lobectomy for seizure control and rendered permanently amnestic.[33,34] Studies of H.M. reveal that since the time of the surgery he has been unable to learn declarative information (information that can be consciously recalled), but procedural memory and learning of new motor tasks are preserved. Beyond a short period of retrograde amnesia that extends to before the surgery, H.M. has retained remote memories for earlier life events. These features, which also occur in memory disorder associated with bilateral temporal lobe dysfunction of other etiologies, characterize the temporal amnestic syndrome.

The temporolimbic amnestic syndrome has resulted from Alzheimer's disease, bitemporal surgery, traumatic brain injury, posterior cerebral artery occlusion and stroke, herpes encephalitis, thalamic and hypothalamic tumors, paraneoplastic syndromes, and Wernicke-Korsakoff syndrome secondary to thiamine deficiency. The syndrome is transient when induced by electroconvulsive therapy.

Delusions and Hallucinations

Delusions are associated with limbic system dysfunction, and most diseases with delusional manifestations involve the temporal lobes or subcortical limbic system structures.[35] Delusions from tumors or epileptic foci involving the temporolimbic system, particularly on the left, have been reported.[36] However, Malloy and Richardson[37] reviewed a number of types of "content-specific" delusions in patients with completed neurological workups and found that the right hemisphere was more commonly involved than the left hemisphere. Davison and Bagley[38] reviewed 77 psychotic patients with brain tumors and found the temporal lobe to be the region of the brain most frequently involved.

About 10% of patients with TLE have an interictal psychosis with schizophrenia-like characteristics.[39] These epileptic patients often have an 11- to 15-year history of poorly controlled complex partial seizures with secondary generalized tonic-clonic seizures prior to developing their delusions.[40] They have prominent paranoid delusions, frequent affective symptoms, relatively preserved affect, normal premorbid personalities, and no family histories of schizophrenia. Some of these epileptic patients develop worsening delusions concomitant with an increase in seizure frequency, and a few others have worsening delusions on control of the seizures.[36,39] The terms *alternating psychosis* and *forced or paradoxical normalization* refer to this antagonism between delusions and ictal activity.

Delusional patients with brain tumors or epileptic foci in the temporolimbic lobe usually have left-sided foci, which may be physiologically active at the time of psychosis.[41] Surgical removal of the tumor or of the epileptic focus may not prevent the development of delusions and psychosis.[42] In contrast, formed auditory, visual, or haptic hallucinations of neurologic origin most often emanate from the right temporolimbic lobe as ictal phenomena.[35]

Recent developments in neuroscience implicate the left temporolimbic lobe in schizophrenia, delusional disorders, and paraphrenia (late-life psychosis). Investigators have found neuropathological changes in the hippocampi of schizophrenic patients, including bilateral disarray of hippocampal pyramidal cells, apparent neuronal loss, and some degree of gliosis.[43] There are abnormal spike discharges in anterior temporal lobes and in orbitofrontal cortex during periods of psychosis.[43] Using quantitative magnetic resonance imaging (MRI) in schizophrenia, Turetsky et al.[44] demonstrated a decrease in brain volume in the left temporal and right frontal regions, and Suddath et al.[45] showed decreased size in the hippocampus and amygdala bilaterally in the affected twin in monozygotic twin pairs discordant for schizophrenia. Moreover, neuropsychological studies indicate that schizophrenic patients have impairments in verbal recognition, which is a left temporolimbic function, and this performance correlates with the severity of the delusions.[46] In the Capgras syndrome, the delusional belief that someone has been replaced by an impostor, Signer[47] found that there is often a lesion in the left temporal or right frontal area. Malloy and Richardson[37] found right hemisphere or bilateral lesions in their review of the Capgras syndrome. In paraphrenia, recent MRI volumetric measurements found smaller left temporal volumes compared with control subjects,[48] and single-photon emission computerized tomography (SPECT) scans revealed left anterior temporal and right frontal hypoperfusion.[49]

Many of these findings illustrate a distributed model involving the left temporolimbic lobe, along with frontal cortex, in the production of delusions. They are in agreement with the theory that the onset of schizophrenia is associated with reactive synaptic regeneration in brain regions that receive degenerating temporal lobe projections, particularly the temporolimbic-prefrontal network.[50] The hippocampus may be a nodal area for delusions given its role in memory, in receiving emotional valencing from the amygdala, and in facilitating perceptual associations.[51] Ultimately, delusions may result from dissonance between disturbances in these functions and disturbed frontal-executive control.[52]

Emotional and Mood Disorders

Depression and Mania: Depression may be the most common neuropsychiatric complication of brain disease.[35] Patients with strokes and brain tumors in the left hemisphere frequently develop depression.[53] Although the localization with these lesions is more frontal or striatal than temporal, localization in the anterior temporal lobe can be the primary finding. Depressive disorders occur in 7.5% to 25% of epileptic patients and do not appear to be explained by a reaction to having a chronic illness.[54,55] These depressed patients tend to have frequent complex partial seizures, particularly when there is a left-sided temporal focus. Furthermore, postictal depression is common, and the risk of completed suicide in epileptic patients is 4 to 5 times greater than among the nonepileptic population.[56]

Mood changes result from disturbances in a temporolimbic-frontal-caudate network. Among bipolar patients, SPECT images show asymmetry in the anterior part of the temporal lobes in both depression and mania; asymmetry disappears with resolution of the mood state.[57] During sad moods, positron emission tomography (PET) shows cerebral blood flow to be increased in the left amygdala and decreased in the right amygdala,[58] and functional MRI shows a significant increase in signal intensity in the left amygdala.[59] These data, as well as information from focal lesion studies, suggest that the temporolimbic lobes, frontal cortex and cingulate gyrus, and caudate nuclei are involved in the production of mood disorders, and that there is reciprocity between the temporolimbic-frontal regulation of emotional experience and frontal lobe hypometabolism.[8,9,60] Far less is understood about mania. However, there is a weaker association of secondary mania with temporobasal or paralimbic lesions, particularly those involving the right side.[35] The same circuits may be implicated in mania as in depression.

Rage and Aggression: The temporolimbic lobes have been implicated in episodes of rage and aggression.[61] This association may occur particularly among patients with temporal lobe epilepsy; other patients with temporal mass lesions have not demonstrated a clear tendency to increased aggression. Episodic dyscontrol or intermittent explosive disorder manifests with prodromal mounting tension and irritability, postictal remorse, and frequent temporal lobe spikes on EEG. Studies have bolstered this association by eliciting aggressive verbalizations with stimulation of the amygdala and by producing defensive rage with hippocampal lesions in cats.[32,35] Among patients in a maximum-security mental hospital, the violent patients had focal temporal slowing or sharp waves on EEG and dilated temporal horns or small temporal lobes on computed tomography (CT) scans.[62] Functional neuroimaging studies with SPECT and PET among aggressive individuals reveal focal abnormalities in the temporal lobes, particularly on the left.[63,64] These results suggest that aggression and violence are associated with abnormalities in the temporal lobes. In addition, patients with left temporal lobe seizure foci have higher scores on hostile feelings than other epileptic patients.

There are other changes in emotion that involve the temporal lobes. These include deficits in the interpretation of emotional components of language (receptive aprosodia), reduction in the frequency and intensity of facial expression, and impairment in the ability to assess mood from facial expressions and emotional situations.[65]

Anxiety and Dissociative Disorders

Focal lesions associated with anxiety disorders usually involve the temporolimbic lobe, particularly on the right.[35,66] Anxiety occurs with poststroke depression in patients with left frontal lesions or right temporal lesions.[53] Subarachnoid hemorrhages and tumors in the vicinity of the third ventricle may produce anxiety. Anxiety and panic disorders occur among epileptic patients, and these disorders must be distinguished from simple partial seizures manifesting as anxiety or panic. Anxiety may be especially associated with right medial temporal lesions.[66] PET and SPECT studies show a relationship between anxiety and regional changes in the right temporo-occipital or temporobasilar regions.[67,68]

There is little information on the localization of dissociative states. Depersonalization and derealization, as well as multiple personality, may occur with right temporal lobe lesions.[35,69] A specific association of epilepsy with multiple personality disorder, depersonalization disorders, possession states, fugue states, and psychogenic amnesia is unresolved. Patients with multiple personality disorder often have EEG changes but rarely have clinical seizures. Among epileptic patients, clinicians must distinguish dissociative states from poriomania (prolonged periods of compulsive wandering resulting from an admixture of ictal and postictal changes) and from periods of amnesia due to prolonged or recurrent partial seizures.

Neurovegetative Disorders

Oral Aggression and Other Hyperoral Behaviors: Oral aggressive behavior can be a consequence of temporolimbic dysfunction.[70] Biting and oral attacks on others rarely occur from brain lesions; however, these behaviors may occur as part of amygdala-related aggression. Oral aggressive behaviors are most frequently self-injurious. Oral behaviors can be a manifestation of both ictal automatisms and the postictal state in patients with partial complex seizures of temporolimbic origin. In addition, a range of miscellaneous oral behaviors occurs in neuropsychiatric disease, such as spitting, yawning, and sucking. The nature and etiology of these behaviors are poorly understood, but they may be related to disorders in temporolimbic structures.

Sexuality Changes: Both hypo- and hypersexuality occur in patients with temporolimbic syndromes.[28,71,72] Hyposexuality is more common, but it often goes undetected because patients and spouses may not complain of it in the course of a visit to a physician. Hypersexuality, although more unusual, is more likely to come to clinical attention because a spouse may experience the sexual demands as unpleasant or abnormal or the patient may run afoul of the authorities for making unwelcome sexual advances to others.

Hypersexuality is a major feature of the Klüver-Bucy syndrome in cats and nonhuman primates, where exaggerated intercourse, self-stimulation, and even cross-species mating behaviors are observed. In humans with the Klüver-Bucy syndrome, altered sexual behavior is more common than frank increases in the amount of sexual activity.[18]

Hyposexuality is characteristic of the Gastaut-Geschwind syndrome, but it is also common in patients with TLE without other features of this distinctive syndrome.[28,71,72] Both men and women experience disturbances of sexual arousal and a lower sexual drive. Possible causes of hyposexuality are temporolimbic physiologic dysfunction, recurrent hyperprolactinemia associated with temporal lobe seizures, or the effects of anticonvulsant medications. Men have an increased risk of erectile dysfunction, suggesting a neurophysiological component, and studies of sex hormones suggest the possibility of a subclinical hypogonadotropic hypogonadism.[71,72]

PATHOLOGIES OF THE TEMPORAL LOBES

The main pathologic conditions associated with temporolimbic neuropsychiatric symptoms are focal lesions such as tumors and strokes, TLE, trauma, and encephalitis. These conditions have been described above and are only briefly recapitulated here.

Cerebral Tumors and Strokes

Neuropsychiatric symptoms can be the sole manifestations of brain tumors or strokes in the temporal lobes.[35,69,73,74] After the frontal lobes, the temporal lobes are the second most common important focus for neuropsychiatric symptoms from tumors and the most common site for delusions.[35,73] There is a significant association between gliomas, hamartomas, and other temporal lobe tumors and delusions.[75] Irritability and depression are frequent, particularly with left temporal lobe tumors,[76,77] and anxiety symptoms and panic attacks can develop, particularly with right temporal lobe tumors.[78,79] In addition, since the middle and posterior cerebral arteries supply regions of the temporal lobe, delusions and psychosis occasionally occur secondary to strokes.[69]

Temporal Lobe Epilepsy

Studies from communities, epilepsy clinics, and psychiatric hospitals demonstrate an increased prevalence of psychiatric problems among TLE patients as compared with nonepileptic patients.[80–82] Although most epileptic patients are normal, one-quarter or more of epileptics have a schizophreniform psychosis, depression, personality changes, or hyposexuality. These behavioral changes are chronic and are present between seizure episodes. Among epileptic patients, much of the psychopathology results from electrophysiological, structural, chemical, or kindled changes in the temporolimbic system.[83] Delusions and depression are particularly correlated with left-sided temporolimbic lesions.[36,55]

Trauma

Head injury produces frontotemporal contusions. In closed head injury with an identifiable posttraumatic amnesia, the anterior temporal and orbitofrontal lobes are the most susceptible sites for neuronal injury because of their proximity to bone in middle and anterior cranial fossae.[36,69] The behavioral and emotional sequelae, which may be severe, reflect temporolimbic system dysfunction; however, because it is common to incur both orbitofrontal and temporal trauma, it is difficult to distinguish the results of temporal from those of orbitofrontal injury. Personality changes, depression, and delusions may occur, particularly in those with left-sided lesions, and there are case reports of posttraumatic mania from right-sided involvement.[35]

Encephalitis

Herpes simplex virus is the prototypical infection of the temporolimbic lobes. Herpes encephalitis is due to herpes simplex (HSV-1) virus, either as a new infection or due to activation of a previously acquired but dormant virus.[84] This encephalitis presents with a sudden change of affect or behavior and evidence of focal CNS involvement. The initial disturbance may be bizarre, with delusions, hallucinations, and other neuropsychiatric symptoms. After the acute period, there is often residual amnesia, irritability and restlessness, distractibility, episodic aggressiveness, depression, or the Klüver-Bucy syndrome. The EEG is markedly abnormal, sometimes showing characteristic repetitive slow wave discharges in a temporal region, and the development of a seizure disorder is common. The most affected parts of the limbic system are in the anterior temporal lobes, notably the uncus; the amygdala; the hippocampus and dentate fascia; the insula; and the parahippocampal, posterior orbital, and cingulate gyri.[85] Persistent behavioral changes may result from impairment of amygdalofrontal pathways, particularly on the left.[86]

CONCLUSIONS

Neuropsychiatric symptoms result from disturbances of an extended neural network with a nidus in the temporolimbic lobe. Although only the Klüver-Bucy and the Gastaut-Geschwind syndromes appear specific to the temporolimbic lobe, a range of other symptoms result from involvement of temporolimbic connections. The hippocampus and portions of the amygdala interface with multiple cortical and subcortical circuits that modulate emotional behavior and affect. Temporolimbic connections especially involve the frontal cortex, caudate nuclei, and cingulate gyri. Further study will help provide a more complete understanding of the regulation of emotional and social behaviors by this extended temporolimbic network.

REFERENCES

1. Hollander B: The Mental Functions of the Brain. London, Grant Richards and Sons, 1901
2. McLean PD: The Triune Brain in Evolution. New York, Plenum, 1990
3. Papez JW: A proposed mechanism of emotion. Archives of Neurology and Psychiatry 1937; 38:725–733
4. Klüver H, Bucy PC: Preliminary analysis of functions of the temporal lobe in monkeys. Archives of Neurology and Psychiatry 1939; 42:979–1000
5. Mega MS, Cummings JL: Frontal-subcortical circuits and neuropsychiatric disorders. J Neuropsychiatry Clin Neurosci 1996; 6:358–370
6. Groenewegen HJ, Berendse HW: The specificity of the "non-specific" midline and intralaminer thalamic nuclei. Trends Neurosci 1994; 17:52–57
7. Mayberg HS: Frontal lobe dysfunction in secondary depression. J Neuropsychiatry Clin Neurosci 1994; 6:428–442
8. Guze BH, Gitlin M: The neuropathologic basis of major affective disorders: neuroanatomic insights. J Neuropsychiatry Clin Neurosci 1994; 6:114–121
9. Cummings JL: The neuroanatomy of depression. J Clin Psychiatry 1993; 54(suppl):14–20
10. Holstege G, Bandler R, Saper CB: The Emotional Motor System. Amsterdam, Elsevier, 1996
11. Livingston KE: Limbic connections: the limbic system as a substrate for epileptic disorder, in Limbic Epilepsy and the Dyscontrol Syndrome, edited by Girgis M, Kiloh LG. New York, Elsevier North-Holland Biomedical, 1979, pp 7–18
12. Alheid GF, Heimer L: Theories of basal forebrain organisation and the "emotional motor system," in The Emotional Motor System, edited by Holstege G, Bandler R, Saper CB. Amsterdam, Elsevier, 1996, pp 461–484
13. Koella W: The functions of the limbic system, in Temporal Lobe Epilepsy, Mania, and Schizophrenia and the Limbic System, edited by Koella W, Trimble MR. Basel, Karger, 1982, pp 12–39
14. LeDoux JE: Emotion and the amygdala, in The Amygdala, edited by Aggleton JP. New York, Wiley-Liss, 1992, pp 339–352
15. Klüver H, Bucy PC: "Psychic blindness" and other symptoms following bilateral temporal lobectomy in rhesus monkeys. Am J Physiol 1937; 119:352–353
16. Marlowe WB, Mancall EL, Thomas JJ: Complete Klüver-Bucy syndrome in man. Cortex 1975; 11:53–59
17. Mendez MF, Foti D: Lethal hyperoral behavior from the Klüver-Bucy syndrome. J Neurol Neurosurg Psychiatry 1997; 62:293–294
18. Lilly R, Cummings JL, Benson DF, et al: Clinical features of the human Klüver-Bucy syndrome. Neurology 1983; 33:1141–1145
19. Gastaut H: Étude électroclinique des épisodes psychotiques survenant en dehors des crises cliniques chez les épileptiques [Electroclinical study of interictal psychotic episodes in epileptics]. Rev Neurol 1956; 94:587–594
20. Waxman SG, Geschwind N: Hypergraphia in temporal lobe epilepsy. Neurology 1974; 24:629–636
21. Bear D: Behavioural changes in temporal lobe epilepsy: conflict, confusion, challenge, in Aspects of Epilepsy and Psychiatry, edited by Trimble MR, Bolwig TG. Chichester, England, Wiley, 1986, pp 19–30
22. Rao SM, Devinsky O, Grafman J, et al: Viscosity and social cohesion in temporal lobe epilepsy. J Neurol Neurosurg Psychiatry 1992; 55:149–152
23. Sachedev HS, Waxman SG: Frequency of hypergraphia in temporal lobe epilepsy: an index of the interictal behaviour syndrome. J Neurol Neurosurg Psychiatry 1981; 44:358–360
24. Hermann BP, Whitman S, Arntson P: Hypergraphia in epilepsy: is there a specificity to temporal lobe epilepsy? J Neurol Neurosurg Psychiatry 1983; 46:848–853
25. Roberts JKA, Robertson MM, Trimble MR: The lateralising significance of hypergraphia in temporal lobe epilepsy. J Neurol Neurosurg Psychiatry 1982; 45:131–138
26. Trimble MR: Hypergraphia, in Aspects of Epilepsy and Psychiatry, edited by Trimble MR, Bolwig TG. Chichester, England, Wiley, 1986, pp 75–88
27. Yamadori A, Mori E, Tabuchi M, et al: Hypergraphia: a right hemisphere syndrome. J Neurol Neurosurg Psychiatry 1986; 49:1160–1164
28. Smith DB, Treiman DM, Trimble MR: Neurobehavioral Problems in Epilepsy. New York, Raven, 1991
29. Mungus D: Interictal behavior abnormality in temporal lobe epilepsy, Arch Gen Psychiatry 1982; 39:108–111

30. Bear D, Fedio P: Quantitative analysis of interictal behavior in temporal lobe epilepsy. Arch Neurol 1977; 34:454–467
31. Nielsen H, Kristensen O: Personality correlates of sphenoidal EEG foci in temporal lobe epilepsy. Acta Neurol Scand 1981; 64:289–300
32. Trimble MR: Biological Psychiatry, 2nd edition. Chichester, England, Wiley, 1996
33. Squire LR: Memory and Brain. New York, Oxford University Press, 1987
34. Zola-Morgan SM, Squire LR: The primate hippocampal formation: evidence for a time-limited role in memory storage. Science 1990; 250:288–290
35. Yudofsky SC, Hales RE (eds): The American Psychiatric Press Textbook of Neuropsychiatry, 2nd edition. Washington, DC, American Psychiatric Press, 1992
36. Mendez MF, Grau R, Doss RC, et al: Schizophrenia in epilepsy: seizure and psychosis variables. Neurology 1993; 43:1073–1077
37. Malloy PF, Richardson ED: The frontal lobes and content-specific delusions. J Neuropsychiatry Clin Neurosci 1994; 6:455–466
38. Davison K, Bagley CR: Schizophrenia-like psychosis associated with organic disorders of the central nervous system. Br J Psychiatry 1969; 4(suppl):113–184
39. Trimble MR: The Psychosis of Epilepsy. New York, Raven, 1991
40. Slater E, Beard A: The schizophrenia-like psychosis of epilepsy: psychiatric aspects. Br J Psychiatry 1963; 109:95–150
41. Jibiki I, Maeda T, Kubota T, et al: ^{123}I-IMP SPECT brain imaging in epileptic psychosis: a study of two cases of temporal lobe epilepsy with schizophrenia-like syndrome. Neuropsychobiology 1993; 28:207–211
42. Krahn LE, Rummans TA, Peterson GC: Psychiatric implications of surgical treatment of epilepsy. Mayo Clin Proc 1996; 71:1201–1204
43. Scheibel AB: Are complex partial seizures a sequela of temporal lobe dysgenesis? Adv Neurol 1991; 55:59–77
44. Turetsky B, Cowell PE, Gur RC, et al: Frontal and temporal lobe brain volumes in schizophrenia: relationship to symptoms and clinical subtype. Arch Gen Psychiatry 1995; 52:1061–1070
45. Suddath RL, Christison GW, Torrey EF, et al: Anatomical abnormalities in the brains of monozygotic twins discordant for schizophrenia. N Engl J Med 1990; 322:789–794
46. Gur RE, Jaggi JL, Shtasel DL, et al: Cerebral blood flow in schizophrenia: effects of memory processing on regional activation. Biol Psychiatry 1994; 35:3–15
47. Signer SF: Localization and lateralization in the delusion of substitution: Capgras symptom and its variants. Psychopathology 1994; 27:168–176
48. Howard R, Mellers J, Petty R, et al: Magnetic resonance imaging volumetric measurements of the superior temporal gyrus, hippocampus, parahippocampal gyrus, frontal and temporal lobes in late paraphrenia. Psychol Med 1995; 25:495–503
49. Cloud BS, Carew TG, Rothenberg H, et al: A case of late-onset psychosis: integrating neuropsychological and SPECT data. J Geriatr Psychiatry Neurol 1996; 9:146–153
50. Ruppin E, Reggia JA, Horn D: Pathogenesis of schizophrenic delusions and hallucinations: a neural model. Schizophr Bull 1996; 22:105–123
51. Fricchione GL, Carbone L, Bennett WI: Psychotic disorder caused by a general medical condition, with delusions: secondary "organic" delusional syndromes. Psychiatr Clin North Am 1995; 18:363–378
52. Mentis MJ, Weinstein EA, Horwitz B, et al: Abnormal brain glucose metabolism in the delusional misidentification syndromes: a positron emission tomography study in Alzheimer disease. Biol Psychiatry 1995; 38:438–449
53. Robinson RG, Kubos KL, Starr LR, et al: Mood disorders in stroke patients: importance of lesion location. Brain 1984; 107:81–93
54. Mendez MF, Cummings JL, Benson DF: Depression in epilepsy: significance and phenomenology. Arch Neurol 1986; 43:766–770
55. Robertson MM, Channon S, Baker J: Depressive symptomatology in a general hospital sample of outpatients with temporal lobe epilepsy: a controlled study. Epilepsia 1994; 35:771–777
56. Mathews WS, Barabas G: Suicide and epilepsy: a review of the literature. Psychosomatics 1981; 22:515–524
57. Gyulai L, Alavi A, Broich K, et al: I-123 iofetamine single-photon computed emission tomography in rapid cycling bipolar disorder: a clinical study. Biol Psychiatry 1997; 41:152–161
58. Schneider F, Gur RE, Mozley LH, et al: Mood effects on limbic blood flow correlate with emotional self-rating: a PET study with oxygen-15 labeled water. Psychiatry Res 1995; 61:265–283
59. Grodd W, Schneider F, Klose U, et al: Functional magnetic resonance tomography of psychological functions exemplified by experimentally induced emotions. Radiologe 1995; 35:283–289
60. George MS, Ketter TA, Post RM: SPECT and PET imaging in mood disorders. J Clin Psychiatry 1993; 54(suppl):6–13
61. Mark VH, Ervin FR: Violence and Epilepsy. New York, Harper and Row, 1970
62. Wong MT, Lumsden J, Fenton GW, et al: Electroencephalography, computed tomography and violence ratings of male patients in a maximum-security mental hospital. Acta Psychiatr Scand 1994; 90:97–101
63. Amen DG, Stubblefield M, Carmichael B, et al: Brain SPECT findings and aggressiveness. Ann Clin Psychiatry 1996; 8:129–137
64. Volkow ND, Tancredi LR, Grant C, et al: Brain glucose metabolism in violent psychiatric patients: a preliminary study. Psychiatry Res 1995; 61:243–253
65. Beeckmans K, Michiels K: Personality, emotions and the temporolimbic system: a neuropsychological approach. Acta Neurol Belg 1996; 96:35–42
66. Ghadirian AM, Gauthier S, Bertrand S: Anxiety attacks in a patient with a right temporal meningioma. J Clin Psychiatry 1986; 47:270–271
67. Benkelfat C, Bradwejn J, Meyer E, et al: Functional neuroanatomy of CCK4-induced anxiety in normal healthy volunteers. Am J Psychiatry 1995; 152:1180–1184
68. O'Carroll RE, Moffoot AP, Van Beck M, et al: The effect of anxiety induction on the regional uptake of ^{99m}Tc-exametazime in simple phobia as shown by single photon emission tomography (SPET). J Affect Disord 1993; 28:203–210
69. Lishman A: Organic Psychiatry, 2nd edition. Oxford, England, Blackwell, 1987
70. Harris JC: Destructive behavior: aggression and self-injury, in Developmental Neuropsychiatry, vol 2. New York, Oxford University Press, 1995, pp 463–484
71. Morrell MJ, Guldner GT: Self-reported sexual function and sexual arousability in women with epilepsy. Epilepsia 1996; 37:1204–1210
72. Murialdo G, Galimberti CA, Fonzi S, et al: Sex hormones and pituitary function in male epileptic patients with altered or normal sexuality. Epilepsia 1995; 36:360–365
73. Keschner M, Bender M, Strauss I: Mental symptoms in cases of tumor of the temporal lobe. Arch Neurol Psychiatry 1936; 35:572–596
74. Malamud N: Organic brain disease mistaken for psychiatric disorder, in Psychiatric Aspects of Neurologic Disease, edited by Benson DF, Blumer D. New York, Grune and Stratton, 1975, pp 287–305
75. Taylor DC: Factors influencing the occurrence of schizophrenia-like psychosis in patients with temporal lobe epilepsy. Psychol Med 1975; 5:249–254
76. Malamud N: Psychiatric disorder with intracranial tumors of limbic system. Arch Neurol 1967; 17:113–123
77. Irle E, Peper MA, Wowra B, et al: Mood changes after surgery for tumors of the cerebral cortex. Arch Neurol 1994; 51:164–174
78. Dietch JT: Cerebral tumor presenting with panic attacks. Psychosomatics 1984; 25:861–863
79. Druback DA, Kelly MP: Panic disorder associated with a right paralimbic lesion. Neuropsychiatry, Neuropsychology, and Behavioral Neurology 1989; 2:282–289

80. Pond DA, Bidwell BH: A survey of epilepsy in 14 general practices, II: social and psychological aspects. Epilepsia 1959/1960; 1:285–299
81. Gudmundsson G: Epilepsy in Iceland: a clinical and epidemiological investigation. Acta Neurol Scand 1966; 25(suppl):1–124
82. Lindsay J, Ounsted C, Richards P: Long-term outcome in children with temporal lobe seizures, III: psychiatric aspects in childhood and adult life. Dev Med Child Neurol 1979; 21:630–636
83. Smith PF, Darlington CL: The development of psychosis in epilepsy: a re-examination of the kindling hypothesis. Behav Brain Res 1996; 75:59–64
84. Longson M: Herpes simplex encephalitis, in Recent Advances in Neurology, edited by Matthews WB, Glaser G. Edinburgh, Churchill Livingstone, 1985, pp 123–140
85. Hierons R, Janota I, Corsellis JAN: The late effects of necrotising encephalitis of the temporal lobes and limbic areas: a clinico-pathological study of ten cases. Psychol Med 1978; 8:21–42
86. Caparros-Lefebvre D, Girard-Buttaz I, Reboul S, et al: Cognitive and psychiatric impairment in herpes simplex virus encephalitis suggest involvement of the amygdalo-frontal pathways. J Neurol 1996; 243:248–256

9

The Neurobiology of Emotional Experience

Kenneth M. Heilman, M.D.

Although almost everyone knows what emotions are, the word *emotion* is a psychological construct that remains difficult to define. Perhaps emotions are difficult to define because they are not objects or things but rather are complex, multicomponent processes.

There are at least three major components of emotion. First, there is emotional behavior, including overt behavior and changes in the viscera and autonomic nervous system. Second, there is emotional communication through word, tone of voice (prosody), and gestures, including facial expressions. Third, there is emotional feeling or experience. Emotional experience, although subjective, is one of the major factors in motivating human behavior. It is not only the direct experience of emotion that motivates behavior, but also the knowledge that certain stimuli, situations, and actions could in the future reproduce emotional states.

In this review, I explore the neural basis of emotional experience. Building models of the neural basis of emotional experience may help us understand abnormal emotional states. Although studies based on the ablation paradigm were used to help form this model, I do not discuss here the psychiatric affective disorders such as panic and depression. I first briefly review the classic feedback and central theories and the more recent revisions of these theories, and I provide evidence that these theories cannot fully explain the neural basis of emotional experience. I then discuss a "distributed modular network theory of emotional experience." The proposed model suggests that emotional experience involves the components of valence, arousal, and motor activation.[1]

FEEDBACK THEORIES

Facial Feedback Hypothesis

Charles Darwin[2] noted that "he who gives violent gesture increases his rage." Darwin also believed that the means by which we express emotions are innate. Izard[3] and Ekman et al.[4] performed cross-cultural studies of facial emotional expression and found that the same seven to nine emotional facial expressions appeared to be universal, thereby providing support for Darwin's hypothesis that emotional expression may be innate. Tomkins[5,6] posited that it was the feedback of these facial

---

This work was supported by the Medical Research Service of the Department of Veterans Affairs.

emotions to the brain that induced emotional feeling. Laird[7] experimentally manipulated facial expressions and found that patients felt emotions, thereby providing some support for the facial feedback hypothesis.

There are, however, many unresolved problems with the facial feedback theory of emotional experience. For example, when one is feeling a strong emotion, the valence of this emotion cannot be changed by altering one's facial expression, and a person can express one emotion while feeling another. It is possible that voluntary facial emotions cannot entirely control emotional experience because the innervatory patterns of voluntary facial gestures are different from those that occur naturally. Patients with pseudobulbar palsy may express strong facial emotions that they are not feeling,[8,9] but it is possible that these patients' brain lesions also interrupt feedback to the brain. Patients with diseases of the facial nerve motor unit, even when the problem is bilateral, do not report a change of emotion. Although it is possible, as Darwin suggested, that facial expressions may embellish emotions, it remains possible that this influence is not from feedback from facial muscles, but rather is mediated totally within the central nervous system by associative networks.

Visceral Feedback Hypothesis

William James[10] proposed that stimuli that provoke emotion induce changes in the viscera and that the self-perception of these visceral changes is what produces emotional experience.

This feedback theory was challenged by Cannon,[11] who argued that the separation of the viscera from the brain, as occurs with cervical spinal cord injuries, does not eliminate emotional experience. However, the Cannon argument was not fully supported by Hohmann,[12] who studied patients with spinal cord injuries and found that patients with either high or low spinal cord transection did experience emotions as predicted by Cannon, but that patients with lower lesions reported stronger emotions than those with higher lesions. Higher cervical lesions would be more likely to affect efferent control of the viscera and injure the autonomic nervous system's efferent output. Therefore, Hohmann's observations provide partial support for the visceral feedback theory. It should also be noted that the vagus nerve is the primary visceral afferent nerve, and cervical spinal cord transection would not interfere with this nerve's ability to provide the brain with feedback from the viscera. However, even if feedback was possible, with a loss of autonomic control there might be little to feed back.

Cannon thought that the viscera have insufficient afferent input to the brain to be important in inducing emotional experience. As mentioned above, however, the vagus nerve is mainly a visceral afferent nerve. In addition, using a heartbeat detection paradigm, Katkin et al.[13] not only found that some normal subjects can accurately detect their heartbeats, but also that the subjects who had the strongest emotional responses to negative slides were the subjects who were able to detect their own heartbeats.[14]

Cannon also thought that because the same visceral responses occur with different emotions, feedback of these visceral responses could not account for the variety of emotions that humans experience. However, Ax[15] demonstrated that different bodily reactions can be associated with different emotions.

Thus, Cannon's arguments fail to refute the visceral feedback theory of emotional experience. However, we have not yet discussed how this feedback system may work. In humans, the neocortex and limbic cortex play a critical role in the analysis and interpretation of various stimuli (see Heilman et al.[16] for review). Because the viscera respond to complex stimuli (such as speech), there must be efferent descending neural systems that enable the cortex to control the viscera via the autonomic nervous system and the endocrine system.

The autonomic nervous system has two components, the sympathetic and the parasympathetic. The descending sympathetic neurons receive projections from the hypothalamus, and the hypothalamus receives projections from many limbic and paralimbic areas, including the amygdala. The most important parasympathetic nerve is the vagus. The vagus originates in the dorsal motor nucleus situated in the brainstem and projects to the viscera, including the heart. The amygdala not only projects to the hypothalamus, but also sends direct projections to the nucleus of the solitary tract and the dorsal motor nucleus of the vagus. In this manner the amygdala may directly influence the parasympathetic system. The amygdala receives neocortical input.

Although the amygdala may be the most important part of the limbic system to influence the autonomic nervous system and viscera, stimulation of other areas, including the insula and orbitofrontal cortex, can also induce autonomic and visceral changes, and these structures also receive input from the neocortex. As mentioned, in regard to feedback the major nerve that carries visceral afferent information back to the brain is the vagus. These vagal afferents terminate primarily in the nucleus of the solitary tract. This nucleus projects to the central nucleus of the amygdala, and the central nucleus of the amygdala projects to other amygdala nuclei and the insula. The amygdala and insula, in turn, project to several neocortical areas, including the temporal, parietal, and frontal lobes.

Luria and Simernitskaya[17] thought that the right

hemisphere might be more important than the left in perceiving visceral changes. To test this postulate, Davidson et al.[18] gave a variety of tapping tests to normal individuals. They found that the left hand was more influenced by heart rate than the right hand, suggesting the right hemisphere may be superior at detecting heartbeat. Unfortunately, this left hand superiority in the detection of heartbeat has not been replicated by other investigators.

There are other observations for which the visceral feedback theory cannot account. For example, Marañon[19] injected epinephrine into subjects and then inquired as to the nature of the emotion felt by these subjects. Epinephrine does not cross the blood–brain barrier, but it affects the autonomic nervous system and viscera. However, Marañon found that injections of epinephrine were not associated with genuine emotional experiences, but rather with "as if" feelings.

Schacter and Singer[20] also injected epinephrine into experimental subjects and reported that pharmacologically induced autonomic and visceral activation did not produce an emotion unless this arousal was accompanied by a cognitive set. Although Schacter and Singer's study suggested that visceral feedback combined with centrally mediated cognition is important for emotional experience, observations in our laboratory did not entirely support these findings.

Recently, we tested the autonomic-visceral feedback theory to learn if the right hemisphere plays a dominant role in perceiving visceral changes. Using a shock anticipation paradigm, we found that when compared with normal control subjects, patients with right hemisphere lesions had a reduced autonomic response. Although their autonomic response was reduced, they showed no differences in the experience of anticipatory anxiety.[21]

In addition, the strongest argument against the cognitive-visceral feedback theory or attribution theory arises from the observations of patients who have strong emotions (such as fear) associated with medial temporal lobe or amygdala seizures. In many of these epileptic patients, the emotional experience is often the first symptom. Therefore, in these patients cognitive set and autonomic/visceral changes come after the experience rather than before the experience. The attribution theory of Schacter and Singer cannot account for these observations.

CENTRAL THEORIES

To account for emotional experience, Cannon[11] proposed that afferent stimuli enter the brain and are transmitted to the thalamus, which activates the hypothalamus. The hypothalamus can in turn activate the endocrine system and the autonomic nervous system, and it is these systems that induce the physiological changes in the viscera. According to Cannon, autonomic and visceral activation are primarily adaptive and aid in the survival of the organism. Emotional experience, according to Cannon, is induced by the hypothalamus feeding back to the cortex. More recently, LeDoux et al.[22] have modified Cannon's thalamic-hypothalamic emotion circuit to include the amygdala in fear conditioning.

Cannon and LeDoux did not propose a critical role for the cortex in the interpretation of stimuli. Although some conditioned stimuli, similar to those used by LeDoux, may induce emotion without cortical interpretation, there is overwhelming evidence that in humans the neocortex is critical for interpreting the meaning of many stimuli, especially stimuli that are complex and rely on past learning. For example, spoken and written stimuli may engender strong emotions. These stimuli need phonological, orthographic, lexical, and semantic processing that is mediated primarily by the left cerebral cortex.

Recently, Ross et al.[23] posited that the right hemisphere is important for primary emotions (such as anger or fear) and the left is important for social emotions (such as embarrassment). They based this theory on the observation that some patients undergoing selective hemispheric anesthesia (the Wada test) changed their emotional response to traumatic events that had occurred earlier in their lives. For example, one of their subjects told of an incident where he was very frightened, but when his right hemisphere was anesthetized, he stated that he felt embarrassed. However, Ross and co-workers did not demonstrate that left hemisphere dysfunction led to a loss of social emotions. In the absence of such evidence, there are alternative explanations of Ross and colleagues' observations, and these will be discussed below.

Cannon's diencephalic-hypothalamic model, LeDoux's diencephalic-limbic (amygdala) model, and Ross and colleagues' left-right social emotional theories not only fail to account for how humans experience emotion in response to complex stimuli, but also fail to explain how humans can experience a variety of emotions.

MODULAR THEORY

There are two ways in which the brain may modulate a variety of emotional experiences.

One possibility is that the brain may contain specialized emotional systems (devoted systems), so that there would be a special system for fear, one for anger, one for happiness, and so forth. This would imply that each emotion was uniquely mediated. A second possibility is that each emotion is not uniquely mediated—that the neural apparatus that mediates one emotion may not only play a role in other emotions, but may also mediate even nonemotional functions (nondevoted systems).

The nondevoted systems view is consistent with the "dimensional" view of emotion favored by many cognitive emotional theorists. Wundt[24] proposed that emotional experiences vary in three dimensions: quality, activity, and excitement (arousal). Similarly, Osgood et al.[25] performed factor analyses on verbal assessments of emotional judgments and found that the variance could be accounted for by three major dimensions: valence (positive/negative, pleasant/unpleasant), arousal (calm/excited), and control or dominance (in control/out of control). Using this type of multidimensional scale, one can define the different emotions. For example, fear would be unpleasant, high arousal, and out of control; sadness could be unpleasant and low arousal. Frijda[26] also explored the cognitive structure of emotion and found that "action readiness" was an important component. Studies using a variety of techniques with normal subjects[27] have supported this dimensional view.

The modular theory posits that the conscious experience of emotion is mediated by anatomically distributed modular networks.[1] By *modular*, we mean that a system works independently. However, this does not mean that modules are encapsulated in such a way that they cannot be influenced by other systems. By *network*, we mean the sum total of modules that are critical in predicating a behavior or experience. If the violinist is the module, the orchestra is the network. The proposed network contains three major modules, one that determines valence (pleasant versus unpleasant), another that controls arousal, and a third that mediates motor activation with approach or avoidance behaviors.

Valence

Goldstein[28] reported that aphasic patients with left hemisphere lesions often appeared anxious, agitated, and sad. He called this behavioral syndrome the "catastrophic reaction." Gainotti[29] studied 160 patients with right or left hemisphere strokes. Like Goldstein, he reported that patients with left hemisphere lesions had a catastrophic reaction, but he thought this was a normal response to serious cognitive and physical deficits such as aphasia or hemiparesis. In contrast, Babinski,[30] Hécean et al.,[31] and Denny-Brown et al.[32] noticed that patients with right hemisphere lesions often appeared either inappropriately indifferent or even euphoric. Gainotti[29] confirmed these observations but suggested that these patients' indifference might be related to the denial or unawareness of illness (anosognosia) that often is associated with right hemisphere lesions.

Although Gainotti proposed that patients' psychological responses to their own illness may account for some of the emotional asymmetries observed between patients with right and left hemisphere lesions, there are additional reports that a psychological reactive theory cannot entirely explain. Terzian[33] and Rossi and Rosadini[34] studied the emotional reactions of patients recovering from selective hemispheric barbiturate-induced anesthesia (the Wada test). These investigators noted that whereas right carotid injections were often associated with euphoria, barbiturate injections into the left carotid artery were often associated with catastrophic reactions. The Wada test is a diagnostic test that causes only transient hemiparesis and aphasia. It therefore seems unlikely that this procedure would cause a reactive depression that occurred during only a part of the test. In addition, we have seen right hemisphere–damaged stroke patients who appear to be emotionally indifferent but do not demonstrate anosognosia or verbally explicit denial of illness.

The catastrophic-depressive reaction associated with left hemisphere lesions is seen most commonly in patients who have anterior (frontal) perisylvian lesions.[35,36] It is possible that the hemispheric emotional asymmetries reported by Gainotti and others may be related to a communication disorder rather than differences in emotional experience. Tucker et al.[37] and Ross and Mesulam[38] reported that right hemisphere–damaged subjects lost their ability to express emotional prosody. Patients with left hemisphere lesions may be aphasic and unable to express themselves using propositional speech. To express feelings, they may rely more on right hemisphere–mediated, nonpropositional affective communication, and therefore these patients may intone their speech more heavily and use more emotional facial expressions. Because these left hemisphere–damaged patients may have a greater need to use these expressive emotional systems, they may appear to be highly emotional. In contrast, patients with right hemisphere lesions may have more difficulty than patients with left hemisphere disease in expressing emotional faces and emotional speech prosody. Therefore, when compared with left hemisphere–damaged patients, those with right hemisphere damage may appear to be indifferent.

Although hemispheric defects of emotional expression may account for some of the behavioral observations by Goldstein,[28] Babinski,[30] and Gainotti,[29] they cannot explain the results of Gasparrini et al.,[39] who administered the Minnesota Multiphasic Personality Inventory (MMPI) to a group of left and right hemisphere–damaged patients. The MMPI is widely used to assess emotional experience. The left hemisphere–damaged patients were not severely aphasic. In addition, the right and left hemisphere patients were balanced for cognitive and motor defects. The MMPI does not require emotionally intoned speech or facial expressions. The findings of Gasparrini et al. supported the hemispheric valence hypothesis. They found that patients with left hemisphere disease showed a marked elevation of the depression scale, whereas patients with right hemisphere disease did not. Therefore, the right–left differences in emotional behavior observed by Gainotti and others cannot be attributed to emotional expressive disorders or to the severity of the motor or cognitive deficit.

Starkstein et al.[40] also studied emotional changes associated with stroke and also found that about one-third of stroke patients had a depression. They not only found that the depression was associated with both left frontal and left caudate lesions, but also that the closer to the frontal pole the lesion was located, the more severe was the depression. Many of the patients with left hemisphere lesions and depression were also anxious. In contrast, patients with right frontal lesions were often indifferent or even euphoric. However, not all investigators agree that after stroke there is more depression with left than right hemisphere lesions. House et al.[41] and Milner[42] could not replicate the emotional asymmetries found in prior reports. In addition, as discussed above, Ross et al.[23] posited that the left hemisphere mediates social emotions and the right hemisphere mediates all primary emotions. However, it is not clear that Ross et al. fully explored this right hemisphere primary emotion hypothesis.

To learn if there are discrete physiological changes associated with depression, several groups of investigators have conducted functional imaging studies of patients with primary depression. Several of these investigations have noted a decrease of activation in the left frontal lobe as well as the left cingulate gyrus.[43,44] However, Drevets and Raichle[45] found increased activity in the left prefrontal cortex, amygdala, basal ganglia, and thalamus.

Davidson et al.[46] and Tucker[47] investigated the hemispheric valence hypothesis by studying normal subjects with electrophysiological techniques. Their studies confirmed the ablation studies, suggesting that the frontal region of each hemisphere makes asymmetric contributions to emotional experience: greater right than left hemisphere activation is associated with negative moods, and greater left than right hemisphere activation is associated with positive moods.

Unfortunately, it is not known how the right and left hemispheres may influence emotional valence. Fox and Davidson[48] suggest that left hemisphere–mediated positive emotions are related to approach and right hemisphere–mediated negative emotions are related to avoidance behaviors. However, recently in our laboratory we studied the relationship of emotions and approach–avoidance behavior. We found that negative emotions can be associated with both approach and avoidance behaviors (G. P. Crucian et al.: Dissociation of behavioral action and emotional valence in the expression of affect. Presented at the annual meeting of the American Academy of Neurology, Boston, MA, April, 1997). For example, fear and anger both have a negative valence, but fear is associated with avoidance and anger is associated with approach. In addition, this approach–avoidance model does not explain how the two hemispheres are differently organized such that they make opposite contributions to mood. This approach–avoidance theory does not explain how other emotions are mediated, and it also does not explain the role of other areas in the brain, such as the limbic system.

Tucker and Williamson[49] think that hemispheric valence asymmetries may be related to asymmetrical control of neuropharmacological systems, the left hemisphere being more cholinergic and dopaminergic and the right hemisphere more noradrenergic. Mayberg et al.[50] reported that pharmacologic changes in the two hemispheres may be different after stroke. Using PET imaging, they reported that whereas strokes in the right hemisphere appear to increase serotonergic receptor binding, left hemisphere strokes lower serotonergic binding. The lower the serotonergic binding, the more severe the depression. Although it is well known from clinical psychiatry that neurotransmitter systems may have a profound influence on mood and that changes in these systems may induce either euphoria or dysphoria, the mechanism by which the pharmacological changes induce mood remains unknown.

The frontal lobes also have strong projections to the limbic system. The dorsolateral and medial frontal lobe project primarily to the cingulate gyrus; the orbitofrontal cortex, via the uncinate fasciculus and anterior temporal lobe, has strong connections to the amygdala. Although limbic structures such as the amygdala are important in negative emotions such as fear, the means by which the right and left frontal lobes selectively influence these limbic structures remain unknown.

Arousal

The term *arousal*, like the terms *attention* and *emotion*, is difficult to define. Behaviorally, whereas an aroused organism is awake, alert, and prepared to process stimuli, an unaroused organism is lethargic to comatose and not prepared to process stimuli. Arousal has several physiologic definitions. In the central nervous system, arousal usually refers to the excitatory state of neurons or the propensity of neurons to discharge when appropriately activated. In functional imaging, arousal is usually measured by increases of blood flow, and electrophysiologically it is measured by desynchronization of the EEG or by the amplitude and latency of evoked potentials. Outside the central nervous system, arousal usually refers to activation of the sympathetic nervous system and the viscera, including the heart.

Arousal and attention are intimately linked and appear to be mediated by a cortical limbic reticular modular network.[51–54] This chapter provides an overview of this modular network, but for details one should refer to the original articles. Much of what we know about the anatomy of this network initially came from studies of monkeys and human patients with discrete brain lesions. However, more recently functional imaging has confirmed much of this ablation research. In humans, lesions of the inferior parietal lobe are most often associated with disorders of attention and arousal,[55,56] and in monkeys temporoparietal ablations are also associated with attentional disorders.[57,58] Physiological recordings from neurons in the parietal lobes of monkeys appear to support the postulate that the parietal lobe is important in attention. Unlike neurons in the primary sensory cortex, these "attentional" neurons in the parietal lobe, a supramodal association cortex, show an association between the rate of firing and the significance of the stimulus to the monkey—the relevant stimuli being associated with higher firing rates than the unimportant stimuli.[58,59]

Sensory information projects to the thalamic relay nuclei. From the thalamus, this modality-specific sensory system projects to the primary sensory cortices. Each of these primary sensory cortices (visual, tactile, auditory) projects only to its association cortex. For example, Brodmann area 17, the primary visual cortex, projects to Brodmann area 18. Subsequently, these modality-specific association areas converge on polymodal areas such as the frontal cortex (periarcuate, prearcuate, and orbitofrontal) and both banks of the superior temporal sulcus.[60] Both of these sensory polymodal convergence areas project to the supramodal inferior parietal lobe.[61]

Although the determination of the novelty of a stimulus may be mediated by modality-specific sensory association cortex, determination of the significance of a stimulus requires knowledge as to both the meaning of the stimulus and the motivational state of the organism. The motivational state is dependent on at least two factors: immediate biological needs and long-term goals. It has been demonstrated that portions of the limbic system, together with the hypothalamus, monitor the internal milieu and develop drive states. Therefore, limbic input into regions important in determining stimulus significance may provide information about immediate biological needs. Regarding long-term goals, the frontal lobe has been demonstrated to play a major role in goal-oriented behavior and set development,[62,63] and frontal input into the attentional-arousal systems may provide information about goals that are not motivated by immediate biological needs. Studies of cortical connectivity in monkeys have demonstrated that the temporoparietal region not only has strong connections with portions of the limbic system (cingulate gyrus) but also with the frontal cortex.

Stimulation of the mesencephalic reticular formation (MRF) in animals induces behavioral and physiological arousal.[64] In contrast, bilateral lesions of the MRF induce coma, and unilateral lesions cause the ipsilateral hemisphere to be behaviorally and physiologically hypo-aroused.[65] The polymodal and supramodal cortex discussed above not only determine stimulus significance, but also modulate arousal by influencing the MRF.[66] The exact means by which these cortical areas influence the MRF and the MRF influences the cortex remain unknown. However, there are at least three possible mechanisms by which the MRF may influence cortical processing. First, Shute and Lewis[67] describe an ascending cholinergic reticular formation. The nucleus basalis, which is in the basal forebrain, receives input from the reticular formation and has cholinergic projections to the entire cortex. These cholinergic projections appear to be important for increasing neuronal responsivity.[68] Second, the MRF may also influence the cortical activity through thalamic projections. Steriade and Glenn[69] demonstrated that nonspecific thalamic nuclei such as centralis lateralis and paracentralis project to widespread cortical regions and these thalamic nuclei can be activated by stimulation of the mesencephalic reticular formation. The third mechanism that may help account for cortical arousal involves the thalamic nucleus reticularis (NR). This thin nucleus envelops the thalamus and projects to all the sensory thalamic relay nuclei. Physiologically, NR inhibits the thalamic relay of sensory information.[70] However, when cortical limbic networks determine that a stimulus is significant or novel, corticofugal projections may inhibit the inhibitory NR, thereby allowing the thalamic sensory nuclei to relay sensory information to the cortex.

The level of activity of the peripheral autonomic nervous system usually mirrors the level of arousal in the central nervous system. One means of measuring peripheral autonomic arousal is to measure hand sweating by using the galvanic skin response. To learn if there were differences in the hemispheric control of sweating, Heilman et al.[71] studied patients with right and left hemisphere damage as well as normal control subjects. These investigators presented these subjects with nociceptive stimuli (electric shock) and demonstrated that patients with right hemisphere lesions had a reduced arousal response when compared with normal subjects and left hemisphere–damaged control subjects. Subsequently, other investigators reported similar findings. For example, Morrow et al.[72] and Schrandt et al.[73] also found that right hemisphere–damaged patients had a reduced skin response to emotional stimuli. However, Heilman and his co-workers[71] reported another interesting finding. When compared with normal subjects, patients with left hemisphere lesions appear to have a greater autonomic response. Using changes in heart rate as a peripheral measure of arousal, Yokoyama et al.[74] obtained results similar to those using skin response. Using physiological imaging, Perani et al.[75] also found that in cases of right hemisphere stroke there was also metabolic depression of the left hemisphere. Unfortunately, left hemisphere–damaged control patients were not reported.

The mechanism underlying the asymmetrical hemispheric control of arousal remains unknown. Because lesions restricted to the right hemisphere could not directly interfere with the left hemisphere's corticofugal projections to the reticular systems or the reticular system's corticipetal influence of the left hemisphere, one would have to propose that the right hemisphere's control of arousal may be related to privileged communication that the right hemisphere has with the reticular activating system. Alternatively, portions of the right hemisphere may play a dominant role in computing stimulus significance. The increased arousal associated with left hemisphere lesions also remains unexplained. Perhaps the left hemisphere maintains some type of inhibitory control over the right hemisphere or the reticular activating system.

Motor Activation and Approach–Avoidance

Some emotions do not call for action (for example, sadness or satisfaction), but others do (anger, fear, joy, surprise). When emotions are associated with action, this action may be toward the stimulus (approach) or away from the stimulus (avoidance) that induces the emotion. Although one would like to avoid emotions that are unpleasant and approach situations that induce pleasant emotions, when we discuss approach and avoidance we are discussing the behavior associated with the emotion and not one's plans for structuring one's behavior in relation to one's environment. For example, whereas one would like to avoid situations that induce anger, when one does become angry, one has a propensity to approach the stimulus that is inducing this emotion. Joy, a positive emotion, is also associated with approach behaviors.

Pribram and McGuiness[76] use the term *activation* to denote the physiological readiness to respond to stimuli. We have posited that motor activation or intention is mediated by a modular network that includes portions of the cerebral cortex, basal ganglia, and limbic system (see Heilman et al.[54] for a detailed review). The dorsolateral frontal lobe appears to be the hub of this motor preparatory network.[52,77] Physiological recording from cells in the dorsolateral frontal lobe reveals neurons that have enhanced activity when the animal is presented with a stimulus that is meaningful and predicts movement.[78] The dorsolateral frontal lobes receive inputs from the cingulate gyrus and from the posterior cortical association areas that are modality-specific, polymodal, and supramodal. Inputs from these posterior neocortical areas may provide the frontal lobes with information about the stimulus, including its meaning and its spatial location. The limbic system (for example, the cingulate gyrus, which is not only part of the Papez circuit but also receives input from Yakolov's basal lateral circuit) may provide information as to the organism's motivational state. The dorsolateral frontal lobe has nonreciprocal connections with the basal ganglia (for example, with the caudate), which in turn projects to the globus pallidus, which projects to the thalamus, which projects back to the frontal cortex.[79] The dorsolateral frontal lobes also have extensive connections with the nonspecific intralaminar nuclei of the thalamus (centromedian and parafascicularis). These intralaminar nuclei may gate motor activation by their influence on the basal ganglia, especially the putamen, or by influencing the thalamic portion of motor circuits (ventralis lateralis pars oralis). Lastly, the dorsolateral frontal lobe has strong input into the premotor areas. The observation that lesions of the dorsolateral frontal lobe, the cingulate gyrus, the basal ganglia, the intralaminar nuclei, and the ventrolateral thalamus may all cause akinesia supports the postulate that this system mediates motor activation.

The right hemisphere appears to play a special role in motor activation or intention. Coslett and Heilman[80] demonstrated that right hemisphere lesions are more likely to be associated with contralateral akinesia than those of the left hemisphere. Howes and Boller[81] measured reaction times (a measure of the time taken to initiate a response) of the hand ipsilateral to a hemispheric lesion and demonstrated that right hemisphere lesions were associated with slower reaction times than left hemisphere lesions. However, as previously discussed, this finding may be related to the important role of the right hemisphere in mediating arousal. Heilman and Van Den Abell[82] measured the reduction of reaction times of normal subjects who received warning stimuli directed to either their right or left hemisphere. They found that, independent of the hand used, warning stimuli delivered to the right hemisphere reduced reaction time to midline stimuli more than did warning stimuli delivered to the left hemisphere.

Some emotions are associated with approach behaviors, others with avoidance behaviors. Unfortunately, the portions of the brain that mediate approach and avoidance behaviors have not been entirely elucidated. Denny-Brown and Chambers[83] suggested that the frontal lobes mediate avoidance behaviors and the parietal lobes mediate approach behaviors. These authors also suggested that approach and avoidance behaviors may be reciprocal, so that a loss of one behavior may release the other behavior.[83] Therefore, because the frontal lobes mediate avoidance behavior, frontal lobe lesions would cause inappropriate approach behaviors, and because the parietal lobes mediate approach behaviors, parietal lesions would induce avoidance. In support of this postulate, one can see patients with frontal lesions who demonstrate a variety of approach behaviors, including manual grasp reflexes, visual grasp reflexes, rooting and sucking responses, magnetic apraxia, utilization behaviors, and defective response inhibition. Unfortunately, which specific area or areas within the frontal lobes, when damaged, cause approach behaviors has not been entirely elucidated. Animals with frontal lesions show an increase of aggressive behavior. Patients with left dorsolateral frontal lesions (which should induce an emotion of negative valence, increased arousal, and approach behaviors) are also prone to hostility and anger.[84] Patients with orbitofrontal lesions are prone to irritability, anger, inappropriate sexual advances, and euphoria-jocularity,[85] and studies of animals with ventral frontal lesions also demonstrate abnormal approach behaviors. The orbitofrontal cortex is closely connected with areas of the limbic system such as the amygdala that are known to be important in emotional experience.

Denny-Brown and Chambers[83] demonstrated that, in contrast to the manual grasp response associated with frontal lesions, patients with parietal lesions may demonstrate avoiding responses. Patients with parietal lesions, especially of the right side, may not only fail to move or show a delay in moving their arms, heads, and eyes toward a part of the space that is opposite the parietal lesion, but these patients may even deviate their eyes, head, and arms toward ipsilateral hemispace. In addition, unlike patients with frontal lesions who cannot withhold their response to stimuli, patients with parietal lesions may be unable to respond to stimuli (neglect). These avoidance responses are more severe with right than left hemisphere lesions.

SUMMARY

In addition to reviewing some of the former feedback and central theories of emotional experience, I have discussed a central modular theory. Although I argue against the postulate that feedback is critical to the experience of emotions, I do suspect that feedback may influence emotions. Emotions may be conditioned and may use thalamic limbic circuits, as proposed by LeDoux. However, most emotional behaviors and experiences are induced by complex stimuli that an isolated thalamus could not interpret. The cerebral cortex of humans has complex modular systems that analyze stimuli, develop percepts, and interpret meaning.

We have proposed that the experience of emotions is dimensional. Almost all primary emotions can be described with two or three factors, including valence, arousal, and motor activation. The determination of valence is based on whether the stimulus is beneficial (positive) or detrimental (negative) to a person's well-being. Whereas the right frontal lobe and its subcortical connections appear to be important in the mediation of emotions with negative valence, the left frontal lobe and its subcortical connections may be important in the mediation of emotions with positive valence. Depending on the nature of the stimulus, some positive and some negative emotions (such as joy and fear) are associated with high arousal, and others (such as satisfaction and sadness) are associated with low arousal. Whereas the right parietal lobe appears to be important in mediating arousal response, the left hemisphere appears to inhibit the arousal response. Some positive and negative emotions (such as anger, fear, and joy) are associated with motor activation, and others (such as sadness) are not. The right frontal lobe appears to be important in motor activation. The motor activation associated with emotions may be either approach or avoidance behaviors. Approach behaviors may be mediated by the parietal lobes; avoidance behaviors may be mediated by the frontal lobes.

The cortical areas we have discussed have rich interconnections. In addition, these neocortical areas have rich connections with the limbic system, basal ganglia, thalamus, and reticular system. Therefore, the anatomic modules that mediate valence, arousal, and activation systems are richly interconnected and form a modular network. Emotional experience depends on the patterns of neural activation of this modular network.

REFERENCES

1. Heilman KM: Emotion and the brain: a distributed modular network mediating emotional experience, in Neuropsychology, edited by Zeidel D. San Diego, CA, Academic Press, 1994, pp 139–158
2. Darwin C: The Expression of Emotion in Man and Animals. London, Murray, 1872
3. Izard CE: Human Emotions. New York, Plenum, 1977
4. Ekman P, Sorenson ER, Freisen WV: Pancultural elements in facial displays of emotions. Science 1969; 164:86–88
5. Tomkins SS: Affect, Imagery, Consciousness, vol 1: The Positive Affects. New York, Springer, 1962
6. Tomkins SS: Affect, Imagery, Consciousness, vol 2: The Negative Affects. New York, Springer, 1963
7. Laird JD: Self-attribution of emotion: the effects of expressive behavior on the quality of emotional experience. J Pers Soc Psychol 1974; 29:475–486
8. Poeck K: Pathophysiology of emotional disorders associated with brain damage, in Handbook of Neurology, vol 3, edited by Vinken PJ, Bruyn GW. New York, Elsevier, 1969
9. Sackeim HA, Greenberg MS, Weiman AL, et al: Hemispheric asymmetry in the expression of positive and negative emotions: neurologic evidence. Arch Neurol 1982; 39:210–218
10. James W: The Principles of Psychology, vol 2 (1890). New York, Dover, 1950
11. Cannon WB: The James-Lange theory of emotion: a critical examination and an alternative theory. Am J Psychol 1927; 39:106–124
12. Hohmann G: Some effects of spinal cord lesions on experimental emotional feelings. Psychophysiology 1966; 3:143–156
13. Katkin ES, Morrell MA, Goldband S, et al: Individual differences in heartbeat discrimination. Psychophysiology 1982; 19:160–166
14. Hantas M, Katkin ES, Blascovich J: Relationship between heartbeat discrimination and subjective experience of affective state (abstract). Psychophysiology 1982; 19:563
15. Ax AF: The physiological differentiation between fear and anger in humans. Psychosom Med 1953; 15:433–442
16. Heilman KM, Bowers D, Valenstein E: Emotional disorders associated with neurological disease, in Clinical Neuropsychology, 3rd edition, edited by Heilman KM, Valenstein E. New York, Oxford University Press, 1993, pp 461–497
17. Luria AR, Simernitskaya EG: Interhemispheric relations and the functions of the minor hemisphere. Neuropsychologia 1977; 15:175–178
18. Davidson RJ, Horowitz ME, Schwartz GE, et al: Lateral differences in the latency between finger tapping and heart beat. Psychophysiology 1981; 18:36–41
19. Marañon G: Contribution à l'étude de l'action emotive de l'adrenaline [Contribution to the study of the emotional action of adrenalin]. Revue Française d'Endocrinologie 1924; 2:301–325
20. Schacter S, Singer JE: Cognitive, social, and physiological determinants of emotional state. Psychol Rev 1962; 69:379–399
21. Slomaine BS: Hemispheric differences in emotional psychophysiology. Doctoral dissertation, University of Florida, 1995
22. LeDoux JE, Cicchetti P, Xagoraris A, et al: The lateral amygdaloid nucleus: sensory interface of the amygdala in fear conditioning. J Neurosci 1990; 10:1062–1069
23. Ross ED, Homan RW, Buck R: Differential hemispheric lateralization of primary and social emotions. Neuropsychiatry, Neuropsychology, and Behavioral Neurology 1994; 7:1–13
24. Wundt W: Grundriss der Psychologie [Outline of psychology]. Stuttgart, Engelmann, 1903
25. Osgood C, Suci G, Tannenbaum P: The Measure of Meaning. Urbana, IL, University of Illinois, 1957
26. Frijda NH: Emotion, cognitive structure, and action tendency. Cognition and Emotion 1987; 1:115–143
27. Greenwald MK, Cook EW, Lang PJ: Affective judgment and psychophysiological response: dimensional co-variation in the evolution of pictorial stimuli. J Psychophysiol 1989; 3:51–64
28. Goldstein K: Language and Language Disturbances. New York, Grune and Stratton, 1948
29. Gainotti G: Emotional behavior and hemispheric side of lesion. Cortex 1972; 8:41–55
30. Babinski J: Contribution à l'étude des troubles mentaux dans l'hemiplegie organique cérébrale (anosognosie) [Contribution to the study of mental problems in organic cerebral hemiplegia (anosognosia)]. Rev Neurol 1914; 27:845–848
31. Hécaen H, Ajuriagerra J, de Massonet J: Les troubles visuoconstuctifs par lésion parieto-occipitale droit [Visuoconstructional problems from right parieto-occipital lesions]. Encephale 1951; 40:122–179
32. Denny-Brown D, Meyers JS, Horenstein S: The significance of perceptual rivalry resulting from parietal lesions. Brain 1952; 75:434–471
33. Terzian H: Behavioral and EEG effects of intracarotid sodium amytal injections. Acta Neurochirurgica 1964; 12:230–240
34. Rossi GS, Rosadini G: Experimental analysis of cerebral dominance in man, in Brain Mechanisms Underlying Speech and Language, edited by Millikan C, Darley FL. New York, Grune and Stratton, 1967
35. Benson DF (ed): Psychiatric aspects of aphasia, in Aphasia, Alexia, and Agraphia. New York, Churchill Livingstone, 1979
36. Robinson RG, Sztela B: Mood change following left hemisphere brain injury. Ann Neurol 1981; 9:447–453
37. Tucker DM, Watson RT, Heilman KM: Affective discrimination and evocation in patients with right parietal disease. Neurology 1977; 17:947–950
38. Ross ED, Mesulam MM: Dominant language functions of the right hemisphere? Prosody and emotional gesturing. Arch Neurol 1979; 36:144–148
39. Gasparrini WG, Satz P, Heilman KM, et al: Hemispheric asymmetries of affective processing as determined by the Minnesota Multiphasic Personality Inventory. J Neurol Neurosurg Psychiatry 1978; 41:470–473
40. Starkstein SE, Robinson RG, Price TR: Comparison of cortical and subcortical lesions in the production of poststroke mood disorders. Brain 1987; 110:1045–1059
41. House A, Dennis M, Warlow C, et al: Mood disorders after stroke and their relation to lesion location. Brain 1990; 113:1113–1129
42. Milner B: Hemispheric specialization: scope and limits, in The Neurosciences: Third Study Program, edited by Schmitt FO, Worden FG. Cambridge, MA, MIT Press, 1974
43. Bench CJ, Friston KJ, Brown RG, et al: The anatomy of melancholia: focal abnormalities of blood flow in major depression. Psychol Med 1992; 22:607–615
44. Phelps ME, Mazziotta JC, Baxter L, et al: Positron emission tomographic study of affective disorders: problems and strategies. Ann Neurol 1984; 15:S149–S156
45. Drevets WC, Raichle ME: Neuroanatomic circuits in depression. Psychopharmacol Bull 1992; 28:261–274
46. Davidson RJ, Schwartz GE, Saron C, et al: Frontal versus parietal EEG asymmetry during positive and negative affect. Psychophysiology 1979; 16:202–203
47. Tucker DM: Lateral brain function, emotion and conceptualization. Psychol Bull 1981; 89:19–46
48. Fox NA, Davidson RJ (eds): Hemispheric substrates for affect: a developmental model, in The Psychobiology of Affective Development. Hillsdale, NJ, Lawrence Erlbaum, 1984
49. Tucker DM, Williamson PA: Asymmetric neural control in human self-regulation. Psychol Rev 1984; 91:185–215

50. Mayberg HS, Robinson RG, Wong DF, et al: PET imaging of cortical S$_2$-serotonin receptors following stroke: lateralized changes and relationship to depression. Am J Psychiatry 1988; 145:937–943
51. Heilman KM: Neglect and related disorders, in Clinical Neuropsychology, edited by Heilman KM, Valenstein E. New York, Oxford University Press, 1979, pp 268–307
52. Watson RT, Valenstein E, Heilman KM: Thalamic neglect: the possible role of the medical thalamus and nucleus reticularis thalami in behavior. Arch Neurol 1981; 38:501–507
53. Mesulam MM: A cortical network for directed attention and unilateral neglect. Ann Neurol 1981; 10:309–325
54. Heilman KM, Watson RT, Valenstein E: Neglect and related disorders, in Clinical Neuropsychology, 3rd edition, edited by Heilman KM, Valenstein E. New York, Oxford University Press, 1993, pp 279–386
55. Critchley M: The Parietal Lobes. New York, Hafner, 1966
56. Heilman KM, Valenstein E, Watson RT: Localization of neglect, in Localization in Neurology, edited by Kertesz A. New York, Academic Press, 1983, pp 471–492
57. Heilman KM, Pandya DN, Geschwind N: Trimodal inattention following parietal lobe ablations. Transactions of the American Neurological Association 1970; 95:259–261
58. Lynch JC: The functional organization of posterior parietal association cortex. Behav Brain Sci 1980; 3:485–534
59. Bushnell MC, Goldberg ME, Robinson DL: Behavioral enhancement of visual responses in monkey cerebral cortex, I: modulation of posterior parietal cortex related to selected visual attention. J Neurophysiol 1981; 46:755–772
60. Pandya DM, Kuypers HGJM: Cortico-cortical connections in the rhesus monkey. Brain Res 1969; 13:13–36
61. Mesulam MM, Van Hesen GW, Pandya DN, et al: Limbic and sensory connections of the inferior parietal lobule (area PG) in the rhesus monkey: a study with a new method for horseradish peroxidase histochemistry. Brain Res 1977; 136:393–414
62. Damasio AR, Anderson SW: The frontal lobes, in Clinical Neuropsychology, 3rd edition, edited by Heilman KM, Valenstein E. New York, Oxford University Press, 1993, pp 410–460
63. Stuss DT, Benson DF: The Frontal Lobes. New York, Raven, 1986
64. Moruzzi G, Magoun HW: Brainstem reticular formation and activation of the EEG. Electroencephalogr Clin Neurophysiol 1949; 1:455–473
65. Watson RT, Heilman KM, Miller BD, et al: Neglect after mesencephalic reticular formation lesions. Neurology 1974; 24:294–298
66. Segundo JP, Naguet R, Buser P: Effects of cortical stimulation on electrocortical activity in monkeys. Neurophysiology 1955; 18:236–245
67. Shute CCD, Lewis PR: The ascending cholinergic reticular system, neocortical olfactory and subcortical projections. Brain 1967; 90:497–520
68. Sato H, Hata Y, Hagihara K, et al: Effects of cholinergic depletion on neuron activities in the cat visual cortex. J Neurophysiol 1987; 58:781–794
69. Steriade M, Glenn L: Neocortical and caudate projections of intralaminar thalamic neurons and their synaptic excitation from the midbrain reticular core. J Neurophysiol 1982; 48:352–370
70. Scheibel ME, Scheibel AB: The organization of the nucleus reticularis thalami: a Golgi study. Brain Res 1966; 1:43–62
71. Heilman KM, Schwartz H, Watson RT: Hypoarousal in patients with the neglect syndrome and emotional indifference. Neurology 1978; 28:229–232
72. Morrow L, Vrtunski PB, Kim Y, et al: Arousal responses to emotional stimuli and laterality of lesions. Neuropsychologia 1981; 19:65–71
73. Schrandt NJ, Tranel D, Damasio H: The effects of total cerebral lesions on skin conductance response to signal stimuli (abstract). Neurology 1989; 39:223
74. Yokoyama K, Jennings R, Ackles P, et al: Lack of heart rate changes during an attention-demanding task after right hemisphere lesions. Neurology 1987; 37:624–630
75. Perani D, Vallar G, Paulesu E, et al: Left and right hemisphere contributions to recovery from neglect after right hemisphere damage. Neuropsychologia 1993; 31:115–125
76. Pribram KH, McGuiness D: Arousal, activation and effort in the control of attention. Psychol Rev 1975; 182:116–149
77. Watson RT, Miller BD, Heilman KM: Nonsensory neglect. Ann Neurol 1978; 3:505–508
78. Goldberg ME, Bushnell BC: Behavioral enhancement of visual responses in monkey cerebral cortex, II: modulation in frontal eye fields specifically to related saccades. J Neurophysiol 1981; 46:773–787
79. Alexander GE, DeLong MR, Strick PL: Parallel organization of functionally segregated circuits linking basal ganglia and cortex. Annu Rev Neurosci 1986; 9:357–381
80. Coslett HB, Heilman KM: Hemihypokinesia after right hemisphere strokes. Brain Cogn 1989; 9:267–278
81. Howes D, Boller F: Evidence for focal impairment from lesions of the right hemisphere. Brain 1975; 98:317–332
82. Heilman KM, Van Den Abell T: Right hemispheric dominance for mediating cerebral activation. Neuropsychologia 1979; 17:315–321
83. Denny-Brown D, Chambers RA: The parietal lobe and behavior. Research Publications, Associations for Research in Nervous and Mental Disease 1958; 36:35–117
84. Grafman J, Vance SC, Weingartner H, et al: The effects of lateralized frontal lesions on mood regulation. Brain 1986; 109:1127–1140
85. Hornak J, Rolls ET, Wade D: Face and voice expression identification in patients with emotional and behavioral changes following ventral frontal damage. Neuropsychologia 1996; 34:247–261

10

The Neurobiology of Recovered Memory

Stuart M. Zola, Ph.D.

During the last two decades, two separate but related developments have led to new insights about memory and about how memory is organized in the brain.

One development has involved the systematic study of memory and memory distortion by cognitive and experimental psychologists. This work addresses questions about whether and how one's memories for events change over time. Interest in this area has been stimulated in part by the increasing number of case reports in which adults appear to uncover long-forgotten memories of sexual abuse and trauma. This phenomenon has been variously referred to as "decades-delayed discovery," "recovery of repressed memories," the "false memory syndrome," and "recovered memory." In this review, I will use the term "recovered memory" to refer to this phenomenon. Recovered memories have been reported to emerge with considerable detail as long as 10, 20, or even 40 years after the alleged abuse or trauma. Individuals identified as the abusers, usually parents or other family members, frequently deny the episodes and say that they have been wrongfully accused; that is, they claim that the recovered memories of abuse are false memories. The recovered memory phenomenon has engendered considerable controversy and debate within the field of psychiatry. Because there is often no physical evidence and no other corroborating evidence to support the claim of abuse, a major part of the controversy has to do with determining the credibility of recovered memories.

The second development that has led to new insights about memory comes from research in neurobiology and neuroscience. New findings from several levels of analysis of brain function have provided insights about how memory is organized in the brain. These have included evidence for ongoing change, reorganization, and plasticity in the brain at the level of cells and synapses; insights about how neural representations of memories in the brain change with time; and other facts, including the discovery that there is more than one memory system in the brain.

This chapter will take into account both of these developments in exploring the issue of recovered memory. It is important at the outset that I make two points. First,

---

The author thanks Larry R. Squire for his helpful comments. Preparation of this manuscript was supported by grants from the Medical Research Service of the Department of Veterans Affairs and the National Institutes of Health. This chapter is based on a presentation at a meeting of the American Neuropsychiatric Association and the Behavioral Neurology Society, October 12–15, 1995.

I am not a clinician or a therapist. I am a neuroscientist who studies how the brain works with regard to memory function. I do not treat patients who have recovered memories of abuse. However, as a result of my interest in the issue of recovered memory, I have had the opportunity to meet with patients who have recovered memories of abuse, with therapists involved in treating patients who have recovered memories of abuse, and with many accused family members. As a result, I have come to understand the enormous impact that recovered memories of abuse can have on patients, therapists, family members, and society as a whole. However, as I maintain below, acceptance at face value of the credibility of all recovered memories is, to my mind, incompatible with the behavioral and biological facts and ideas about memory derived from scientific research. Second, the discussion that follows is not meant to imply that childhood sexual abuse does not occur, or that in a particular case traumatic events did not happen. Indeed, a tragic consequence of the recovered memory controversy is that the existence of false memories in some cases could cast doubt on the validity of true recollections in other cases.[1,2]

My intent in this chapter is to show that many biological and behavioral facts and ideas about memory are relevant to the issue of recovered memory, and that the credibility of recovered memories will be usefully served when considered in the light of these facts and ideas. In the sections that follow, I first review a paradigmatic case of recovered memory, the well-publicized case of Eileen Franklin, and underscore several characteristics that the case shares with other cases of recovered memory. I then describe recent findings and ideas about how memory is organized in the brain. Next, I turn to behavioral evidence related to two questions about memory that have become a focus of debate in the recovered memory controversy: whether memories for traumatic events change over time, and whether memories can be created for traumatic events that did not actually happen. Finally, I discuss the Eileen Franklin case as well as other cases of recovered memory in light of information about the biological and behavioral bases of memory.

Therapists and lawyers involved in cases of recovered memory face the dilemma of having to distinguish between *narrative truth* (how much a person believes that what he or she is reporting is true) and *historical truth* (how accurate the report is, whether the event actually happened, and whether it happened in the way it was described).[3] In the context of the present discussion, the question can be reformulated as follows: Is it possible that an event that is narrated to a therapist with considerable detail, with strongly felt emotion, with a high level of confidence, and with a strong belief by the individual holding the memory that it is real and accurate, nevertheless did not happen historically the way it was described, or might not have happened at all?

THE CASE OF EILEEN FRANKLIN

Ten weeks after 9-year-old Susan Nason disappeared from her home in Foster City, California, in September 1969 her body was found in a ravine off a highway not far from her home. The crime remained unsolved for 20 years. Then, in November 1989, a 35-year-old woman named Eileen Franklin told prosecutors in San Mateo County, California, that she had recovered a long-repressed memory of having been present at the scene of the crime. Moreover, Eileen Franklin identified her own father as the person who had first molested and then murdered her childhood friend.[4,5]

Based on Eileen Franklin's information, prosecutors charged her father, George Franklin, with first-degree murder. A jury became convinced of the veracity of Eileen Franklin's story and concluded that she knew facts that only a witness to the crime could have known. Throughout the trial, George Franklin proclaimed his innocence, but in 1990 he was convicted of first-degree murder and sentenced to a life term in prison. A United States district judge recently overturned the conviction, ruling that George Franklin's rights were violated during the trial (for reasons described later in this chapter). In the summer of 1996, the San Mateo County district attorney decided not to retry the case, and George Franklin was released from custody.

This case is especially informative for three reasons. Reason 1: This case has key aspects that characterize many other cases of recovered memory. The essence of many cases of recovered memory is a claim by the victim (or in this case, an alleged eyewitness) that certain things were done, and a claim by the alleged perpetrator that these things were not done. As is also true for many cases of recovered memory of abuse, Eileen Franklin's story of the murder was not corroborated by any physical evidence or other eyewitness evidence that placed her or her father at the scene of the crime. Thus, the degree to which the narrative truth and the historical truth overlap is unknowable, and the outcome of the case depended, for the most part, on the statements of the two adversaries. Reason 2: The process of a jury trial necessarily made public two additional pieces of information that are ordinarily not available in discussions of recovered memory: how Eileen Franklin came to recover her memory of the episode, and what aspects of her testimony convinced the jurors that her memory for the events of the murder were accurate. Reason 3: The evidence at the murder scene was well documented by the police and by the media at the time of discovery 20 years earlier, and this documentation provided a way of evaluating the accuracy of Eileen Franklin's recollections. I will return to these points in the last section of this chapter.

HOW MEMORY IS ORGANIZED IN THE BRAIN

The Static View of Memory

There are two popular views about how the brain stores memories. The first view, referred to here as the *static view of memory*, is that the brain stores the memory of an event like a time capsule or a videotape. When the event is later reflected upon, whether it is a few minutes later or even 20 or 30 years later, all the components of the recalled event are potentially available to recollection, unchanged from their original occurrence, as would be the case if the time capsule were opened or the videotape simply replayed. If this were the case, then recall of events would typically be quite accurate, and the likelihood of generating or recalling a false memory would be rather small.

The static view of memory is a popularly held view. One's own experience of recollecting some apparently forgotten detail from the remote past reinforces the idea that memory can remain stored, unchanged, for long periods of time. Seventeen years ago Elizabeth and Geoffrey Loftus[6] polled 169 lay people and persons with graduate training in psychology, asking them to choose which of the following two statements matched their view of how memory worked:

Statement 1: Everything we learn is permanently stored in the mind, although sometimes particular details are not accessible. With hypnosis, or other special techniques, these inaccessible details could eventually be recovered.

Statement 2: Some details that we learn may be permanently lost from memory. Such details would never be able to be recovered by hypnosis, or any other special techniques, because these details are simply no longer there.

The majority of individuals chose Statement 1 (84% of the psychology trainees and 69% of the lay people). In a recent informal survey (Zola and Squire, unpublished data), 645 health care professionals (including nurses, social workers, and psychologists) attending workshops on memory and the brain were polled using the same two questions. Nearly two-thirds of the respondents (62%) chose Statement 1.

Additional information about the widespread appeal of the static view of memory comes from work by the clinical psychologist Michael Yapko, who gathered information from 864 therapists around the country about their ideas and practices regarding the role of suggestion and memory in therapy.[7] One issue that Yapko's survey addressed directly was whether or not therapists view memory as objective and infallible. Approximately 41% of all respondents believed (strongly or slightly) that "early memories, even from the first year of life, are accurately stored and retrievable." About 24% of respondents agreed (strongly or slightly) with the statement, "One's level of certainty about a memory is strongly positively correlated with that memory's accuracy." Yapko additionally obtained information about beliefs in the use of hypnosis and its relationship to memory. Approximately 31% of respondents agreed (strongly or slightly) with the statement, "When someone has a memory of a trauma while in hypnosis, it objectively must have actually occurred." And about 54% of respondents agreed (strongly or slightly) with the statement, "Hypnosis can be used to recover memories of actual events as far back as birth." Thus, many lay people and practicing therapists share a belief in the static view of memory, the belief that everything is retained and that forgetting simply reflects a loss in accessibility.

The Dynamic View of Memory

An alternative view, referred to here as the *dynamic view of memory*, is that memory storage in the brain changes over time, and that memory storage can also be affected by such things as the intrusion of new events and by the act of retrieval itself. In this view, when an earlier event is recalled (whether it is an event from yesterday or an event from 20 years ago), some components of the recalled event may be different than the original occurrence, and some of the information may have been entirely lost. Thus, the recollection of an earlier event could be incomplete and it could contain distortions or false memories, or the recollection for the whole event could be a false memory.

It turns out that biological facts about how memory is organized in the brain, views about brain function derived from connectionist models, and experimental and behavioral studies of how memory operates all support the dynamic view of memory. Biological-minded scientists of today seem to have accepted this view. In 1996, at a scientific meeting about the nervous system, 48 scientists with advanced degrees in biology or neuroscience were asked to choose between the two statements that appear above.[8] Eighty-seven percent of them chose

Statement 2, the statement that implies that memory changes with the passage of time and that some aspects of memory may be permanently lost.

Biologically based information so far available on this point comes mostly from studies in animals with simple nervous systems.[9,10] The results of these studies suggest that forgetting, whether it occurs in a few hours or many days, reflects in part an actual loss of information from storage. Moreover, there is evidence that the neural connectivity and patterns of neural activity that originally represented the stored information change as time passes. For example, in *Aplysia*, long-term facilitation of the gill-withdrawal reflex involves the growth of new synaptic terminals; that is, there are morphological changes of the synapses that contribute to the storage of information. The decay of these structural changes parallels the decay of behavioral memory.[9,10]

Studies in animals with more complex nervous systems, together with studies in humans, have shown that the brain is organized in a highly specialized way. That is, different regions of neocortex are specialized for processing and storing particular kinds of information: for example, visual cortex for visual stimuli, auditory cortex for sounds. This circumstance has led to the idea that memory for an event is represented in the brain in a "distributed" fashion; that is, bits and pieces of an experience are parceled out to different regions of the brain. Memories of the visual components of events settle into the visual cortex, and memories of sounds settle into the auditory cortex. In this sense, ensembles of neurons in multiple, specialized brain regions each represent a little bit of the memory. All of the scattered fragments remain physically linked, however. A region of the brain thought to be responsible for assembling and binding the distributed components of a memory is the medial temporal lobe.[11] This region of the brain encompasses several areas, including the hippocampal region and the entorhinal, perirhinal, and parahippocampal cortices.[12] The medial temporal lobe region binds the disparate aspects of a memory from the separate specialized regions distributed in the cortex and helps make them into a cohesive whole again at the time of recollection.

Recent ideas from the fields of cognitive psychology and computational neuroscience are consistent with the dynamic view of memory storage. Moreover, these ideas argue against the static view of memory. For example, cognitive psychologist Daniel Schacter[13] has argued that the very nature of distributed representations precludes any simple notion of a stored snapshot of an event: memories are stored as patterns of activation across numerous units and connections that are involved with the storage of many different memories.

Consistent with this idea are suggestions from work on parallel distributed processing models. For example, McClelland[14] has suggested that remembering involves a system of interconnected processing units. The units that participate in representing one episode or event also participate in representing other episodes. Accordingly, the output that a connectionist model produces as a memory of a particular event may be affected by many other experiences. These ideas are meaningful to the discussion of recovered memories because they correspond to our personal experiences with regard to recalling past events, in which we sometimes inadvertently combine details from different episodes.

It is useful to describe two additional points in the context of the dynamic view of memory storage. The first is that the role in memory of the medial temporal lobe structures is temporally limited. Eventually, remembering a particular event or experience becomes possible without the participation of the medial temporal lobe structures. That is, during a process of consolidation, which can be quite lengthy, the distributed components of a memory are bound together by processes that no longer include the medial temporal lobe structures and that presumably take place in neocortex.[11,12] The process of consolidation further underscores the idea that memory representations in the nervous system change and become reorganized with the passage of time.

The second point is that new information about how the brain may represent events has been uncovered by recent work using functional brain imaging techniques (both positron emission tomography and functional magnetic resonance imaging). The findings suggest that visual imagery relies on activation of some of the same neural substrates in visual cortex as the perceptual images generated during normal perception.[15] A similar picture appears to hold for the motor system, where the same regions of the supplementary motor area were activated when human subjects imagined a motor movement and when they actually carried out the movement.[16]

The precise relationship between the neural activation that accompanies imagery and the neural activation that accompanies real perception and actual motor behavior is not well understood. However, the fact that similar brain areas can sometimes be activated for real and for imagined events suggests the possibility that, at the time of recollection, memory for an imagined event could become confused with memory of a real event. This point is relevant to the issue of recovered memory because when patients initially report no actual memory of abuse or of a traumatic event, they are sometimes encouraged to imagine what it would have been like.[1,7]

Before we discuss behavioral evidence concerned

with the issue of memory distortion, three additional facts should be considered with respect to the biology of memory: the phenomenon of source amnesia, the phenomenon of infantile amnesia, and the fact that there is more than one kind of memory.

Source Amnesia

In everyday remembering, one can often recall a fact or an idea but cannot recall the source of the information (where or when the fact was learned). This phenomenon, referred to as *source amnesia*,[17,18] is commonly observed in normal adults when newly learned information is tested after a delay of several weeks.[19] Additionally, source memory and source amnesia have been linked to the frontal lobes.[18,20] Lesions that are restricted to the frontal lobes do not ordinarily produce a severe amnesic syndrome of the kind that is observed after medial temporal lobe damage, but they can impair recollection for temporal order and related aspects of memory.[20] The frontal lobes are slow to mature and they are especially vulnerable to the effects of normal aging. Correspondingly, source memory errors are common in children[21] and in the elderly.[22,23]

Source amnesia is relevant to the issue of recovered memory because when people fail to recollect the source of their knowledge they become susceptible to various kinds of memory distortions. For example, they confuse real and imagined events, or they fail to remember whether something actually happened or was only suggested. Different memories can become confused because people forget the sources of their memories.[24] Source amnesia is observed more frequently after long delays from the time of learning than after short delays from the time of learning,[17,25] and memory distortions that are attributable to source amnesia are observed most readily after long delays.[13,21,26]

Infantile Amnesia

This phenomenon refers to the normal inability to consciously recollect events or experiences from the first 2 to 3 years of our lives.[8] One explanation for this phenomenon is that the cortical areas that eventually become the repository for permanent memory storage appear to mature slowly after birth. During the first 2 to 3 years they are undergoing a process of maturation, during which they may not be available for processing and storing the kind of information needed for reconstruction of conscious memories of events at a later time.[8]

Infantile amnesia provides a limiting factor on the issue of recovered memories. Specifically, it would be unusual, perhaps even impossible, to personally recollect, especially in detail, an event that had occurred prior to the age of two. Yet recovered memories are sometimes described from before the age of two, and often with rich detail.[27]

Multiple Memory Systems

One of the most penetrating insights about memory to emerge in the past decade is that there is more than one kind of memory and that different kinds of memory depend on different brain systems.[12,28] In particular, considerable research has centered on the distinction between declarative or explicit memory and nondeclarative or implicit memory.[28,29] The fact that there is more than one kind of memory means that memory is not a unitary entity and that it is necessary to qualify statements about memory in the context of discussions of "recovered memory."[13]

Declarative memory refers to the capacity for having conscious recollections of recently occurring facts and events. In contrast, nondeclarative memory refers to a heterogeneous collection of abilities, all of which are independent of the structures damaged in amnesia. Nondeclarative memory is nonconscious. Information is acquired as changes within perceptual or response systems, or as changes encapsulated as habits, skills, or conditioned responses, and these changes are expressed through performance without any sense of memory being involved—without any sense of "pastness," of having learned the activity or practiced it before.[19]

The distinction between declarative and nondeclarative memory has become important to discussions of the recovered memory issue. Specifically, it has been supposed by some researchers that nondeclarative memory (for example, priming or perceptual fluency, or emotional feelings or sensations) might facilitate declarative memory by providing a cue for its retrieval.[30-32] In this view, declarative memory of a traumatic event may be stored so weakly that the person has no conscious memory of the original painful event. The same event might also be captured by the nondeclarative system through physiological sensations or gestures. Perhaps later, the nondeclarative system might provide clues such as physical sensations that might help stir the recall of the weak declarative memories.[30]

An alternative view is that although nonconscious memory processes do have the potential to evoke conscious mental states, one must then interpret any conscious mental content that is produced. It is then a separate question whether the interpretation refers to a declarative memory that is correctly associated with the sensation or emotional response and whether the memory is accurate or inaccurate.[19] A recent study asked whether in ordinary recognition memory tasks the recognition responses (declarative memory for words) were supported in part by priming (nondeclarative memory). The results showed that priming did not cause words to be consciously recollected.[33]

Organization of Memory in the Brain: Summary

There is substantial evidence that the memory systems of the brain are organized in ways that can be characterized as dynamic—that is, that neural representations of events are continually being modified and reorganized with time. In addition, there is evidence that at least some brain regions represent real and imagined events by similar patterns of neural activity. Other important biological features of memory and memory impairment—source amnesia, infantile amnesia, and the fact that there is more than one kind of memory system in the brain—also affect how memory may be expressed at the time of recollection. This dynamic view of memory, rooted in biology, suggests that there ought to be behavioral examples in which the accuracy of memory is found to be poor, or in which there is confusion between different actual events, or in which there is confusion between real and imagined events. It turns out that there are many points of contact between pertinent behavioral findings and what I have described as the dynamic, biological basis of memory.

BEHAVIORAL EVIDENCE CONCERNING TRAUMATIC MEMORY

In this section, I discuss behavioral evidence related to two questions about memory that have become a focus of debate in the recovered memory controversy: whether memories for traumatic events change over time, and whether memories can be created for traumatic events that did not actually happen.

Do Memories for Traumatic Events Change With Time?

A distinction is often made by therapists, patients, and some researchers between the processes that govern how ordinary memories change with the passage of time and the processes that govern how traumatic memories change with the passage of time. This distinction is based on the idea that traumatic memories are protected from forgetting because they have different operating rules than ordinary memory.[32,34]

Although it seems clear that emotional memories are often well retained, there is little basis for supposing that they are less susceptible to distortion than are nonemotional memories. Moreover, there is evidence that when a memory is vivid and detailed, it need not necessarily be accurate, and this evidence holds even in the case of emotionally traumatic memories.[13,21,35-37]

For ethical reasons, it has not been possible to evaluate systematically the hypothesis that emotionally traumatic memories are relatively immune to decay and distortion. Nevertheless, there is useful anecdotal evidence that addresses this issue,[1,21] and some systematic experimental evidence has begun to accumulate as well.[36,38] I next describe examples from both of these categories of evidence.

Anecdotal Evidence: In 1967, Tony Conigliaro, who was a star hitter and fielder with the Boston Red Sox, had his career prematurely ended when he was hit in the head with a fastball by California Angels pitcher Jack Hamilton. Although Conigliaro survived the injury, it impaired his vision and he could no longer play baseball. On the occasion of Conigliaro's death in 1990, *New York Times* sports writer Dave Anderson interviewed Jack Hamilton, the pitcher whose fastball had nearly killed Conigliaro in an evening game at Boston's Fenway Park.[21,39] Hamilton expressed regret over the incident:

> I know in my heart I wasn't trying to hit him. . . . It was like the sixth inning when it happened. I think the score was 2–1, and he [Conigliaro] was the eighth hitter in their batting order. . . . I had no reason to throw at him. . . . I tried to go see him in the hospital late that afternoon or early that evening but they were just letting his family in.[21,39]

Hamilton is remembered for hitting Conigliaro. Much of what Hamilton remembers, however, is in error. It was not the sixth inning, it was only the fourth inning, and Conigliaro was batting sixth, not eighth. Moreover, Conigliaro was a strong hitter. He already had 20 home runs and 67 runs batted in that season, so it would not be surprising for Hamilton to have tried to brush him back from the plate.[21] In addition, because it was an evening game, Hamilton actually did not visit the hospital until the following day. Thus, despite Hamilton's claim that the event was emotionally traumatic for him and that he had often relived the event in his mind, Hamilton's memory had become distorted over the years, even for key aspects of the event (such as the hospital visit). Although anecdotal, this example is interesting because, unlike in many cases of recovered memory, there was an accurate record available of the earlier events against which Hamilton's recollections could be compared.

Experimental Evidence: Flashbulb memories are said to be associated with the occurrence of an event that is so surprising or emotionally intense that it is described as leaving an indelible picture in the mind that does not change with time. A common example of a flashbulb event is the assassination of President John F. Kennedy. People who were old enough to be emotionally affected by the event often claim, more than 30 years later, to remember in full detail the circumstances under which they learned about the tragedy. Questions about the accuracy of flashbulb memories have rarely been asked because we take their accuracy for granted; we do this partly because, in the absence of independent records, it is hard to resist the confidence of the subjects.[36,38] However, there is some evidence that personal consequentiality of events can affect the accuracy of recollections. For example, the circumstances surrounding Margaret Thatcher's resignation were remembered better by those who rated it as important than by those who did not.[38]

The question of whether flashbulb memories are immune to forgetting has begun to be examined systematically.[36,38] For example, consider the following written descriptions of two memories. These two recollections were actual written responses to the question, "How did you first hear of the news of the Challenger disaster?" The event, which was one of the most catastrophic in the history of the United States space program, was seen live on television by millions of people on the morning of January 28, 1986.[36]

> *Description 1:* When I first heard about the explosion I was sitting in my freshman dorm room with my roommate and we were watching TV. It came on a news flash and we were both totally shocked. I was really upset and I went upstairs to talk to a friend of mine and then I called my parents.

> *Description 2:* I was in my religion class and some people walked in and started talking about it. I didn't know any details except that it had exploded and the schoolteacher's students had all been watching, which I thought was so sad. Then after class I went to my room and watched the TV program talking about it and I got all the details from that.

Forty-four students at Emory University in Atlanta, Georgia, were asked on the morning after the event to recall the event, and they were asked again in the fall of 1988, 2½ years later.[36] Description 1 was given by a student 2½ years after the event. Description 2 comes from the same student and is her answer to the same question 24 hours after the explosion. About one-third of the students in this study showed substantial changes in their recollection of the event. Moreover, the overall correlation between accuracy (how closely the second description matched the first description) and level of confidence ratings obtained at the 2½-year timepoint was 0.29, a nonsignificant correlation indicating little relationship between the two variables. In particular, there were many cases in which students rated their confidence in their descriptions as quite high, even though the accuracy of their descriptions was quite low.

Additional evidence from flashbulb events comes from descriptions by military personnel recorded during or shortly after their firsthand experiences of the attack on Pearl Harbor. In some cases, their original accounts were found to differ considerably from their subsequent memoirs written many years later. (For a discussion of these accounts, see Loftus.[1]) This is an important point because these observations challenge the view that memories for events that have actually happened to a person (in the case of Pearl Harbor) are not as susceptible to distortion as events a person has simply witnessed on television (as in the case of the Challenger disaster).[40]

The findings described thus far make two important points. First, memories for traumatic events personally experienced (the anecdotal evidence) and memories for flashbulb events, whether experienced as an observer or personally experienced, do appear to change with the passage of time. Second, confidence may not always serve as a reliable measure of accuracy.

Can False Traumatic Memories Be Created?

Direct experimental demonstrations of whether people can falsely create an entire history of traumatic sexual abuse when none occurred are precluded because of ethical considerations. Nevertheless, researchers have successfully developed several useful paradigms, within the bounds of ethical considerations, in which subjects have been induced to create false memories of emotionally laden events. When hypnotized subjects were regressed to a "past life" and then given suggestions that they were abused in their past life, many of them subsequently "remembered" the abuse.[13,41] In addition, there is both anecdotal and experimental evidence relevant to this question. Examples from both categories will be described next.

Anecdotal Evidence: One of the best-known examples comes from developmental psychologist Jean Piaget.[42] This is the episode of his childhood memory of an attempted kidnapping:

One of my first memories would date, if it were true, from my second year. I can still see most clearly the following scene, in which I believed until I was about fifteen. I was sitting in my pram, which my nurse was pushing in the Champs Elysées, when a man tried to kidnap me. I was held in by the strap fastened around me while my nurse bravely tried to stand between me and the thief. She received various scratches and I can still see those vaguely on her face.... When I was about fifteen, my parents received a letter from my former nurse.... She wanted to confess to her past faults, and in particular to return the watch she had been given as a reward.... She had made up the whole story.... I therefore must have heard as a child the account of the story, which my parents believed, and projected it into the past in the form of a visual memory. (p. 117)

Experimental Evidence: Recent experimental work has involved attempts by researchers to create in subjects memories of full events, and in some cases events that could be considered traumatic. In a recent study by Loftus and Coan,[43] adults were asked about childhood events; one of these was an event that had never occurred, namely getting lost in a shopping mall. The subjects were "reminded" of the event by other family members during the first interview session. On subsequent interviews, 6 of the 24 individuals in the study (25%) "remembered" the false event as a real episode and provided additional details about the episode.

In an extension of this research, Hyman et al.[44] undertook a study in which college students were asked about various childhood events that never happened, including an overnight hospitalization for an ear infection and an episode in which they released the parking brake on a car after being left alone in it. At the end of the first interview, in which no subjects "recalled" the false events, subjects were encouraged to continue thinking about the events, to try to remember more information about them before the next interview, and to not discuss the events with their parents. During the second interview, 1 to 7 days after the first, 4 of the 20 subjects remembered detailed information about the false events.

In a clever follow-up study using the same paradigm, Hyman and his colleagues[44] attempted to inject false memories that were alleged to have happened at one of three ages: when the subjects were 2, 6, or 10 years of age. No differences were found in the rate of false recollections based on the supposed age of the subject when the event happened. Despite the phenomenon of infantile amnesia described earlier—the normal inability to consciously recollect events or experiences from the first 2 to 3 years of life—subjects in this study were just as likely to endorse a false memory from 2 years of age as a false memory from 10 years of age. This finding is important for the discussion of recovered memories because it demonstrates that, despite the limiting biological factor of infantile amnesia, false memories for events purported to have taken place during infancy can be created in some individuals.

In related work, Ceci and colleagues[21,45] have compellingly shown that some children of preschool age also can be convinced that somewhat traumatic events have happened to them even though these events did not actually occur. In one study,[45] on 12 separate occasions spaced approximately 1 week apart, 40 preschool children were individually interviewed by a trained adult who read the child a set of events, including actual events that the child had experienced (such as birthday parties) and fictitious events that the child never experienced. The fictitious events included negative events (such as falling off a bicycle and getting stitches in the leg), positive events (such as going for a ride in a hot air balloon with classmates), and neutral events (such as waiting for a bus).

At each session, the child was told that all the things on the list had happened to him or her. The interviewer read the list of events, and for each event the child was asked first to visualize the event and then to recollect it. ("After you make a picture of it in your head, and think real hard about each thing for a minute, I want you to tell me if you can remember it or not, OK?")[45] Ceci et al. found that a surprisingly high percentage of children (approximately 34% of 3- to 4-year-olds and 29% of 5- to 6-year-olds) assented to the fictitious events on the very first interview. Moreover, as the children participated in more interviews the overall assent rate grew higher: approximately 45% of the younger children and 40% of the older children assented by the twelfth session. The pattern of results was the same when data for the negative fictitious events were considered separately (assent rate for younger children: 17% for the first session, 28% for the second session; assent rate for older children: 10% for the first session, 23% for the twelfth session). Psychologists who specialized in interviewing children were shown videotaped interviews of the children describing both false events and actual events. The psychologists were not reliably better than chance at detecting which events were real.[45]

SUMMARY

In the preceding sections I have attempted to provide insights into the malleable nature of memory. If one emphasizes the dynamic aspects of how memory is organized in the brain—including the idea that memory is physically distributed, that a process of consolidation continues with the passage of time, that different structures are critical for memory at different points in time from the occurrence of an event, and that some brain regions are involved in representing both real and imagined events—then one begins to provide a biological basis for the behavioral phenomena of memory and memory distortion, and for why memories can be endorsed as real even though they are false.

The behavioral evidence, both anecdotal and experimental, suggests that memories for personally experienced traumatic events can be altered by new experiences. Moreover, research has suggested that entire events that never happened can enter into memory, and that individuals might not always be able to distinguish between a real memory and a false memory. Even when memories are vivid and subjectively compelling, there is still no guarantee that they are accurate. Memory function is imperfect; it is subject to error and distortion and to dissociation between confidence and accuracy. Moreover, memory distortions may increase in severity as the time between an actual event and the recall of the event increases, and individuals in a wide range of age categories, from preschool to adult, are susceptible to memory distortion.

THE EILEEN FRANKLIN CASE REVISITED

I now return briefly to the Eileen Franklin case. In her own account, Eileen Franklin claimed that her repressed memory for the event surfaced after her daughter, who was then just about Susan Nason's age, had glanced at her with a certain facial expression that triggered memories of Susan Nason at the murder scene.[4] However, there are several different versions of how Eileen Franklin actually came to recover her memory. One version in particular is germane to this discussion because it describes a process shared with many other cases of recovered memory. Testimony by Eileen Franklin's therapist, Kirk Barrett, indicated that her memories developed during the course of therapy that involved visualization and hypnosis. According to this account, Eileen Franklin initially visualized the crime scene in June 1989, and eventually visualized the murder but could not make out who the murderer was. According to Barrett, it required several additional sessions before Eileen Franklin was able to visualize that the murderer was her father.[5]

Post-trial discussion with jurors indicated that many had concluded that the only way Eileen Franklin could have known the facts to which she testified was by being an eyewitness. This impression by the jurors was based in part on several features of Eileen Franklin's testimony.[5] First, she provided considerable detailed information that, according to the investigating detectives, was consistent with the facts of the case. For example, she described how a silver ring (which was found crushed on Susan Nason's right hand at the crime scene) had been crushed as Susan Nason tried to block her face from the impending blow of a large rock used as the murder weapon by George Franklin. Second, Eileen Franklin described events with appropriate affect during her testimony. For example, at one point in the questioning process she became very excited and childlike in describing how, while driving in the van with her father, she unexpectedly saw her friend Susan Nason across the street and convinced her father to pick her up. Dramatic pauses and heavy emotions then punctuated her testimony as she described the van pulling over in a wooded area. Finally, she answered key questions with confidence and firmness, such that the jury had the impression that she was sure of her statements.

As it turned out, everything that Eileen Franklin testified to was either unverifiable (such as comments made by Susan Nason at the scene) or available to her through means other than having personally experienced the event. The killing of Susan Nason was a significant event for the town, and it received substantial news coverage on television and in the print media. All of the facts described by Eileen Franklin had appeared in the early news accounts. However, the jurors were not allowed to read the newspaper articles about the murder to determine whether Franklin could have used them to piece together what she knew about the crime.

There was also evidence that Eileen Franklin knew the reported facts well enough a few years after the crime to describe some of the very same details to her sister that she described years later to the jury as part of her recovered memory.[2] Significantly, some of the reports that appeared in the newspaper at the time of the incident contained factual inaccuracies, and Eileen Franklin's first statements, when she spoke 20 years later to the detectives who were assigned the case, contained the same errors. For example, Susan Nason was wearing two rings when her body was discovered, a silver Indian brocade on her right hand (which was found crushed on her finger) and a gold ring with a topaz stone on her left hand. A local newspaper confused the rings, reporting that the silver ring had the stone. In her initial statements to the detectives, Eileen Franklin recalled how her father had crushed Susan Nason's head and that in the process, he crushed a ring containing a small stone on Susan Nason's right hand.[5]

CONCLUSIONS

The biological and behavioral findings that I have described cannot help us to know with certainty where the narrative truth ends and the historical truth begins in the Eileen Franklin case or in other individual cases or claims of recovered memory. But the facts about the way memory works suggest that misremembering and retrospective reworking and reconstruction of the past might not be uncommon. Accordingly, it seems unlikely that there could be perfectly detailed memory, flawlessly preserved through decades, sitting unchanged as if in a time capsule. The dynamic change and reorganization of memory in the brain with the passage of time provide a biological basis for errors in retrieving information about the past. To further clarify the issue of recovered memory, it will be important to determine precisely how such biological processes are related to cognitive and behavioral evidence of accuracy and distortion in human memory.

REFERENCES

1. Loftus EF: The reality of repressed memories. Am Psychol 1993; 48:518–537
2. Schacter DL (ed): Memory Distortion: How Minds, Brains, and Societies Reconstruct the Past. Cambridge, MA, Harvard University Press, 1995
3. Gutheil T: Therapists leave their chair and don the mantle of investigator. Psychiatric Annals 1993; 23:527–531
4. Franklin E, Wright W: Sins of the Father. New York, Fawcett Crest, 1991
5. MacLean HN: Once Upon a Time: A True Story of Memory, Murder, and the Law. New York, Harper Collins, 1993
6. Loftus EF, Loftus GR: On the permanence of stored information in the human brain. Am Psychol 1980; 35:49–72
7. Yapko M: Suggestions of Abuse. New York, Simon and Schuster, 1994
8. Squire LR: Memory. Encyclopedia Americana (in press)
9. Bailey CH, Chen M: Time course of structural changes at identified sensory neuron synapses during long-term sensitization in *Aplysia*. J Neurosci 1989; 9:1774–1780
10. Bailey CH, Kandel ER: Structural changes accompanying memory storage. Annu Rev Physiol 1993; 55:397–426
11. Alvarez P, Squire LR: Memory consolidation and the medial temporal lobe: a simple network model. Proc Natl Acad Sci USA 1994; 91:7041–7045
12. Zola-Morgan S, Squire LR: Neuroanatomy of amnesia. Annu Rev Neurosci 1993; 16:547–563
13. Schacter DL: Memory distortion: history and current status, in Memory Distortion: How Minds, Brains, and Societies Reconstruct the Past, edited by Schacter DL. Cambridge, MA, Harvard University Press, 1995, pp 1–43
14. McClelland JL: Constructive memory and memory distortions: a parallel-distributed processing approach, in Memory Distortion: How Minds, Brains, and Societies Reconstruct the Past, edited by Schacter DL. Cambridge, MA, Harvard University Press, 1995, pp 69–90
15. Kosslyn SM: Image and Brain. Cambridge, MA, Harvard University Press, 1994
16. Jeanerod M: Visual imagery: metric properties of a scene or object are retained in mental representations. Behav Brain Sci 1994; 17:187–245
17. Schacter DL, Harbluk JL, MacLachlan DR: Retrieval without recollection: an experimental analysis of source amnesia. Journal of Verbal Learning and Verbal Behavior 1984; 23:593–611
18. Shimamura AP, Squire LR: A neuropsychological study of fact memory and source amnesia. J Exp Psychol Learn Mem Cogn 1987; 13:464–473
19. Squire LR: Biological foundations of accuracy and inaccuracy in memory, in Memory Distortion: How Minds, Brains, and Societies Reconstruct the Past, edited by Schacter DL. Cambridge, MA, Harvard University Press, 1995, pp 197–225
20. Janowsky JS, Shimamura AP, Squire LR: Source memory impairment in patients with frontal lobe lesions. Neuropsychologia 1989; 27:1043–1056
21. Ceci SJ: False beliefs: some developmental and clinical considerations, in Memory Distortion: How Minds, Brains, and Societies Reconstruct the Past, edited by Schacter DL. Cambridge, MA, Harvard University Press, 1995, pp 91–125
22. Craik FIM, Morris LW, Morris RG, et al: Relations between source amnesia and frontal lobe functioning in older adults. Psychol Aging 1990; 5:148–151
23. Schacter DL, Kaszniak AK, Kihlstrom JF, et al: The relation between source memory and aging. Psychol Aging 1991; 6:559–568
24. Lindsay DS, Johnson MK: The eyewitness suggestibility effect and memory for source. Memory and Cognition 1989; 17:349–358
25. Shimamura AP, Squire LR: The relationship between fact and source memory: findings from amnesic patients and normal subjects. Psychobiology 1991; 19:1–10
26. Lindsay DS: Misleading suggestions can impair eyewitnesses' ability to remember event details. J Exp Psychol Learn Mem Cogn 1990; 16:1077–1083
27. Loftus E, Ketcham K: The Myth of Repressed Memory: False Memories and Allegations of Sexual Abuse. New York, St. Martin's, 1994
28. Squire LR: Mechanisms of memory. Science 1986; 232:1612–1619
29. Schacter DL: Implicit memory: history and current status. J Exp Psychol Learn Mem Cogn 1987; 13:501–518
30. Kandel M, Kandel ER: Flights of memory. Discover 1994; 15:32–38
31. Jacoby L, Whitehouse K: An illusion of memory: false recognition influenced by unconscious perception. J Exp Psychol Gen 1989; 118:126–135
32. van der Kolk BA: The body keeps the score: memory and the evolving psychobiology of PTSD. Harvard Review of Psychiatry 1994; 1:253–265
33. Haist F, Shimamura AP, Squire LR: On the relationship between recall and recognition memory. J Exp Psychol Learn Mem Cogn 1992; 18:492–508
34. Terr L: Unchained Memories. New York, Basic Books, 1994
35. Ofshe R, Watters E: Making monsters: false memories, psychotherapy, and sexual hysteria. New York, Charles Scribner's Sons, 1994
36. Neisser U, Harsch N: Phantom flashbulbs: false recollections of hearing the news about *Challenger*, in Affect and Accuracy in Recall: Studies of Flashbulb Memories, edited by Winograd E, Neisser U. Cambridge, MA, Cambridge University Press, 1992, pp 9–31
37. Spiegel D: Hypnosis and suggestion, in Memory Distortion: How Minds, Brains, and Societies Reconstruct the Past, edited by Schacter DL. Cambridge, MA, Harvard University Press, 1995, pp 129–149
38. Conway MA, Anderson SJ, Larsen SF, et al: The formation of flashbulb memories. Memory and Cognition 1994; 22:326–343
39. Anderson D: New York Times, February 27, 1990, p B-9
40. Yuille JC: We must study forensic eyewitnesses to know about them. Am Psychol 1993; 48:572–573
41. Spanos N, Menary E, Gabora N, et al: Secondary identity enactments

during hypnotic past-life regression: a sociocognitive perspective. J Pers Soc Psychol 1991; 61:308–320
42. Piaget J: The child's construction of reality. London, Routledge and Kegan Paul, 1954
43. Loftus EF, Coan D: The construction of childhood memories, in The Child Witness in Context: Cognitive, Social, and Legal Perspectives, edited by Peters D. New York, Kluwer (in press)
44. Hyman IE, Husband TH, Billings FJ: False memories of childhood experiences. Applied Cognitive Psychology 1995; 9:181–197
45. Ceci SJ, Loftus E, Leichtman M, et al: The possible role of source misattributions in the creation of false beliefs among preschoolers. Int J Clin Exp Hypn 1994; 47:304–320

11

The Medial Temporal Lobe in Schizophrenia

Steven E. Arnold, M.D.

The temporal lobe has been a region of special interest in schizophrenia ever since the concept of the illness first crystallized. Kraepelin[1] speculated that the thought disorder and hallucinations of dementia praecox were due to irritative damage of the temporal lobes. As research in schizophrenia ensued in the early and middle twentieth century, data continued to accumulate in support of the role of the temporal lobe and the strongly interconnected limbic subcortical and frontal brain regions as the neuroanatomic substrates of schizophrenia. A number of investigators highlighted the similar clinical profiles between schizophrenia and known organic schizophrenia-like psychoses due to temporal lobe lesions, including those caused by temporal lobe epilepsy, herpetic encephalitides, temporal lobe tumors, temporal lobe cerebrovascular lesions, and neurodegenerative diseases such as Alzheimer's and Pick's diseases, which have a particular predilection for the temporal lobe.[2-6] Primitive imaging measures also implicated the temporal lobes with pneumoencephalographic evidence of enlarged temporal horns in schizophrenia,[7] and EEG studies described frequent temporal lobe abnormalities.[8]

With these early findings setting the stage, advances in cellular and molecular neurobiology, as well as the advent of modern in vivo neuroimaging modalities, ushered in a new era of intensive examination of the temporal lobe in schizophrenia. Indeed, the medial temporal lobe has been the most frequently studied brain region in both clinical and postmortem neurobiological studies of schizophrenia.[9,10] This review will focus on the medial temporal lobe, including the hippocampus, amygdala, and entorhinal and posterior parahippocampal cortices. These regions are components of a highly integrated neural system with strong connections to other parts of the limbic system as well as association cortices in all four lobes. By dint of these extensive connections, medial temporal lobe structures play critical roles in a variety of higher cognitive and emotional functions. The medial temporal lobe is especially closely interconnected with the prefrontal cortices, the diencephalon, and the basal ganglia. There is substantial evidence that these regions also show abnormalities in structure and function in schizophrenia, and several recent reviews have highlighted their involvement in the illness.[9-12]

Work from the authors' laboratories described here has been supported by grants from the National Institute of Mental Health, the National Alliance for Research in Schizophrenia and Depression, and the Stanley Foundation.

NEUROPSYCHOLOGICAL STUDIES

Neuropsychology associates particular domains of cognition and emotion with their neuroanatomical substrates. The neural systems supporting different aspects of cognition have been well specified after decades of observations in patient populations with frank structural lesions.[13] In schizophrenia, neuropsychological instruments have been used extensively to delineate which neural systems are dysfunctional. Although there has been substantial controversy over the years, evidence has accumulated that within the context of generalized deficits, there are also differential deficits in memory that are attributable to medial temporal lobe dysfunction. In examining the neuropsychological profile of schizophrenia with a battery of standardized tests administered to unmedicated patients and well-matched control subjects, Saykin et al.[14,15] found that the patients showed generalized impairments in all major cognitive domains (executive, attentional, language, sensorimotor, and memory) relative to control subjects. However, patients with schizophrenia had more severe deficits in verbal learning, verbal memory, and visual memory, and these findings were interpreted as indicating selective temporohippocampal region involvement in schizophrenia. Follow-up studies from this group have found these deficits to remain stable over 2 to 4 years even after the more florid psychotic symptoms attenuate with treatment.[16] Similarly, Gruzelier et al.[17] administered a more focused battery to putatively test the integrity of frontohippocampal and temporohippocampal functions in schizophrenia, affective disorders, and normal control samples. Again, patients with schizophrenia showed generalized deficits, but the data were especially notable for these patients' frequent and severe deficits in memory-related tasks involving the hippocampus.

There appear to be both similarities and differences between the memory impairment in schizophrenia and that of neurologic patients with medial temporal lobe lesions and focal amnesia.[18-20] Like focal neurologic amnesia patients, schizophrenic patients show impaired episodic memory but intact procedural ("motor") learning. Unlike focal amnesia patients, however, schizophrenic patients have been found to have deficits in working memory, in addition to episodic memory. Working memory has been more closely associated with prefrontal neocortices and is usually spared in medial temporal lobe amnesia. These findings, in addition to still other cognitive deficits, raise the issue of the anatomic specificity of the memory impairment in schizophrenia[21] and indicate that the pathophysiological substrates of schizophrenia also extend beyond the hippocampal region to involve other brain regions—most notably, the frontal lobes.

NEUROIMAGING

As with the neuropsychological studies of schizophrenia, structural and functional neuroimaging studies have described generalized and diverse, as well as specific, abnormalities in patients with schizophrenia compared with control subjects. Again, the structures in the medial temporal lobe stand out as being particularly affected.

Structural Neuroimaging

Early structural neuroimaging studies with X-ray computed tomography and magnetic resonance imaging described generalized brain abnormalities in many patients with schizophrenia.[22-24] These included ventricular enlargement and increased sulcal cerebrospinal fluid volumes. Advances in MRI technology and resolution have brought increased ability to measure individual anatomic structures. Quantitative high-resolution MRI studies have identified relatively selective volumetric deficits in frontal and temporal lobes, and—even more selectively—the hippocampus, amygdala, and parahippocampal gyrus; however, some controversy exists, with other investigators reporting that the changes are less selective.[25-31]

To better understand the relationship of volumetric measures to schizophrenia, a few studies have attempted to correlate structure size with clinical symptoms. In a study of monozygotic twins discordant for schizophrenia, Goldberg et al.[32] reported that left anterior hippocampal volume correlated with verbal memory scores. Shenton and colleagues[29,33] reported that there were significant associations between poor scores on tests of verbal memory, abstraction, and categorization and reduced volume in temporal lobe structures, including the parahippocampal gyrus and posterior superior temporal gyrus. Bogerts et al.[34] found significantly reduced medial temporal lobe tissue volumes that inversely correlated with severity of psychotic symptoms but not with negative symptoms. In contrast, although they did not isolate the medial temporal lobe structures, Turetsky et al.[35] found that left temporal volume reduction correlated with severity of negative symptoms in schizophrenia. Another study found that selective decreases in the volume in the anterior medial temporal lobe were associated with lower scores on neuropsychological measures of executive and motor functions usually considered sensitive to the integrity of frontal lobe systems. Measures of other neuropsychological functions and of global intellectual ability were not related to medial temporal lobe volumes.[36] This finding is especially interesting vis-à-vis hippocampal neurodevelopmental connectivity hypotheses of schizophrenia, in which a primary abnormality in the hippocampal region occurring early in brain development is proposed to alter the maturation and function of other brain regions with which it is strongly connected.[37–39] It is obvious that structure-function relationships are complex in schizophrenia and further study is much needed.

Functional Neuroimaging

Regional cerebral blood flow (rCBF), single-photon emission computed tomography (SPECT), and positron emission tomography (PET) studies make use of radioactive compounds to image cerebral blood flow, glucose metabolism, or neurotransmitter receptor binding in schizophrenia either at rest or during activation with particular cognitive or emotional tasks. Tamminga et al.[40] found that fluorodeoxyglucose metabolic activity was diminished in the hippocampus and anterior cingulate, but not in basal ganglia or neocortices, in schizophrenic subjects compared with control subjects. In a complex canonical analysis of summary psychopathology scores and regional cerebral blood flow using PET, Friston et al.[41] found abnormalities, and the strongest relationships between brain and behavioral measures, in the left parahippocampal region in individuals with schizophrenia. Other regions, including left superior temporal, insular, cingulate cortices and parietal association cortices, striatum, and midbrain, also showed correlations, but these were less robust. In subtyping symptom profiles of schizophrenic patients into particular syndromes, Schroeder et al.[42] found that the delusional syndrome was associated with decreased glucose utilization in hippocampus as well as medial frontal regions. Finally, Silbersweig et al.[43] have found that a small number of schizophrenic patients showed abnormal activations in the hippocampus and other limbic and paralimbic regions while experiencing classic auditory hallucinations.

Activation studies using xenon-133 rCBF or SPECT scanning of patients while they are performing cognitive tasks have also had success in delineating abnormalities in regional brain metabolic activity in schizophrenia. Most have focused on the response of the frontal lobes to putative frontal activation tasks.[44,45] Interestingly, reduced medial temporal lobe volumes on MRI have been found to correlate with dorsolateral prefrontal rCBF activation during frontal activation cognitive tasks[46]—again indicating a relationship between dysfunction in an interconnected frontal brain region and a structural abnormality in the medial temporal lobe.

Functional magnetic resonance imaging (fMRI) is another relatively new, noninvasive technique to measure brain function by utilizing changes in blood oxygenation in the brain to map cerebral activation.[47] It is a promising imaging modality in schizophrenia because of its ease of use and superior spatial and temporal resolution compared with PET. fMRI has already been used to identify abnormal medial temporal lobe activity with a memory activation task in schizophrenic patients in a preliminary study.[48]

An in vivo imaging technique that gives information about the biochemical status of particular brain regions is proton magnetic resonance spectroscopy. In schizophrenia, this has been used to measure various neuronal constituents, neurotransmitters, and metabolites. There has been much recent interest in *N*-acetyl aspartate (NAA); the function of this putatively neuron-specific chemical is not fully understood, but it is considered to be a marker of neuron vitality. Several groups have reported significant decreases in NAA in medial temporal lobe structures[49–51] early in the course of the illness.[52] Studies are just beginning to correlate findings with symptoms ratings.

Another functional imaging modality that implicates medial temporal lobe structures in schizophrenia is electrophysiology. An increased frequency of abnormal EEG recordings with nonspecific findings (such as slowing or sharp waves) localized to the temporal lobes has been recognized in schizophrenia for decades.[8,53] In the 1950s, Heath[54] and Sem-Jacobsen et al.[55] found that patients with schizophrenia frequently exhibited abnormal paroxysmal and spike activity in the temporal lobes (and to a lesser degree in the frontal and parietal lobes) on depth electrode recordings, even when the routine scalp EEG was normal. It was further noted that the abnormal activity correlated with hallucinatory and delusional behaviors.

Of more interest in recent years have been event-related potentials (ERPs, also known as evoked potentials), which average the scalp EEG responses to auditory, visual, or somatosensory stimulation. The amplitudes and latencies of a number of middle and late components of the ERP waveform reflect higher order information processing and are seen to be altered in schizophrenia.[56] In particular, the P300 component has been found to be smaller in numerous studies of schizophrenia.[57,58] This component is thought to be generated by the hippocampus, inferior parietal and frontal cortices, and thalamus. In addition, earlier ERP components, including mismatch negativity (MMN) and the N200 component (which themselves are composites of several underlying waveforms that are largely generated by medial temporal lobe, frontotemporal, and temporoparietal association cortices), have a lower amplitude in schizophrenia.[57,59] In developing the importance of ERPs in schizophrenia vis-à-vis temporal lobe pathology, Egan et al.[59] found that there were significant correlations between the hippocampal area on MRI scan and the N200 and P300 potentials and that auditory P300 amplitude inversely correlated with the positive symptoms of schizophrenia. These findings were interpreted as reflecting a relationship between temporal lobe pathology and these ERP components in schizophrenia.

POSTMORTEM STUDIES

For over a century, clinicians and investigators have recognized schizophrenia as a brain disease, and for almost as long, neuropathologists have sought to define the neuropathology of the disorder. Early efforts identified various abnormalities such as cell loss in particular cortical layers, disorganization of neuronal processes, and occasional gliosis.[60] However, the field was fraught with inconsistencies that we now realize were due to diagnostic uncertainties of the time as well as poorly controlled tissue handling procedures and the limitations of the histological stains then in use. With modern diagnostic standards and the growing array of quantitative neuroanatomic and molecular biological methods, postmortem research now has a much greater potential for identifying abnormalities in schizophrenia at the cellular and molecular levels, and numerous leads have been proffered.

The first step in defining a neuropathology of schizophrenia is to determine the frequency of recognized pathologic entities in the brains of patients with schizophrenia. Several extensive surveys of multiple regions in the brains of patients with schizophrenia have found only nonspecific or assorted identifiable neuropathologic abnormalities in some cases, even in elderly, severely ill patients.[61–64] Given the lack of obvious neuropathological lesions in the brains of most patients with schizophrenia, more experimental approaches have been applied to delineate subtle anatomic, cellular, and molecular changes. The medial temporal lobe has been a central focus of many of these studies. In reviewing these studies, we will divide them into several broad categories of interest: morphometry, cytoarchitecture, synapse and connectional neuroanatomy, neurochemistry and neuroreceptors, and those that examine neurodevelopmental and neurodegenerative markers. Although abnormal findings have been described in each of these categories, considerable controversy exists with some, and others await confirmation.

Macroscopic and Microscopic Morphometric Changes in the Medial Temporal Lobe

Planimetric and volumetric procedures similar to those used in in vivo neuroimaging have been applied to postmortem schizophrenia specimens. It is interesting that whereas accumulating evidence from in vivo MRI findings suggests that medial temporal lobe structures are smaller in schizophrenic patients compared with control subjects, findings are less consistent in postmortem studies. Decreases in hippocampal area and volume, parahippocampal area and volume, or parahippocampal gyrus cortical thickness have been reported.[65–71] However, other studies have not been able to confirm some of these findings.[61,67,70,72,73] To date, only two studies have examined amygdala size, and no differences were seen between schizophrenic and control subjects.[72,74]

Neuron number, density, and size in the medial temporal lobes of patients with schizophrenia have been the focus of a number of investigations. Decreases in neuron density have been observed in the hippocampus and entorhinal cortex,[68,69,75,76] although other investigators have reported no change in neuron density in these areas.[73,77–79] Only one study has estimated the number of neurons in the amygdala, and no difference from control subjects was observed in the basolateral nucleus.[74] In considering these data, we must appreciate that determining neuron number accurately is methodologically arduous. Nonetheless, when we view these microscopic morphometric neuron density data in the light of macroscopic postmortem and in vivo neuroimaging data indicating smaller medial temporal lobe structures, it appears likely that patients with schizophrenia do have some reduction in the normal complement of neurons within the medial temporal lobe. Additional studies employing modern stereologic counting methods will be helpful in confirming this.

The size of neurons has also been of interest in microscopic morphometric analyses. Benes et al.[78] examined the hippocampal subfields CA1–CA4 and found a decrease in mean neuron size; the greatest difference from control subjects was in CA1. Arnold et al.[79] also assessed neuron size in CA1–CA4, as well as the subiculum, three layers of the entorhinal cortex, and two neocortical "control" regions. We found that neurons were significantly smaller in the subiculum, CA1, layer II of the entorhinal cortex, and, to a lesser degree, in the other hippocampal region subfields, whereas no size differences were seen in the primary visual and motor neocortical control regions. This finding indicates some degree of specificity of the abnormalities for the hippocampal region, although a preliminary report indicated that neuron size may be smaller in dorsolateral prefrontal cortices as well.[80]

The neurobiological basis for smaller neurons in schizophrenia is uncertain. The size of neuronal somata throughout the brain is determined by the neuronal cytoskeletal protein elements that give rise to the cells' morphology, and size also generally correlates with the lengths of axons and extent of dendritic arbors that must be metabolically supported by those cells. Alterations in these components have been reported and are described below.

Cytoarchitectural Changes in the Medial Temporal Lobe

Cytoarchitecture refers to the spatial arrangement of neurons within a given region of the brain. Although relatively few in number, descriptions of cytoarchitectural abnormalities in the hippocampal region in schizophrenia have been very influential in discussions of the neurodevelopmental aspects of the disorder. Scheibel and Kovelman[77,81] reported that the angles of orientation of pyramidal cells in the rostral and mid-hippocampus were altered in schizophrenia. This cell disarray was most prominent at the interfaces between subfields, especially CA1/prosubiculum and CA1/CA2, and was worse in more severely ill patients. They proposed that such a pattern would most likely arise during the period of cell migration in embryogenesis. Although this group later confirmed their initial findings in another sample using a new measuring technique,[82] other investigators have had difficulty confirming the findings using the same and different methods.[78,79,83,84]

The entorhinal cortex has also been a focus of cytoarchitectural studies. Jakob and Beckmann[85–87] first described cytoarchitectural abnormalities in up to two-thirds of a series of 66 patients with schizophrenia; abnormalities consisted of a disordered arrangement of neurons in layer II, apparent heterotopic displacement of layer II–type neurons into layer III, and poorly developed layers III and IV. Arnold et al.[37] later described assorted cytoarchitectural abnormalities in the rostral portion of the entorhinal cortex in all 6 brains of a series of carefully screened cases from the Yakovlev Collection. These included bizarre invaginations of the normally smooth entorhinal cortical surface, a decrease in the density of neurons, apparent heterotopic positioning of layer II–type neurons in layer III, and attenuation of deeper layers. These findings were interpreted as most likely due to abnormal neuron migration during brain development. Furthermore, because of the pivotal role the entorhinal cortex plays in neural systems mediating cognition,[13,88] it was proposed that a defect in this region would have far-reaching neuropsychological consequences that could be part of the clinical profile of schizophrenia.

More recently, investigators have tried to apply more rigorous quantitative methods to assess cytoarchitecture in the entorhinal cortex in schizophrenia. This effort is especially important given the qualitative nature of the earlier descriptions. Using complex spatial point pattern analyses, our group has found an abnormal spatial arrangement of neurons in the rostral entorhinal cortex: indices of dispersion of neurons and effective radii surrounding neurons were significantly different in superficial layers in schizophrenic subjects compared with normal subjects.[89] These findings were concordant with those of the previous qualitative studies. In another preliminary study using different methods, Krimer et al.[76] noted decreases in mean neuronal densities in the entorhinal cortices of schizophrenic subjects, but no significant differences in the volumes or positions of individual layers relative to the pial surface.

The pathophysiological basis for the cytoarchitectural abnormalities described in schizophrenia can be the subject of much debate. Cytoarchitecture is determined by a complex orchestration of developmental and degenerative neurobiological processes,[90–92] and a defect in any one or more of these processes could affect the final anatomy. Migration, neuronal enlargement and differentiation under the influence of intrinsic and extrinsic trophic factors, programmed cell death, age- and disease-related cell death, and the amount of space and composition of the neuropil surrounding neurons all determine the cytoarchitectural appearance at the time of postmortem examination. Which of these factors are responsible for the differences evident in at least some cases of schizophrenia is not yet known, and they represent potentially fertile territories for research.

Synaptic and Connectional Neuroanatomy

It has been hypothesized that unusual thought processes and cognitive deficits in schizophrenia may arise from aberrant patterns of neuroanatomical connections, or "miswiring," in the brain.[93–96] Studies of axons, dendritic arbors and spines, synapses, and synapse-related proteins have been conducted to explore this hypothesis. Most have been conducted in frontal cortices (see review by Arnold and Trojanowski[10]), although several reports have focused on the medial temporal lobe. As with most postmortem research in schizophrenia, results are far from conclusive at this point.

To explore possible abnormalities of synapse in the hippocampal region in schizophrenia, Eastwood et al.[97,98] used in situ hybridization and immunoautoradiography to assess synapse-related proteins in the hippocampus in schizophrenia. They reported that both messenger ribonucleic acid (mRNA) and protein levels for synaptophysin, an important synaptic vesicle protein, are decreased in various hippocampal and parahippocampal subfields in schizophrenic subjects. Since synaptophysin is considered a general marker of synapses, decreases in its expression were interpreted as reflecting an overall decrease in synaptic density in the medial temporal lobe in schizophrenia. Another study found no change in synaptophysin immunoreactivity but did find a decrease in synapsin, another synaptic vesicle protein, in the hippocampus.[99] Alternative strategies have been employed in other brain regions to assess if there are synaptic changes. Preliminary Golgi studies have found normal dendritic arborization but a decrease in spine density numbers in the prefrontal cortex,[100,101] and one ultrastructural case study found synaptic vesicle and synaptic active zone changes in the temporal pole.[102] Whether the changes in synapse-related proteins reported in the medial temporal lobe reflect a change in synaptic density, a down- or up-regulation of their expression, or alterations in their metabolic turnover awaits clarification.

In a preliminary study assessing innervation of the entorhinal cortex, Longson et al.[103] recently reported an increased density of small-caliber glutamatergic vertical axons in a small cohort of schizophrenic subjects compared with control subjects. On the basis of analogy to known connectional neuroanatomic circuits in nonhuman species, the authors proposed that increases in glutamatergic axons from the amygdala may be responsible for these changes.

Neurochemical and Receptor Changes in the Medial Temporal Lobe

Since the late 1970s, there has been considerable interest in possible derangements of neurotransmitter systems in schizophrenia, and various neurochemical levels, receptors, and uptake sites have been studied (see reviews[9,104–107]). Most investigations of monoamines have been conducted in regions other than the medial temporal lobe. Studies of the medial temporal lobe have reported increased dopamine in the left amygdala[108] and normal dopamine levels in the hippocampus.[109]

More recently, a preliminary study found that there was a decrease in the density of tyrosine hydroxylase immunoreactive axons of the entorhinal cortex in schizophrenia.[110] Serotonin (5-HT) and 5-HIAA levels were found to be lower in the hippocampus in schizophrenia;[109] subsequent studies found $5-HT_{1A}$ and $5-HT_2$ receptors to be elevated, with no change in the number of 5-HT uptake sites.[111] Norepinephrine levels have been reported as normal in the hippocampus,[109] but an altered pattern of noradrenergic β_2 receptor expression and a lack of the normal β_2 receptor asymmetry in the hippocampus has been noted.[112]

A theoretically encompassing model of schizophrenia that is gaining increasing attention is the excitatory amino acid receptor dysfunction hypothesis.[113–115] It is postulated in this model that aberrant glutamatergic neurotransmission results in both the cognitive and the psychiatric manifestations of the disorder. It has long been recognized that a schizophrenia-like psychosis can be induced by phencyclidine (PCP), a noncompetitive N-methyl-D-aspartate (NMDA) receptor ion channel blocker. Not only does PCP cause hallucinations and delusions, but it also causes an associated apathetic state and a type of formal thought disorder that are more distinctive features of schizophrenia. These features are not induced by other psychotogenic drugs such as amphetamines. In addition, the model is appealing from a neurobiological perspective in that glutamate has both neurotrophic and neurotoxic properties;[116] thus, a glutamatergic dysfunction model potentially can help explain both neurodevelopmental and neurodegenerative aspects of schizophrenia.

Postmortem studies of glutamatergic mechanisms in schizophrenia so far have yielded complex results in the medial temporal lobe as well as the rest of the brain. Decreases in non-NMDA receptor binding and receptor mRNA transcripts have been found in portions of the hippocampus and parahippocampal gyrus.[97,117] Increased receptor binding of phencyclidine (a noncompetitive NMDA receptor ion channel blocker) has been reported in the hippocampus as well as neocortex.[118] In contrast, Deakin et al.[119] found no significant differences between schizophrenic and control subjects in glutamate uptake and non-NMDA receptor binding in homogenates of hippocampus and amygdala. Finally, changes in the levels of aspartate, glutamate, N-acetylaspartylglutamate, and N-acetyl-α-linked acidic dipeptidase have been noted in the hippocampus and other regions.[120] Although the precise nature of glutamatergic dysfunction in schizophrenia is not known yet, the strength of the hypothesis and these early findings indicate that it is another promising avenue of research.

Studies of Neurodevelopment, Neurodegeneration, and Neural Injury

There have been two principal neurobiological hypotheses about schizophrenia since the time of Kraepelin and Bleuler. One is that schizophrenia arises from an injurious or degenerative process occurring near the time of the onset of symptoms, which persists throughout the illness and may culminate in a severe, "burnt-out" state of dementia in late life. The other hypothesis holds that schizophrenia represents a more static brain disorder. Increasing data on neurodevelopmental abnormalities in individuals who later present with schizophrenia are interpreted as more compatible with this view. In this model, embryogenetic or early neurodevelopmental processes result in a static encephalopathy that manifests in psychosis when maturational factors in adolescence or early adulthood allow its full expression.[38]

One of the chief findings from postmortem investigations that has been used to support the neurodevelopmental model of schizophrenia is the relative absence of gliosis in cortical and limbic brain regions in schizophrenia. In the postnatal and mature brain, reactive gliosis occurs as an acute, nonspecific reaction in the setting of assorted injurious processes such as infection, trauma, autoimmune disorders, ischemia, toxic insults, and neurodegenerative diseases.[121] In contrast, because of the immaturity of immune mechanisms until late in gestation, injury occurring during fetal brain development does not result in an inflammatory response and leaves no trace of glial reaction. Investigation of gliosis in medial temporal lobe structures using both traditional stains and immunohistochemical methods have generally failed to identify gliosis in the medial temporal lobe or elsewhere in schizophrenia.[68,74,75,78,122–126]

Other studies have assessed the frequency of neurodegenerative disease pathology among elderly patients with schizophrenia, many of whom show severe deterioration consistent with a coexisting degenerative-like dementia. Although there has been some controversy, recent evidence from prospectively accrued samples of clinically well characterized patients has failed to identify excess senile plaques, neurofibrillary tangles, Lewy bodies, ubiquitinated neurons, or decreased cholinergic activity. These are all markers that are found in commonly occurring neurodegenerative diseases.[124,127–129] Thus, evidence for postmaturational injury or neurodegeneration is generally absent in schizophrenia, even in elderly, severely ill patients. This absence provides indirect evidence in favor of alternative hypotheses of the pathophysiology of schizophrenia, such as the neurodevelopmental model.

Given the substantial additional clinical evidence supporting the developmental model, it is at first surprising that there have been so few postmortem studies to directly examine developmental markers in schizophrenia. However, it is not easy to detect residual evidence of abnormal brain development that would have occurred decades before postmortem examination. The cytoarchitectural studies of the hippocampus and entorhinal cortex described above have provided the most compelling data in this regard, but they yield little in terms of delineating particular mechanisms involved. In investigating one candidate protein that is involved in neuronal migration, Barbeau et al.[130] reported a decrease in the expression of the embryonic polysialylated neural cell adhesion molecule (N-CAM) isoform in neurons in the hilum of the hippocampus. N-CAMs are a family of proteins that are important in neuronal migration and that may be related to ongoing synaptic plasticity. A deficit could result in altered plasticity, and it could also provide a molecular abnormality that might have caused aberrant cell migration during fetal development. An interesting piece of information from epidemiological studies of schizophrenia is the increased risk for schizophrenia among offspring of mothers infected with the influenza virus during their second trimester of pregnancy.[131,132] Influenza viruses are among the few viruses that express capsular neuraminidase, a protein that alters the normal function of N-CAMs.

Developmentally regulated cytoskeletal proteins have also been examined in the hippocampal region in schizophrenia in order to investigate possible molecular correlates of abnormalities in neuron size, shape, or polarity. Although results are preliminary, two groups have described deficits in the expression of microtubule-associated proteins (MAP2 and MAP1b) in the subiculum and entorhinal cortex in schizophrenia.[133,134] These cytoskeletal elements are important for microtubule stabilization and the establishment of dendritic shape and stability during development and in maturity. Diminished cytoskeletal protein expression could result in changes in cell size and dendritic architecture, with functional implications for the fidelity of neural circuitry supported by those neurons.

ANIMAL STUDIES RELEVANT TO SCHIZOPHRENIA

Two recently described animal models are relevant to the medial temporal lobe and developmental and glutamatergic hypotheses of schizophrenia. In examining the sequelae of injury to the hippocampal region during development in rats, Lipska and colleagues[39,135,136] have conducted an important series of experiments in which an excitotoxic lesion with ibotenic acids was placed in the ventral hippocampus of neonatal and young adult rats. Compared with sham operated rats, those that received lesions as neonates behaved normally when tested shortly before puberty under various conditions such as environmental stress and pharmacological manipulations. After puberty, however, these rats were hyperresponsive to environmental stress and dopaminergic challenges. In contrast, the rats that received lesions as young adults showed a different profile of behavioral abnormalities and no delay in onset. Furthermore, it was recently reported that the neonatally lesioned animals also displayed abnormal prepulse inhibition of the acoustic startle response. This psychophysiologic sensorimotor gating response, which is related to attention and information processing, is abnormal in schizophrenic individuals. Altogether, these data highlight the importance of the timing of the lesion in relation to the onset of symptoms and provide supportive data for the hypothesized role of developmental medial temporal lobe abnormalities in schizophrenia.

In another model, Bardgett et al.[137] examined the consequences of a more generalized excitotoxic disturbance in rats. After administering intracerebroventricular injection of kainic acid, they found a dose-dependent, regionally specific loss of neurons in the ventral and dorsal hippocampus, piriform cortex, and thalamus, along with persistent behavioral changes. They also observed an increase in dopamine receptor binding in the nucleus accumbens, which they interpreted as a neurochemical response to denervation from limbic afferents. By creating a diffusely excitotoxic state, the investigators induced specific hippocampal region neuropathology and neurochemical alteration similar to that seen in human postmortem studies.

CONCLUDING REMARKS

Neuroimaging and neuropathological research have identified numerous abnormalities in the brains of patients with schizophrenia, and the medial temporal lobe appears to be a principal site for many. Although the abnormalities are diverse in nature (and some remain controversial), their common location within the medial temporal lobe provides an explanation for many of the clinical and neuropsychological features of schizophrenia when viewed from a neural systems perspective. The medial temporal lobe is a critical region for learning and memory, one of the most affected neuropsychological functions in schizophrenia. Beyond this, it is strongly interconnected with widespread cortical and subcortical regions throughout the telencephalon. Consequently, one might expect related anatomic, physiologic, or cognitive changes in those regions as they try to interact with a defective medial temporal lobe system. Such a phenomenon has been demonstrated in anatomic-physiologic activation correlation studies.[46] Future work using experimental cognitive and emotional activation paradigms with neuroimaging modalities should continue to be fruitful in this regard and may also be useful in assessing treatment responses.

Postmortem research is now entering a phase where more reliable tissue resources are becoming available that have the benefit of improved antemortem diagnostic information and sound tissue processing methods. Although most postmortem avenues of research are still young, growing evidence supports theories of aberrant development in the medial temporal lobe in schizophrenia.

The particular genetic, molecular, and cellular nature of these abnormalities as yet remains unclear. Given that schizophrenia is most likely a heterogeneous disorder, it is also likely that the neuropathological "lesions" will be heterogeneous. The factors that allow diverse cellular and molecular mechanisms to manifest in a common syndrome may be more related to location and timing than to the specific signature of the lesion. Hemiparesis, for example, results from damage to primary motor cortices or tracts, whether due to stroke, tumor, trauma, or any of a number of pathological processes. Similarly, it is proposed that an anatomic lesion involving the medial temporal lobe and/or strongly interconnected limbic region results in a schizophrenic psychosis, especially if it occurs during brain development. Such a possibility can be successfully explored in animal models at the same time that genetic, molecular, neuropathologic, and epidemiologic research determine specific etiologic factors.

REFERENCES

1. Kraepelin E: Dementia Praecox and Paraphrenia. Edinburgh, Livingstone, 1919
2. Kleist K: Schizophrenic symptoms and cerebral pathology. J Ment Sci 1960; 106:246–255
3. Davison K, Bagley CR: Schizophrenia-like psychoses associated with organic disorders of the CNS: a review of the literature, in Current Problems in Neuropsychiatry, part II, edited by Herrington P. Br J Psychiatry Special Publication No 4. Ashford, UK, Headley Bros, 1969, pp 113–184
4. Davison K: Schizophrenia-like psychoses associated with organic cerebral disorders: a review. Psychiatric Development 1983; 1:1–34
5. Cummings JL: Organic psychosis. Psychosomatics 1988; 29:16–26

6. Fricchione GL, Carbone L, Bennet WI: Psychotic disorder caused by a general medical condition, with delusions. Psychiatr Clin North Am 1995; 18:363–378
7. Jacobi W, Winkler H: Untersuchungen des Liquor cerebrospinalis mit dem zeisschen Spektographan für Chemiker. Deutch Ztschr f Nervenh 1929; 111:5–18
8. Ellingson RJ: The incidence of EEG abnormality among patients with mental disorders of apparently nonorganic origin: a critical review. Am J Psychiatry 1954; 111:263–285
9. Shapiro RM: Regional neuropathology in schizophrenia: Where are we? Where are we going? Schizophr Res 1993; 10:187–239
10. Arnold SE, Trojanowski JQ: Recent advances in defining the neuropathology of schizophrenia. Acta Neuropathol 1996; 42:217–231
11. Goldman-Rakic PS: Working memory dysfunction in schizophrenia. J Neuropsychiatry Clin Neurosci 1994; 6:348–357
12. Sedvall G, Farde L: Chemical brain anatomy in schizophrenia. Lancet 1995; 346:743–749
13. Damasio H, Damasio AR: Lesion Analysis in Neuropsychology. New York, Oxford University Press, 1989
14. Saykin AJ, Gur RC, Gur RE, et al: Neuropsychological function in schizophrenia: selective impairment in memory and learning. Arch Gen Psychiatry 1991; 48:618–624
15. Saykin AJ, Shtasel DL, Gur RE, et al: Neuropsychological deficits in neuroleptic naive patients with first-episode schizophrenia. Arch Gen Psychiatry 1994; 51:124–131
16. Gur RE, Cowell P, Turetsky BI, et al: A follow-up study of neuroanatomical, clinical, and neurobehavioral measures in schizophrenia. Arch Gen Psychiatry (in press)
17. Gruzelier J, Seymour K, Wilson J: Impairments on neuropsychological tests of temporohippocampal and frontohippocampal functions and word fluency in remitting schizophrenic and affective disorders. Arch Gen Psychiatry 1988; 45:623–629
18. McKenna P, Tamlyn D, Lund C, et al: Amnesic syndrome in schizophrenia. Psychol Med 1990; 20:967–972
19. Gold JM, Randolph C, Carpenter CJ, et al: Forms of memory failure in schizophrenia. J Abnorm Psychol 1992; 101:487–494
20. Tamlyn D, McKenna PJ, Mortimer AM, et al: Memory impairment in schizophrenia: its extent, affiliations and neuropsychological character. Psychol Med 1992; 22:101–115
21. Gold JM, Weinberger DR: Cognitive deficits and the neurobiology of schizophrenia. Curr Opin Neurobiol 1995; 5:225–230
22. Kelsoe JR, Cadet JL, Pickar D, et al: Quantitative neuroanatomy in schizophrenia: a controlled magnetic resonance imaging study. Arch Gen Psychiatry 1988; 45:533–541
23. Pearlson GD, Kim WS, Kubos KL, et al: Ventricle-brain ratio, computed tomographic density, and brain area in 50 schizophrenics. Arch Gen Psychiatry 1989; 46:690–697
24. Gur RE, Mozley PD, Resnick SM, et al: Magnetic resonance imaging in schizophrenia, I: volumetric analysis of brain and cerebrospinal fluid. Arch Gen Psychiatry 1991; 48:407–412
25. Suddath RL, Casanova MF, Goldberg TE, et al: Temporal lobe pathology in schizophrenia: a quantitative magnetic resonance imaging study. Am J Psychiatry 1989; 146:464–472
26. Suddath RL, Christison GW, Torrey EF, et al: Cerebral anatomic abnormalities in monozygotic twins discordant for schizophrenia. N Engl J Med 1990; 322:789–794
27. Bogerts B, Ashtari M, Degreef G, et al: Reduced temporal limbic structure volumes on magnetic resonance images in first episode schizophrenia. Psychiatry Res Neuroimaging 1990; 35:1–13
28. Breier A, Buchanan RW, Elkashef A, et al: Brain morphology and schizophrenia: a magnetic resonance imaging study of limbic, prefrontal cortex, and caudate structures. Arch Gen Psychiatry 1992; 49:921–926
29. Shenton ME, Kikinis R, Jolesz FA: Abnormalities of the left temporal lobe and thought disorder in schizophrenia: a quantitative magnetic resonance imaging study. N Engl J Med 1992; 327:604–612
30. Zipursky RB, Lim KO, Sullivan EV, et al: Widespread cerebral gray matter volume deficits in schizophrenia. Arch Gen Psychiatry 1992; 49:195–205
31. Zipursky RB, Marsh L, Lim KO: Volumetric MRI assessment of temporal lobe structures in schizophrenia. Biol Psychiatry 1994; 35:501–516
32. Goldberg TE, Torrey EF, Berman KF, et al: Relations between neuropsychological performance and brain morphological and physiological measures in monozygotic twins discordant for schizophrenia. Psychiatry Res 1994; 55:51–61
33. Nestor PG, Shenton ME, McCarley RW, et al: Neuropsychological correlates of MRI temporal lobe abnormalities in schizophrenia. Am J Psychiatry 1993; 150:1849–1855
34. Bogerts B, Lieberman JA, Ashtari M, et al: Hippocampus-amygdala volumes and psychopathology in chronic schizophrenia. Biol Psychiatry 1993; 33:236–246
35. Turetsky B, Cowell PE, Gur RC, et al: Frontal and temporal brain volumes in schizophrenia. Arch Gen Psychiatry 1995; 52:1061–1070
36. Bilder RM, Bogerts B, Ashtari M, et al: Anterior hippocampal volume reductions predict frontal lobe dysfunction in first episode schizophrenia. Schizophr Res 1995; 17:47–58
37. Arnold SE, Hyman BT, Hoesen GWV, et al: Some cytoarchitectural abnormalities of the entorhinal cortex in schizophrenia. Arch Gen Psychiatry 1991; 48:625–632
38. Weinberger DR: From neuropathology to neurodevelopment. Lancet 1995; 346:552–557
39. Lipska BK, Jaskiw GE, Weinberger DR: Postpubertal emergence of hyperresponsiveness to stress and to amphetamine after neonatal excitotoxic hippocampal damage: a potential animal model of schizophrenia. Neuropsychopharmacology 1993; 9:67–75
40. Tamminga CA, Thaker GK, Buchanan R, et al: Limbic system abnormalities identified in schizophrenia using positron emission tomography with fluorodeoxyglucose and neocortical alterations with deficit syndrome. Arch Gen Psychiatry 1992; 49:522–530
41. Friston KJ, Liddle PF, Frith CD, et al: The left medial temporal region and schizophrenia: a PET study. Brain 1992; 115:367–382
42. Schroeder J, Buchsbaum MS, Siegel BV, et al: Structural and functional correlates of subsyndromes in chronic schizophrenia. Psychopathology 1995; 28:38–45
43. Silbersweig DA, Stern E, Frith C, et al: A functional neuroanatomy of hallucinations in schizophrenia. Nature 1995; 378:176–179
44. Weinberger DR, Berman KF, Zec RF: Physiologic dysfunction of dorsolateral prefrontal cortex in schizophrenia, I: regional cerebral blood flow evidence. Arch Gen Psychiatry 1986; 43:11–19
45. Berman KF, Torrey EF, Daniel DG, et al: Regional cerebral blood flow in monozygotic twins discordant and concordant for schizophrenia. Arch Gen Psychiatry 1992; 49:927–934
46. Weinberger DR, Berman KF, Suddath R, et al: Evidence of dysfunction of a prefrontal-limbic network in schizophrenia: a magnetic resonance imaging and regional cerebral blood flow study of discordant monozygotic twins. Am J Psychiatry 1992; 149:890–897
47. Humberstone MR, Sawle GV: Functional magnetic resonance imaging in clinical neurology. Eur Neurol 1996; 26:117–124
48. Saykin AJ, Riordan HJ, Weaver JB, et al: Memory activation in schizophrenia: preliminary observations using functional magnetic resonance imaging (FMRI) (abstract). Schizophr Res 1995; 15:97
49. Nasrallah HA, Skinner TE, Schmalbrock P, et al: Proton magnetic resonance spectroscopy (1H MRS) of the hippocampal formation in schizophrenia: a pilot study. Br J Psychiatry 1994; 165:481–485
50. Fukuzako H, Takeuchi K, Hokazono Y, et al: Proton magnetic-resonance spectroscopy of the left medial temporal and frontal lobes in chronic schizophrenia: a preliminary report. Psychiatry Res 1995; 61:193–200
51. Maier M, Ron MA, Barker GJ, et al: Proton magnetic spectroscopy: an in vivo method of estimating hippocampal neuronal depletion in schizophrenia. Psychol Med 1995; 25:1201–1209

52. Renshaw PF, Yurgelun-Todd DA, Tohen M, et al: Temporal-lobe proton magnetic-resonance spectroscopy of patients with first episode psychosis. Am J Psychiatry 1995; 153:444–446
53. Abrams R, Taylor MA: Differential EEG patterns in affective disorder and schizophrenia. Arch Gen Psychiatry 1979; 36:1355–1361
54. Heath RG: Correlation of electrical recordings from cortical and subcortical regions of the brain with abnormal behavior in human subjects. Confin Neurol 1958; 18:305–315
55. Sem-Jacobsen CW, Petersen MC, Lazarte J, et al: Intracerebral electrographic recordings from psychotic patients during hallucinations and agitation. Am J Psychiatry 1955; 112:278–288
56. Baribeau-Braun J, Picton TW, Gosselin J-Y: Schizophrenia: a neurophysiological evaluation of abnormal information processing. Science 1983; 219:874–876
57. Pritchard WS: Cognitive event-related potential correlates of schizophrenia. Psychol Bull 1986; 100:43–66
58. Duncan CC, Perlstein WM, Morihisa JM: The P300 metric in schizophrenia: effects of probability and modality, in Current Trends in Event Related Potential Research (EEG Supplement 40), edited by Johnson JR, Rohrbaugh JW, Parasurman R. Amsterdam, Elsevier Science, 1987, pp 670–674
59. Egan MF, Duncan CC, Suddath RL, et al: Event-related potential abnormalities correlate with structural brain alterations and clinical features in patients with chronic schizophrenia. Schizophr Res 1994; 11:259–271
60. Proceedings of the First International Congress of Neuropathology, vol 1. Turin, Italy, Rosenberg and Sellier, 1952
61. Bruton CJ, Crow TJ, Frith CD, et al: Schizophrenia and the brain: a prospective clinico-neuropathological study. Psychol Med 1990; 20:285–304
62. Stevens JR: Neuropathology of schizophrenia. Arch Gen Psychiatry 1982; 39:1131–1139
63. Arnold SE, Gur RE, Shapiro RM, et al: Prospective clinicopathological studies of schizophrenia: accrual and assessment. Am J Psychiatry 1995; 152:731–737
64. Golier JA, Davidson M, Haroutunian V, et al: Neuropathological study of 101 elderly schizophrenics: preliminary findings. Schizophr Res 1995; 15:120
65. Bogerts B, Meertz E, Schonfeldt-Bausch R: Basal ganglia and limbic system pathology in schizophrenia. Arch Gen Psychiatry 1985; 42:784–791
66. Brown R, Colter N, Corsellis JAN, et al: Postmortem evidence of structural brain changes in schizophrenia. Arch Gen Psychiatry 1986; 43:36–42
67. Colter N, Battal S, Crow TJ, et al: White matter reduction in the parahippocampal gyrus of patients with schizophrenia. Arch Gen Psychiatry 1987; 44:1023–1029
68. Falkai P, Bogerts B, Rozumek M: Limbic pathology in schizophrenia: the entorhinal region: a morphometric study. Biol Psychiatry 1988; 24:515–521
69. Jeste DV, Lohr JB: Hippocampal pathologic findings in schizophrenia: a morphometric study. Arch Gen Psychiatry 1989; 46:1019–1024
70. Altshuler LL, Casanova MF, Goldberg TE, et al: The hippocampus and parahippocampus in schizophrenic, suicide, and control brains. Arch Gen Psychiatry 1990; 47:1029–1034
71. Bogerts B, Falkai P, Haupts M, et al: Post-mortem volume measurements of limbic system and basal ganglia structures in chronic schizophrenics: initial results from a new brain collection. Schizophr Res 1990; 3:295–301
72. Heckers S, Heinsen H, Heinsen YC, et al: Limbic structures and lateral ventricle in schizophrenia: a quantitative postmortem study. Arch Gen Psychiatry 1990; 47:1016–1022
73. Heckers S, Heinsen H, Geiger B, et al: Hippocampal neuron number in schizophrenia: a stereological study. Arch Gen Psychiatry 1991; 48:1002–1008
74. Pakkenberg B: Pronounced reduction of total neuron number in mediodorsal thalamic nucleus and nucleus accumbens in schizophrenia. Arch Gen Psychiatry 1990; 47:1023–1028
75. Falkai P, Bogerts B: Cell loss in the hippocampus of schizophrenics. Eur Arch Psychiatry Neurol Sci 1986; 236:154–161
76. Krimer LS, Herman MM, Saunders RC, et al: Qualitative and quantitative analysis of the entorhinal cortex cytoarchitectural organization in schizophrenia (abstract). Society for Neuroscience Abstracts 1995; 21:239
77. Kovelman JA, Scheibel AB: A neurohistological correlate of schizophrenia. Biol Psychiatry 1984; 19:1601–1621
78. Benes FM, Sorensen I, Bird ED: Reduced neuronal size in posterior hippocampus of schizophrenic patients. Schizophr Bull 1991; 17:597–608
79. Arnold SE, Franz BR, Gur RC, et al: Smaller neuron size in schizophrenia in hippocampal subfields that mediate cortical-hippocampal interactions. Am J Psychiatry 1995; 152:738–748
80. Rajkowska G, Selemon LD, Goldman-Rakic PS: Reduction in neuronal sizes in prefrontal cortex of schizophrenics and Huntington patients (abstract). Society for Neuroscience Abstracts 1994; 20:620
81. Scheibel AB, Kovelman JB: Disorientation of the hippocampal pyramidal cell and its processes in the schizophrenic patient. Biol Psychiatry 1981; 16:101–102
82. Conrad AJ, Abebe T, Austin R, et al: Hippocampal pyramidal cell disarray in schizophrenia as a bilateral phenomenon. Arch Gen Psychiatry 1991; 48:413–417
83. Altshuler LL, Conrad A, Kovelman JA, et al: Hippocampal pyramidal cell orientation in schizophrenia. Arch Gen Psychiatry 1987; 44:1094–1098
84. Christison GW, Casanova MF, Weinberger DR, et al: A quantitative investigation of hippocampal pyramidal cell size, shape, and variability of orientation in schizophrenia. Arch Gen Psychiatry 1989; 46:1027–1032
85. Jakob H, Beckmann H: Prenatal developmental disturbances in the limbic allocortex in schizophrenics. J Neural Transm 1986; 65:303–326
86. Jakob H, Beckmann H: Gross and histological criteria for developmental disorders in brains of schizophrenics. J R Soc Med 1989; 82:466–469
87. Jakob H, Beckmann H: Circumscribed malformation and nerve cell alterations in the entorhinal cortex of schizophrenics: pathogenetic and clinical aspects. J Neural Transm 1994; 98:83–106
88. Van Hoesen GW: The parahippocampal gyrus: new observations regarding its cortical connections in the monkey. Trends Neurosci 1982; 5:345–350
89. Arnold SE, Rusceinsky DR, Han L-Y: Further evidence of abnormal cytoarchitecture of the entorhinal cortex in schizophrenia using spatial point pattern analyses. Biol Psychiatry 1997 (in press)
90. Nowakowski RS, Rakic P: The site of origin and route and rate of migration of neurons to the hippocampal region in the rhesus monkey. J Comp Neurol 1981; 196:129–154
91. Arnold SE, Trojanowski JQ: Human fetal hippocampal development, I: cytoarchitecture, myeloarchitecture and neuronal morphology. J Comp Neurol 1996; 367:274–292
92. Arnold SE, Trojanowski JQ: Human fetal hippocampal development, II: the neuronal cytoskeleton. J Comp Neurol 1996; 367:293–307
93. Frith CD, Done DJ: Towards a neuropsychology of schizophrenia. Br J Psychiatry 1988; 1988:437–443
94. Gray JA, Feldon J, Rawlins JNP, et al: The neuropsychology of schizophrenia. Behav Brain Sci 1991; 14:1–84
95. Benes FM: Neurobiological investigations in cingulate cortex of schizophrenic brain. Schizophr Bull 1993; 19:537–549
96. Stevens JR: Abnormal synaptic reinnervation as the basis of schizophrenia: a hypothesis. Arch Gen Psychiatry 1992; 49:238–243
97. Eastwood SL, McDonald B, Burnet PW, et al: Decreased expression of mRNAs encoding non-NMDA glutamate receptors GluR1 and

GluR2 in medial temporal lobe neurons in schizophrenia. Brain Res Molec Brain Res 1995; 29:211–223
98. Eastwood SL, Burnet PWJ, Harrison PJ: Altered synaptophysin expression as a marker of synaptic pathology in schizophrenia. Neurosci 1995; 66:309–319
99. Browning MD, Dudek EM, Rapier JL, et al: Significant reductions in synapsin but not synaptophysin specific activity in the brains of some schizophrenics. Biol Psychiatry 1993; 34:529–535
100. Glantz LA, Lewis DA: Assessment of spine density on layer III pyramidal cells in the prefrontal cortex of schizophrenic subjects (abstract). Society for Neuroscience Abstracts 1995; 21:239
101. Glantz LA, Lewis DA: Specificity of decreased spine density on layer III pyramidal cells in schizophrenia (abstract). Society for Neuroscience Abstracts 1996; 22:1679
102. Ong WY, Garey LJ: Ultrastructural features of biopsied temporopolar cortex (area 38) in a case of schizophrenia. Schizophr Res 1993; 10:15–27
103. Longson D, Deakin JFW, Benes FM: Increased density of entorhinal glutamate-immunoreactive vertical fibers in schizophrenia. J Neural Transm 1996; 103:503–507
104. Kleinman JE, Casanova MF, Jaskiw GE: The neuropathology of schizophrenia. Schizophr Bull 1988; 14:209–216
105. Seeman P, Niznik HB: Dopamine receptors and transporters in Parkinson's disease and schizophrenia. FASEB 1990; 4:2737–2744
106. Reynolds GP: Beyond the dopamine hypothesis: the neurochemical pathology of schizophrenia. Br J Psychiatry 1989; 155:305–316
107. Joyce JN: The dopamine hypothesis of schizophrenia: limbic interactions with serotonin and norepinephrine. Psychopharmacol (Berl) 1993; 112:S16–S34
108. Reynolds GP: Increased concentrations and lateral asymmetry of amygdala dopamine in schizophrenia. Nature 1983; 305:527–529
109. Winblad B, Blucht G, Gotfries CG, et al: Monoamines and monoamine metabolites in brains from demented schizophrenics. Acta Psychiatr Scand 1979; 60:17–28
110. Akil M, Lewis DA: The catecholaminergic innervation of the human entorhinal cortex: alterations in schizophrenia (abstract). Society for Neuroscience Abstracts 1995; 21:238
111. Joyce JN, Shane A, Lexow N, et al: Serotonin uptake sites and serotonin receptors are altered in the limbic system of schizophrenics. Neuropsychopharmacology 1993; 8:315–336
112. Joyce JN, Lexow N, Kim SJ, et al: Distribution of beta-adrenergic receptor subtypes in human post-mortem brains: alterations in limbic regions of schizophrenics. Synapse 1992; 10:228–246
113. Kim JS, Kornhuber HH, Schmid-Burgk W, et al: Low cerebrospinal fluid glutamate in schizophrenic patients and a new hypothesis on schizophrenia. Neurosci Lett 1980; 30:379–382
114. Olney JW, Farber NB: Glutamate receptor dysfunction and schizophrenia. Arch Gen Psychiatry 1995; 52:998–1007
115. Javitt DC, Zukin SR: Recent advances in the phencyclidine model of schizophrenia. Am J Psychiatry 1991; 148:1301–1308
116. Mattson MP: Cellular signaling mechanisms common to the development and degeneration of neuroarchitecture: a review. Mech Ageing Dev 1989; 50:103–157
117. Kerwin RW, Beats BC: Increased forskolin binding in the left parahippocampal gyrus and CA 1 region in post mortem schizophrenic brain determined by quantitative autoradiography. Neurosci Lett 1990; 118:164–168
118. Simpson MDC, Royston MC, Slater P, et al: Phencyclidine and sigma receptor abnormalities in schizophrenic post-mortem brain (abstract). Schizophr Res 1990; 3:32
119. Deakin JFW, Slater P, Simpson MDC, et al: Frontal cortical and left temporal glutamatergic dysfunction in schizophrenia. J Neurochem 1989; 52:1781–1786
120. Tsai G, Passani LA, Slusher BS, et al: Abnormal excitatory neurotransmitter metabolism in schizophrenic brains. Arch Gen Psychiatry 1995; 52:829–836
121. Norton WT, Aquino DA, Hozumi I, et al: Quantitative aspects of reactive gliosis: a review. Neurochem Res 1992; 17:877–885
122. Stevens CD, Altshuler LL, Bogerts B, et al: Quantitative study of gliosis in schizophrenia and Huntington's chorea. Biol Psychiatry 1988; 24:697–700
123. Crow TJ, Ball J, Bloom SR, et al: Schizophrenia as an anomaly of development of cerebral asymmetry: a postmortem study and a proposal concerning the genetic basis of the disease. Arch Gen Psychiatry 1989; 46:1145–1150
124. Arnold SE, Trojanowski JQ, Gur RE, et al: Investigations of neurodegeneration and neural injury in the brains of elderly patients with schizophrenia. Society for Neuroscience Abstracts 1995
125. Benes FM, Davidson J, Bird ED: Quantitative cytoarchitectural studies of the cerebral cortex of schizophrenics. Arch Gen Psychiatry 1986; 43:31–35
126. Selemon LD, Rajkowska G, Goldman-Rakic PS: Abnormally high neuronal density in the schizophrenic cortex: a morphometric analysis of prefrontal area 9 and occipital area 17. Arch Gen Psychiatry 1995; 52:805–818
127. Purohit DP, Davidson M, Perl DP, et al: Severe cognitive impairment in elderly schizophrenic patients: a clinicopathological study. Biol Psychiatry 1993; 33:255–260
128. Powchik P, Davidson M, Nemeroff CB, et al: Alzheimer's-disease-related protein in geriatric schizophrenic patients with cognitive impairment. Am J Psychiatry 1993; 150:1726–1727
129. Arnold SE, Franz BR, Trojanowski JQ: Elderly patients with schizophrenia exhibit infrequent neurodegenerative lesions. Neurobiol Aging 1994; 15:299–303
130. Barbeau D, Liang JJ, Robitaille Y, et al: Decreased expression of the embryonic form of the neural cell adhesion molecule in schizophrenic brains. Proc Natl Acad Sci USA 1995; 92:2785–2789
131. Mednick SA, Machon RA, Huttunen MO, et al: Adult schizophrenia following prenatal exposure to an influenza epidemic. Arch Gen Psychiatry 1988; 45:189–192
132. Wright P, Takei N, Rifkin L, et al: Maternal influenza, obstetric complications, and schizophrenia. Am J Psychiatry 1995; 152:1714–1720
133. Arnold SE, Lee VMY, Gur RE, et al: Abnormal expression of two microtubule-associated proteins (MAP2 and MAP5) in specific subfields of the hippocampal formation in schizophrenia. Proc Natl Acad Sci USA 1991; 88:10850–10854
134. Rosoklija G, Kaufman MA, Liu D, et al: Subicular MAP-2 immunoreactivity in schizophrenia (abstract). Society for Neuroscience Abstracts 1995; 21:2126
135. Lipska BK, Weinberger DR: Behavioral effects of subchronic treatment with haloperidol or clozapine in rats with neonatal excitotoxic hippocampal damage. Neuropsychopharmacology 1994; 10:199–205
136. Lipska BK, Swerdlow NR, Geyer MA, et al: Neonatal excitotoxic hippocampal damage in rats causes post-pubertal changes in prepulse inhibition of startle and its disruption by apomorphine. Psychopharmacol 1995; 122:35–43
137. Bardgett ME, Jackson JL, Taylor GT, et al: Kainic acid decreases hippocampal neuron number and increases dopamine receptor binding in the nucleus accumbens: an animal model of schizophrenia. Behav Brain Res 1995; 70:153–164

12

Limbic-Cortical Dysregulation

A Proposed Model of Depression

Helen S. Mayberg, M.D.

A critical role for limbic structures in the regulation of mood and affect is now considered almost axiomatic. As first articulated by Broca,[1] and later Papez,[2] Yakovlev,[3] and MacLean,[4] these regions are centrally involved in integrating exteroceptive and interoceptive inputs required for widespread motor, cognitive, and autonomic processes.[5–7] The neurobiological substrate for this integration has been further substantiated by comparative cytoarchitectural, connectivity, and neurochemical studies. These studies have delineated reciprocal pathways linking midline limbic structures (cingulate, hypothalamus, hippocampus, and amygdala) with widely distributed brainstem, striatal, paralimbic, and neocortical sites.[6–18] While there is little debate that "limbic" brain is critically involved in various aspects of motivational, affective, and emotional behaviors,[4–7,19–24] the full role of these regions in the pathogenesis of depressive illness is not known.

New strategies for testing limbic hypotheses in depressed patients have emerged with the development of in vivo structural and functional imaging techniques. To date, imaging studies have identified regional abnormalities that appear to both support and contradict the involvement of limbic structures in this disorder. Anatomical studies of patients with major depression have not demonstrated consistent changes in primary limbic regions, but frontal and striatal abnormalities have been repeatedly demonstrated.[25,26] Functional imaging studies, on the other hand, report involvement of limbic as well as frontal, striatal, and paralimbic sites, although there is tremendous variability among published reports.[27–30] Depression likely involves the disruption of a widely distributed and functionally interactive network of cortical-striatal and cortical-limbic pathways that is critical to the integrated regulation of mood and associated motor, cognitive, and somatic behaviors.

The working model of depression formulated in this chapter attempts to both consolidate these diverse experimental observations and accommodate the various symptoms that characterize the clinical syndrome (Figure 12–1). This model has evolved from an earlier prototype[30] developed to interpret a series of positron emission tomographic (PET) studies of patients with major depression associated with specific neurological disorders.[30–37] This current, expanded version now includes data from a more recent series of experiments

---

The author thanks her collaborators at the Research Imaging Center in San Antonio, Stephen Brannan, M.D., Roderick Mahurin, Ph.D., Mario Liotti, M.D., Ph.D., Scott McGinnis, B.S., and Peter Fox, M.D., for their significant contributions to the collective body of research discussed in this chapter and their insightful criticism of the manuscript. This work was supported by National Institute of Mental Health Grant MH49553, an Independent Investigator's Award from the National Alliance for Research on Schizophrenia and Depression (NARSAD), a Clinical Hypothesis Award from the Charles A. Dana Foundation, and a grant from Eli Lilly and Company.

examining 1) blood flow changes with induced sadness in healthy subjects, 2) resting state patterns of regional metabolism in patients with primary and secondary depression, and 3) changes in metabolism with antidepressant treatment.[38-43] The convergence of findings from these experiments and other clinical, anatomical, neurochemical, and functional imaging studies of depression is the basis for the model presented below.

For this discussion, the model will be limited to the syndrome of major depression, clinically defined as the presence of a persistent negative mood state occurring in conjunction with an array of core behavioral symptoms, including disturbances of attention, motivation, motor and mental speed, sleep, appetite, and libido as well as anhedonia, anxiety, guilt, and recurrent thoughts of death with or without suicidal ideations or attempts.[44] All clinical and biological features of this syndrome cannot be fully accounted for by this or any model at our present stage of knowledge. Rather, this formulation is offered as an evolving and adaptable framework to facilitate the integration of clinical functional imaging findings with complementary basic human and animal research in the study of the pathogenesis of primary major depression and other affective disorders.

DEPRESSION MODEL

The proposed model has three main components, each composed of brain regions previously identified in PET studies of depression.

The *dorsal compartment* (Figure 12–1, red boxes) includes both neocortical and midline limbic elements, and it is postulated to be principally involved with attentional and cognitive features of the illness.[45-48] Depression symptoms such as apathy, psychomotor slowing, and impaired performance on tasks of selective and directed attention and executive function are hypothesized to localize to anterior and posterior aspects of the dorsal components of the model, specifically dorsolateral prefrontal cortex (dFr 9/46), dorsal anterior cingulate (dCg 24b), inferior parietal cortex (inf Par 40), and striatum (BG). This hypothesis is based on complementary structural and functional lesion–deficit correlational studies in patients with both discrete brain lesions and other neurological syndromes (with and without depression)[18,22,24,49-51] and functional activation studies designed to specifically map these cognitive domains.[52-55] The grouping of individual regions into this dorsal compartment is based on the previous delineation of reciprocal connections of these regions with one another[5,16-18,56-59] and their communication with regions of the ventral compartment through the rostral and dorsal anterior cingulate, caudate-putamen, mediodorsal thalamus, and posterior cingulate.[6-7,10-18,59]

The *ventral compartment* (Figure 12–1, blue boxes) is composed of paralimbic cortical, subcortical, and brainstem regions, and it is hypothesized to mediate the vegetative and somatic aspects of the illness.[4,60,61] Sleep, appetite, libido, and endocrine disturbances reflect dysregulation of predominantly paralimbic and subcortical components of the compartment, specifically the hypothalamic-pituitary-adrenal axis (Hth), insula (vIns), subgenual cingulate (Cg 25), and brainstem (mb-p). This hypothesis is based primarily on clinical, biochemical, and electrophysiological evidence and related animal studies.[19-22,60-63] Like those in the dorsal compartment, the individual members of the ventral compartment have known reciprocal connections with one another,[4,8-12,62-64] as well as links to the dorsal compartment via the rostral cingulate, ventral striatum, anterior thalamus, hippocampus, and posterior cingulate.[7,10-15,17-19,59]

As illustrated in the model schematic, the *rostral cingulate* (Figure 12–1, yellow box) is isolated from both the ventral and dorsal compartments on the basis of its cytoarchitectural characteristics,[8-10,17] its reciprocal connections to both dorsal and ventral anterior cingulate,[10,12,17,59] and the recent PET finding that metabolism in this region uniquely predicts antidepressant response in acutely depressed patients.[39-40] These anatomical and clinical distinctions suggest that the rostral anterior cingulate may serve an important regulatory role in the overall network by facilitating the interactions between the dorsal and ventral compartments. Dysfunction in this area thus could have significant impact on remote brain regions regulating a variety of behaviors, including the interaction among mood, cognitive, somatic, and autonomic responses.

It is clear that depression involves many different behaviors, none of which localizes to any single brain region. In this model, it is proposed that these behaviors are modulated by specific subsets of regions that group predominantly to either the dorsal or the ventral compartment. Interactions among these regions and compartments are necessary for the normal regulation of mood and associated motor, cognitive, and vegetative processes. Depression is not simply dysfunction of one or another of these components, but is the failure of the coordinated interactions between the subcomponents of either compartment and between the two compartments. Support for this hypothesis is presented below.

LITERATURE BASIS FOR MODEL

Depression in Neurological Disease

Classical lesion–deficit studies have consistently reported a strong association between frontal, temporal, and basal ganglia lesions and the development of secondary depression,[25,65–76] although the issue of lesion laterality is still debated.[66,69–70,73] It is in considering the potential common link to depression in different groups of neurological patients that the limitations of the lesion–deficit approach become most apparent. Lesion–behavior correlations fail to identify the uninjured components of the overall network that regulates mood symptoms or their functional organization. Functional imaging, on the other hand, is able to delineate the

FIGURE 12–1. Depression model. Brain regions consistently identified in PET studies of depression are represented in this schematic model. Regions with known anatomical interconnections that also show synchronized changes (using PET) in three behavioral states—normal transient sadness (control subjects), baseline depressed (patients), and post-fluoxetine treatment (patients)—are grouped into three main compartments: dorsal *(red)*, ventral *(blue)*, and rostral *(yellow)*. The dorsal-ventral segregation additionally identifies those brain regions where an inverse relationship is seen across the different PET paradigms. Sadness and depressive illness are both associated with decreases in dorsal limbic and neocortical regions *(red areas)* and relative increases in ventral paralimbic areas *(blue areas)*; with successful treatment, there is a reversal of these findings. The model proposes that illness remission occurs when there is inhibition of the overactive ventral regions and activation of the previously hypofunctioning dorsal areas *(solid black arrows)*, an effect facilitated by fluoxetine action in dorsal raphe and its projection sites *(dotted lines)*. Integrity of the rostral cingulate *(yellow)*, with its direct anatomical connections to both the dorsal and ventral compartments, is postulated to be additionally required for the occurrence of these adaptive changes, since pretreatment metabolism in this region uniquely predicts antidepressant treatment response.

White regions delineate brain regions potentially critical to the evolution of the model but where changes have not been consistently identified across PET studies. *Colored arrows* identify segregated ventral and dorsal compartment afferents and efferents to and from the striatum (caudate, putamen, nucleus accumbens) and thalamus (predominantly mediodorsal and anterior thalamus), although individual cortical-striatal-thalamic pathways are not delineated. *Black arrows* indicate reciprocal connections through the anterior and posterior cingulate linking the dorsal and ventral compartments. *Dotted lines* indicate serotonergic projections to limbic, paralimbic, subcortical, and cortical regions in both compartments. Red: dFr = dorsolateral prefrontal; inf Par = inferior parietal; dCg = dorsal anterior cingulate; pCg = posterior cingulate. Blue: Cg 25 = subgenual (infralimbic) cingulate; vIns = ventral anterior insula, Hc = hippocampus; vFr = ventral frontal; Hth = hypothalamus. Yellow: rCg = rostral anterior cingulate. White: mb-p = midbrain-pons; BG = basal ganglia; Th = thalamus; Am = amygdala. Numbers are Brodmann designations.

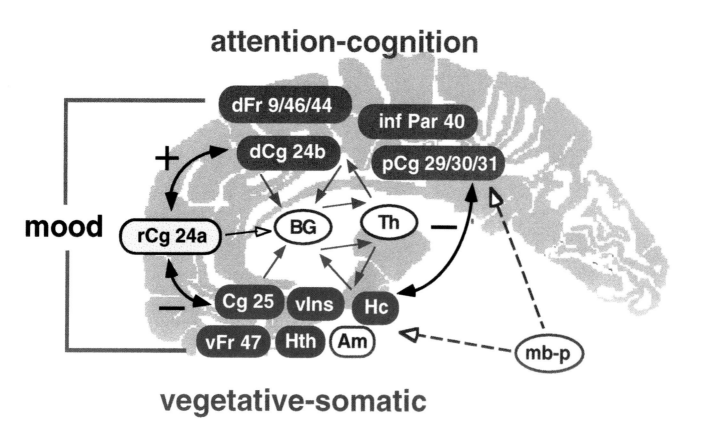

consequences of anatomic, chemical, or degenerative lesions for global and regional brain function and to identify common patterns across patient groups.

Resting state measures of brain function in secondary depressions have confirmed the anatomical observations of the lesion–deficit studies and have added the dimension of connectivity. Fluorodeoxyglucose (FDG) PET studies of depressed patients with degenerative and focal lesions have consistently identified ventral prefrontal and anterior temporal metabolic abnormalities independent of disease etiology, suggesting a critical role for these paralimbic and neocortical pathways in the regulation of mood and associated cognitive deficits.[30,33–34,77–81] Disease-specific disruption of converging pathways to these regions best explains the presence of similar depressive symptoms in patients with distinctly different disease pathologies.[30]

Proposed mechanisms for common paralimbic hypometabolism in depression associated with three basal ganglia disorders—Parkinson's disease, Huntington's disease, and caudate strokes—include anterograde or retrograde disruption of corticobasal ganglia circuits from striatal degeneration or injury, degeneration of mesencephalic monoamine neurons and their cortical projections, involvement of serotonergic neurons via disruption of orbital frontal outflow to the dorsal raphe, and remote changes in basotemporal limbic regions, with or without involvement of the amygdala.[15,48,66,76,82–84] All of these possibilities are consistent with and supportive of the model.

Primary Depression

Unlike the findings in neurological depressions, disease-specific structural changes in limbic, paralimbic, or neocortical regions have not been consistently identified in primary unipolar depressed patients, although nonspecific changes in ventricular size and T_2-weighted MRI changes in subcortical gray and periventricular white matter have been reported, particularly in late-onset patients.[26,85,86]

Resting state functional imaging studies in primary depression, on the other hand, have repeatedly reported the involvement of frontal (dorsal and ventral) and, less commonly, temporal and cingulate cortex, consistent with the general pattern seen in neurological depressions.[27–29,36,87–91] A critical issue is whether these functional regional abnormalities are disease markers or, alternatively, reflect the presence of specific depressive symptoms such as apathy, anxiety, psychomotor slowing, and executive cognitive dysfunction that are variably expressed in individual depressed patients. The latter theory might actually explain the variability in the pattern of regional changes reported in the literature. The most consistent finding is an inverse relationship between depression severity and frontal metabolism or blood flow, which has been replicated by a number of investigators (reviewed in Ketter et al.[28]). These same regions also have been found to correlate with psychomotor speed,[36,92] as well as with other unrelated cognitive measures not usually associated with depression but seen in other neurological and psychiatric diseases.[50–51,55,77,93] The presence of regional overlaps cautions against definitive conclusions regarding the role of any one brain area in regulating particular behaviors in depressive illness and suggests a more complex relationship between regional metabolic or blood flow defects and individual symptoms.

Neurochemical Markers

Evidence of neurochemical mechanisms that would account for the limbic, paralimbic, and neocortical metabolic abnormalities is compelling but circumstantial. No single neurotransmitter abnormality can fully explain the pathophysiology of depression or the associated constellation of mood, motor, cognitive, and somatic symptoms.[94] Moreover, when a peripheral chemical marker is identified, it still must be interpreted in the context of multiple neuroreceptor subtypes, second messenger effects, and regionally specific regulatory mechanisms.[95–97] Despite these caveats, a large literature exists to support changes in a number of different monoamines and peptides in depression.[98–103] However, to date there has been little direct focus on the target regions identified in published imaging studies.

Serotonergic and noradrenergic mechanisms have dominated the neurochemical literature on depression because most typical antidepressant drugs affect synaptic concentrations of these two transmitters.[94,104] Changes in both serotonergic and noradrenergic metabolites have been reported in subsets of depressed patients, but the relationship of these peripheral measures to changes in brainstem nuclei or their cortical projections is unknown. Postmortem studies of brains of depressed suicide victims have reported changes in serotonergic and noradrenergic receptors.[105] S_2 serotonin receptor changes measured with PET have been described in the temporal cortex of depressed stroke patients,[31–32] but these measures have not yet been characterized in depressed patients who are not neurologically impaired. There is, however, clear evidence of both direct and indirect monoaminergic modulation of intrinsic cingulate, hippocampal, amygdala, thalamic, and hypothalamic neurons and their afferent and efferent projections that may have direct implications for expanding the working depression model (Figure 12–1, dotted lines).[106–109]

Dopaminergic projections from the ventral tegmental area (VTA) show regional specificity for the orbital/ventral prefrontal cortex and anterior cingulate,[110,111] a finding also of relevance in validating the model. A dopamine hypothesis is appealing, given the mood-enhancing properties of methylphenidate in treating some depressed patients.[48,83,102,112–114] However, dopaminergic stimulation alone is clearly inadequate in treating the full depression syndrome, and degeneration of VTA neurons or their projections has not been demonstrated.[115]

Increasing attention has focused on other transmitter and peptide systems, particularly those with known monoaminergic interactions.[61,99–100,116–119] Unfortunately, functional imaging ligands for many of the systems of greatest interest have not yet been developed. Increases in paralimbic mu-opiate receptors have been demonstrated with PET in unipolar depressed patients[37] (a finding consistent with autoradiography studies in depressed suicide victims[117]) and also in brain regions critical to the proposed depression model. The relationship of this finding to regional metabolic and perfusion changes awaits further investigation.

TESTING THE MODEL

Transient Sadness in Healthy Subjects

Mood induction experiments in healthy subjects have shown involvement of many of the same regions identified in depressed patients, but with some critical differences. Induction of transient sadness results in a combination of cortical and limbic increases and decreases in regional cerebral blood flow.[120–122,38] The exact pattern appears to be highly dependent on the provocation strategy used to elicit the mood state, the timing of the scan acquisition relative to the induction of the desired mood, and the data analysis methods used to interpret the results. However, despite technical differences among studies, the limbic, paralimbic, and neocortical components of the proposed model are repeatedly identified in all reports. Increases in the ventromedial frontal cortex and anterior cingulate are the most consistent finding.

In our experiment,[38] the goal was to separate the neural systems for dysphoria from those for attention and cognition in order to better interpret the results of our ongoing FDG studies of primary and secondary depression.[30,39–40] Surprisingly, the results of this study suggested that in healthy subjects, these behaviors were inseparable: the entire limbic-cortical depression network was simultaneously activated. Specifically, when a sad mood state was induced and maintained, blood flow increased in ventral paralimbic regions (anterior insula, subgenual cingulate) and decreased in dorsal neocortical and limbic regions (prefrontal, inferior parietal, dorsal anterior cingulate, posterior cingulate) (Figure 12–2). The localization of the dorsal decreases overlaps both resting state abnormalities seen in depressed patients[27–28,30,39] and sites of increased blood flow seen in studies of normal selective and directed attention.[52–55] Of relevance to the model is the finding that the normal experience of sadness appears to affect widespread cortical systems that control selective cognitive behaviors, in a pattern similar to that seen in depressed patients.

Resting State Patterns in Depressed Patients: Unique Role of the Rostral Cingulate

Despite the general consensus as to the regional localization of functional changes across imaging studies of depressed patients, there are some troubling discrepan-

FIGURE 12–2. Reciprocal changes in cortical and paralimbic function with manipulation of mood state. Left images: Z-score maps demonstrating changes in regional glucose metabolism (fluorodeoxyglucose PET) in depressed patients following 6 weeks of fluoxetine treatment. Right images: changes in regional blood flow (oxygen-15 water PET) in healthy volunteers 10 minutes after induction of acute sadness. Depression recovery and induced sadness involve changes in identical dorsal frontal and ventral paralimbic brain regions. Depression recovery is associated with increases in dorsal regions and decreases in ventral regions. The reverse is seen with induced sadness, where dorsal areas decrease and ventral areas increase with change in mood state. F = frontal; cd = caudate; ins = anterior insula; Cg 25 = subgenual cingulate; Hth = hypothalamus; pCg 31 = posterior cingulate. Color scale: *red* = increases, *green* = decreases in flow or metabolism.

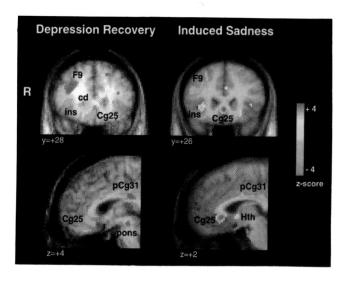

cies. These include contradictory reports as to whether depression is characterized by frontal and cingulate hypo- or hyperfunctioning.[28-29,90,101] One view maintains that this variability is somehow related to the heterogeneity of clinical symptoms such as inattention, apathy, psychomotor slowing, or cognitive impairment, and several studies support this argument.[92,93] Other explanations implicate medication status (drug naive versus drug washouts of varying duration), patient selection (familial versus random), severity of the illness, and transient fluctuations in mood at the time of the PET study.[28,36,90-91]

Findings from our own studies suggest an alternative explanation. We tested the hypothesis that specific metabolic patterns could predict the responsiveness of depressed patients to antidepressant medication.[39,40] In both treatment responders and nonresponders, decreases were found in frontal, parietal, dorsal cingulate, and insular cortex, consistent with previous reports. In contrast, metabolism in the rostral anterior cingulate uniquely distinguished the two groups. Patients with high pretreatment rostral anterior cingulate metabolism went on to show a good response, whereas those with low metabolism remained significantly depressed after 6 weeks of treatment (Figure 12–3). Metabolism in no other region discriminated the two groups, nor did associated demographic, clinical, or behavioral measures, including motor speed, cognitive performance, depression severity, or illness chronicity.

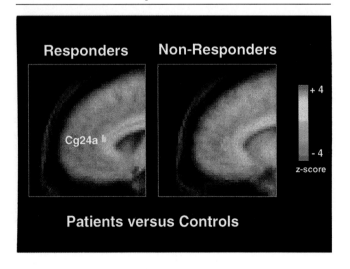

FIGURE 12–3. Predictive value of rostral cingulate metabolism in depression for 6-week treatment response. Z-score maps demonstrating differences in the direction of changes seen in pretreatment rostral cingulate glucose metabolism (Brodmann area 24a) in two groups of depressed patients compared with healthy control subjects. Rostral cingulate hypometabolism (right image, negative z-score shown in *green*) characterizes the eventual nonresponder group, in contrast to hypermetabolism (left image, positive z-score shown in *yellow*) seen in the eventual treatment responders.

This variation in rostral anterior cingulate activity (Brodmann area 24a) is of particular relevance to the model because this region has reciprocal connections with dorsal anterior cingulate, as well as with a number of ventral paralimbic regions (insula, basal frontal, hippocampus, subgenual cingulate, amygdala).[8,10,12,17,59] These rostral cingulate projection sites are the same areas where metabolic changes were seen in this study across the entire depressed patient group.[39] The finding that metabolic activity in the rostral cingulate discriminates eventual responders from nonresponders suggests that this area may function as a bridge linking dorsal and ventral pathways necessary for the normal integrative processing of mood, motor, autonomic, and cognitive behaviors—all of which are disrupted in depression (see Figure 12–1).

The fact that responders and nonresponders show an inverse pattern compared with control subjects further suggests that an adaptive hypermetabolic change in the rostral cingulate may be required to facilitate response to treatment—a compensatory response not present in the nonresponder group. In this context, a central role for rostral cingulate in the depression model is reinforced because integrity of this region appears necessary for the normalization of cortical and paralimbic dysfunction that accompanies recovery from depression. Rostral cingulate function may be a marker of potential network plasticity. Functional failure of this region, as indexed by hypometabolism, appears to be predictive of poor adaptive potential and eventual poor outcome. Presence of this metabolic signature in individual patients may be clinically useful in identifying those at risk for a difficult disease course.

Treatment Effects

A final element in testing the model is to explicitly examine whether the regional abnormalities are static or dynamic. State-trait studies of acutely ill and remitted patient groups are one approach. An alternative with direct implications for the model is to examine how dorsal and ventral regions change with different treatments and how these changes reflect overall and symptom-specific improvement.

Published studies demonstrate that recovery from depression is associated with normalization of certain regional abnormalities. Increases in dorsal frontal and dorsal anterior cingulate hypometabolism and hypoperfusion have been reported with drug therapy, suggesting that these defects are state markers of the illness.[27,29,123-125] On the other hand, studies of sleep deprivation and electroconvulsive therapy (ECT), as well as group comparisons of ill and remitted patients, suggest additional state and trait changes involving limbic and paralimbic regions including the anterior cingulate, ventral frontal cortex, subgenual cingulate, caudate, and amygdala.[29,90,101,126,127] However, the regions affected and the direction of the changes reported are variable across studies and across treatment modalities. This is particularly true of changes seen in the anterior cingulate, where differences in study design, the size and location of cingulate sampling, and data analysis strategies preclude making direct comparisons across experiments.[29,126,127]

In a recent study, we tested the validity of the model by using a pharmacological treatment trial in acutely ill depressed patients.[41-43] Fluoxetine treatment resulted in regional changes in both the dorsal compartment (prefrontal, premotor, dorsal anterior cingulate, and posterior cingulate) and the ventral compartment (subgenual cingulate area 25, anterior insula, hippocampus, and ventral frontal cortex). Clinical response was reflected in the direction of changes in dorsal neocortical and ventral paralimbic regions: dorsal frontal cortex increases were seen only in responders, and this increase was a normalization of the pretreatment hypometabolic pattern (Figure 12–2); ventral paralimbic areas showed decreases, and, unlike changes seen in dorsal neocortex, these were not due to the normalization of an abnormal metabolic pattern. Pretreatment metabolism in these regions was normal to slightly elevated, and symptom improvement was actually associated with new hypometabolism of these ventral paralimbic regions. In contrast, nonresponders with identical treatment showed an increase in metabolism in these same ventral regions. These divergent effects in responders and nonresponders suggest that there are differences in the adaptation of target regions to chronic serotonergic modulation in different patient groups. Studies using varying doses and different drugs are needed to validate this hypothesis.

Interpretation of Findings

These findings indicate that recovery requires both the inhibition of overactive paralimbic regions and the normalization of hypofunctioning dorsal neocortical sites (Figure 12–1, black arrows). A further inference is that, via the rostral and posterior cingulate, suppression of ventral paralimbic activity results in the disinhibition of the dorsal compartment. This theory is supported by the strong correlation of mood improvement with both dorsal compartment increases (dFr 46/9) and ventral compartment decreases (Cg 25, Hc). Remission of sleep and vegetative disturbances, on the other hand, correlated most significantly with ventral paralimbic suppression, and improved cognitive performance tracked primarily with normalization of dorsal prefrontal hypometabolism.[42,43]

Consistent with its role as a trait marker, rostral cingulate metabolism showed no change with treatment in either the responder or nonresponder group and did not correlate with improvement in any of the measured behaviors. As previously noted, the direct anatomical connections of this region to dorsal and ventral regions showing metabolic changes with treatment further supports its key regulatory role in modulating the interaction between the dorsal and ventral compartments.

Similarities Between Depression Recovery and Induced Sadness

The localization of the regional changes seen in patients with a good response to fluoxetine treatment is identical to that demonstrated in the induced sadness experiment (Figure 12–2). Recovery from depression is associated with decreases in ventral paralimbic areas and increases in the dorsal limbic and neocortical regions. Induction of transient sadness shows this identical pattern, but in reverse: increases in ventral paralimbic regions and decreases in dorsal neocortex. These shifts in the relative relationship between the ventral and dorsal compartments as a function of varying the overall mood state provide strong evidence for a reciprocal interaction among these regions in both health and disease (Figure 12–1).

Psychosurgical Parallels

The postulate that poor response reflects the inability to suppress ventral paralimbic regions (required for the release and normalization of dorsal neocortical areas) is further supported by the use of destructive limbic-paralimbic lesions (anterior leukotomy, subcallosal or superior cingulotomy) to alleviate severe refractory depression.[128-131] The model maintains that recovery requires the disinhibition or release of the abnormally functioning dorsal compartment by suppression or disconnection of ventral paralimbic inputs, an effect facilitated by these specific neurosurgical lesions.

CONCLUSIONS

Although many unanswered questions remain, this series of experiments provides strong support for the proposed componential model of depression. Sadness and depressive illness are both associated with decreases in dorsal limbic (anterior and posterior cingulate) and neocortical regions (prefrontal, premotor, parietal cortex) and relative increases in ventral paralimbic areas (subgenual cingulate, anterior insula, hypothalamus, caudate). While additional experimental studies are clearly needed, it is postulated that illness remission, whether facilitated by psychotherapy, medication, ECT, or surgery, requires the inhibition of overactive ventral areas with resulting disinhibition of underactive dorsal regions. Integrity of the rostral cingulate, with its direct anatomical connections to both compartments, appears to be additionally required for the occurrence of these adaptive responses. In the course of further testing of these hypotheses, this model will certainly evolve. For now, it provides a useful framework for facilitating the continued integration of functional imaging findings with ongoing basic neuroanatomical, neurochemical, electrophysiological, and developmental studies.[95–97,132–135] It is hoped that these strategies will contribute to the development of new system- and symptom-specific treatments and the elucidation of the pathogenesis of depression and related affective disorders.

REFERENCES

1. Broca P: Anatomie comparé des circonvolutions cérèbrales: le grant lobe limbique et la scissure limbique dans la série des mammifères [Comparative anatomy of cerebral circumvulations: the great limbic lobe and the limbic fissure in the mammalian series]. Rev Anthropol 1878; 1:385–498
2. Papez JW: A proposed mechanism of emotion. Archives of Neurology and Psychiatry 1937; 38:725–743
3. Yakovlev PI: Motility, behavior, and the brain: stereodynamic organization and neural coordinates of behavior. J Nerv Ment Dis 1948; 107:313–335
4. MacLean PD: The Triune Brain in Evolution: Role in Paleocerebral Function. New York, Plenum, 1990, pp 247–268
5. Mesulam M-M: Large scale neurocognitive networks and distributed processing for attention, language and memory. Ann Neurol 1990; 28:587–613
6. Petrides M, Pandya D: Comparative architectonic analysis of the human and macaque frontal cortex, in Handbook of Neuropsychology, edited by Grafman J, Boller F. Amsterdam, Elsevier Science, 1994, pp 17–58
7. Nauta WJH: Circuitous connections linking cerebral cortex, limbic system, and corpus striatum, in The Limbic System: Functional Organization and Clinical Disorders, edited by Doane BK, Livingston KE. New York, Raven, 1986, pp 43–54
8. Vogt BA, Nimchinsky EA, Vogt LJ, et al: Human cingulate cortex: surface features, flat maps, and cytoarchitecture. J Comp Neurol 1995; 359:490–506
9. Carmichael ST, Price JL: Architectonic subdivision of the orbital and medial prefrontal cortex in the macaque monkey. J Comp Neurol 1994; 346:366–402
10. Morecraft RJ, Geula C, Mesulam M-M: Cytoarchitecture and neural afferents of orbitofrontal cortex in the brain of the monkey. J Comp Neurol 1992; 323:341–358
11. Mesulam M-M, Mufson EJ: Insula of the old world monkey, I, II, III. J Comp Neurol 1992; 212:1–52
12. Kunishio K, Haber SN: Primate cingulostriatal projection: limbic striatal versus sensorimotor striatal input. J Comp Neurol 1994; 350:337–356
13. Alexander GE, Crutcher MD, De Long MR: Basal ganglia-thalamocortical circuits: parallel substrates for motor, oculomotor, "prefrontal" and "limbic" functions. Prog Brain Res 1990; 85:119–146
14. Goldman-Rakic PS, Selemon LD: Topography of corticostriatal projections in nonhuman primates and implications for functional parcellation of the neostriatum, in Cerebral Cortex, edited by Jones EG, Peters A. New York, Plenum, 1984, pp 447–466
15. Nauta WJH: The problem of the frontal lobe: a reinterpretation. Journal of Psychiatric Research 1971; 8:167–187
16. Baleydier C, Mauguiere F: The duality of the cingulate gyrus in the rhesus monkey: neuroanatomical study and functional hypotheses. Brain 1980; 103:525–554
17. Vogt BA, Pandya DN: Cingulate cortex of the rhesus monkey, II: cortical afferents. J Comp Neurol 1987; 262:271–289
18. Morecraft RJ, Geula C, Mesulam MM: Architecture of connectivity within a cingulo-fronto-parietal neurocognitive network for directed attention. Arch Neurol 1993; 50:279–284
19. Rolls ET: Connections, functions and dysfunctions of limbic structures, the prefrontal cortex and hypothalamus, in The Scientific Basis of Clinical Neurology, edited by Swash M, Kennard C. London, Churchill Livingstone, 1985, pp 201–213
20. Damasio AR: Descartes' Error. New York, Putnam, 1994
21. LeDoux J: The Emotional Brain. New York, Simon and Schuster, 1996
22. Devinsky O, Morrell MJ, Vogt BA: Contributions of anterior cingulate cortex to behavior. Brain 1995; 118:279–306
23. Davidson RJ, Sutton SK: Affective neuroscience: the emergence of a discipline. Curr Opin Neurobiol 1995; 5:217–224
24. Liotti M, Tucker DM: Emotions in asymmetric corticolimbic networks, in Brain Asymmetry, edited by Davidson RJ, Hugdahl K. Cambridge, MA, MIT Press, 1995, pp 389–423
25. Starkstein SE, Robinson RG (eds): Depression in Neurologic Diseases. Baltimore, Johns Hopkins University Press, 1993
26. Coffey CE, Wilkinson WE, Weiner RD, et al: Quantitative cerebral anatomy in depression. Arch Gen Psychiatry 1993; 50:7–16
27. Baxter LR, Schwartz JM, Phelps ME, et al: Reduction of prefrontal cortex glucose metabolism common to three types of depression. Arch Gen Psychiatry 1989; 46:243–250
28. Ketter TA, George MS, Kimbrell TA, et al: Functional brain imaging, limbic function, and affective disorders. The Neuroscientist 1996; 2:55–65
29. Ebert D, Ebmeier K: Role of the cingulate gyrus in depression: from functional anatomy to depression. Biol Psychiatry 1996; 39:1044–1050
30. Mayberg HS: Frontal lobe dysfunction in secondary depression. J Neuropsychiatry Clin Neurosci 1994; 6:428–442
31. Mayberg HS, Robinson RG, Wong DF, et al: PET imaging of cortical S_2-serotonin receptors after stroke: lateralized changes and relationship to depression. Am J Psychiatry 1988; 145:937–943
32. Mayberg HS, Parikh RM, Morris PLP, et al: Spontaneous remission of post-stroke depression and temporal changes in cortical S_2-serotonin receptors. J Neuropsychiatry Clin Neurosci 1991; 3:80–83
33. Mayberg HS, Starkstein SE, Sadzot B, et al: Selective hypometabolism in the inferior frontal lobe in depressed patients with Parkinson's disease. Ann Neurol 1990; 28:57–64
34. Mayberg HS, Starkstein SE, Peyser CE, et al: Paralimbic frontal lobe hypometabolism in depression associated with Huntington's disease. Neurology 1992; 42:1791–1797
35. Starkstein SE, Mayberg HS, Berthier ML, et al: Mania after brain injury: neuroradiological and metabolic findings. Ann Neurol 1990; 27:652–659
36. Mayberg HS, Lewis PJ, Regenold W, et al: Paralimbic hypoperfusion in unipolar depression. J Nucl Med 1994; 35:929–934

37. Mayberg HS, Ross CA, Dannals RF, et al: Elevated mu opiate receptors measured by PET in patients with depression (abstract). J Cereb Blood Flow Metab 1991; 11(suppl):821
38. Mayberg HS, Liotti M, Jerabek PA, et al: Induced sadness: a PET model of depression (abstract). Human Brain Mapping 1995(suppl 1):396
39. Mayberg HS, Brannan SK, Mahurin RK, et al: Cingulate function in depression: a potential predictor of treatment response. Neuroreport 1997; 8:1057–1061
40. Mayberg HS, Brannan SK, Mahurin RK, et al: Anterior cingulate function and mood: evidence from FDG PET studies of primary and secondary depression (abstract). Neurology 1996; 46:A327
41. Mayberg HS, Mahurin RK, Brannan SK, et al: Parkinson's depression: discrimination of mood-sensitive and mood-insensitive cognitive deficits using fluoxetine and FDG PET (abstract). Neurology 1995; 45(suppl):A166
42. Mayberg HS, Brannan SK, Mahurin RK, et al: Functional correlates of mood and cognitive recovery in depression: an FDG PET study (abstract). Human Brain Mapping 1995; Suppl 1:428
43. Mayberg HS, Brannan SK, Liotti M, et al: The role of the cingulate in mood homeostasis (abstract). Society for Neuroscience Abstracts 1996; 22:267
44. American Psychiatric Association: Diagnostic and Statistical Manual of Mental Disorders, 4th edition. Washington, DC, American Psychiatric Association, 1994
45. Weingartner H, Cohen RM, Murphy DL, et al: Cognitive processes in depression. Arch Gen Psychiatry 1981; 38:42–47
46. Flint AJ, Black SE, Campbell-Taylor I, et al: Abnormal speech articulation, psychomotor retardation, and subcortical dysfunction in major depression. J Psychiatr Res 1993; 27:285–287
47. Taylor AE, Saint-Cyr JA, Lange AE: Idiopathic Parkinson's disease: revised concepts of cognitive and affective status. Can J Neurol Sci 1988;15:106–113
48. Rogers D, Lees AJ, Smith E, et al: Bradyphrenia in Parkinson's disease and psychomotor retardation in depressive illness: an experimental study. Brain 1987; 110:761–776
49. Mesulam MM: Patterns in behavioral neuroanatomy: association areas, the limbic system, and hemispheric specialization, in Principles of Behavioral Neurology, edited by Mesulam MM. Philadelphia, FA Davis, 1985, pp 1–70
50. Dolan RJ, Bench CJ, Liddle PF, et al: Dorsolateral prefrontal cortex dysfunction in the major psychoses: symptom or disease specificity? J Neurol Neurosurg Psychiatry 1993; 56:1290–1294
51. Mayberg HS: Clinical correlates of PET and SPECT defects in dementia. J Clin Psychiatry 1994; 55:12–21
52. Pardo JV, Reichle ME, Fox PT: Localization of a human system for sustained attention by positron emission tomography. Nature 1991; 349:61–63
53. Corbetta M, Miezin FM, Shulman GL, et al: A PET study of visuospatial attention. J Neurosci 1993; 13:1020–1226
54. Posner MI, Petersen SE: The attention system of the human brain. Annu Rev Neurosci 1990; 13:25–42
55. Paus T, Petrides M, Evans AC, et al: Role of the human anterior cingulate cortex in the control of oculomotor, manual and speech responses: a PET study. J Neurophysiol 1993; 70:453–469
56. Pandya DN, Kuypers HGJM: Cortico-cortical connections in the rhesus monkey. Brain Res 1969; 13:13–36
57. Chavis DA, Pandya DN: Further observations on corticofrontal connections in the rhesus monkey. Brain Res 1976; 117:369–386
58. Petrides M, Pandya DN: Projections to the frontal cortex from the posterior parietal region in the rhesus monkey. J Comp Neurol 1984; 228:105–116
59. Van Hoesen GW, Morecraft RJ, Vogt BA: Connections of the monkey cingulate cortex, in The Neurobiology of Cingulate Cortex and Limbic Thalamus: A Comprehensive Handbook, edited by Vogt BA, Gabriel M. Boston, Birkhauser, 1993, pp 249–284
60. Cowan PJ: Biological markers of depression. Psychol Med 1991; 21:831–836
61. Nemeroff CB, Ranga K, Krishnan R: Neuroendocrine alterations in psychiatric disorders, in Neuroendocrinology, edited by Nemeroff CB. Boca Raton, FL, CRC press, 1992, pp 413–441
62. Neafsey EJ: Prefrontal cortical control of the autonomic nervous system: anatomical and physiological observations, in Progress in Brain Research, vol 85, edited by Uylings HBM, Van Eden CG, DeBruin JPC, et al. New York, Elsevier, 1990, pp 147–166
63. Augustine JR: The insular lobe in primates including humans. Neurol Res 1985; 7:2–10
64. Porrino LJ, Crane AM, Goldman-Rakic PS: Direct and indirect pathways from the amygdala to the frontal lobe in rhesus monkeys. J Comp Neurol 1981; 198:121–136
65. Kleist K: Bericht über die Gehirnpathologie in ihrer Bedeutung für Neurologie und Psychiatrie [Report on the pathology of the brain and its significance for neurology and psychiatry]. Zeitschrift für des Gesamte Neurologie und Psychiatrie 1937; 158:159–193
66. Robinson RG, Kubos KL, Starr LB, et al: Mood disorders in stroke patients: importance of location of lesion. Brain 1984; 107:81–93
67. Stuss DT, Benson DF: The Frontal Lobes. New York, Raven, 1986, pp 121–138
68. Damasio AR, Van Hoesen GW: Emotional disturbances associated with focal lesions of the limbic frontal lobe, in Neuropsychology of Human Emotion, edited by Heilman KM, Satz P. New York, Guilford, 1983, pp 85–110
69. Grafman J, Vance SC, Weingartner H, et al: The effects of lateralized frontal lesions on mood regulation. Brain 1986; 109:1127–1148
70. Ross ED, Rush AJ: Diagnosis and neuroanatomical correlates of depression in brain-damaged patients. Arch Gen Psychiatry 1981; 39:1344–1354
71. Altshuler LL, Devinsky O, Post RM, et al: Depression, anxiety, and temporal lobe epilepsy: laterality of focus and symptoms. Arch Neurol 1990; 47:284–288
72. Honer WG, Hurwitz T, Li DKB, et al: Temporal lobe involvement in multiple sclerosis patients with psychiatric disorders. Arch Neurol 1987; 44:187–190
73. Starkstein SE, Robinson RG, Price TR: Comparison of cortical and subcortical lesions in the production of post-stroke mood disorders. Brain 1987; 110:1045–1059
74. Mendez MF, Adams NL, Lewandowski KS: Neurobehavioral changes associated with caudate lesions. Neurology 1989; 39:349–354
75. Mayeux R: Emotional changes associated with basal ganglia disorders, in Neuropsychology of Human Emotion, edited by Heilman KM, Satz P. New York, Guilford, 1983, pp 141–164
76. McHugh PR: The basal ganglia: the region, the integration of its systems and implications for psychiatry and neurology, in Function and Dysfunction of the Basal Ganglia, edited by Franks AJ, Ironside JW, Mindham RHS, et al. Manchester, England, Manchester University Press, 1990, pp 259–269
77. Jagust WJ, Reed BR, Martin EM, et al: Cognitive function and regional cerebral blood flow in Parkinson's disease. Brain 1992; 115:521–537
78. Ring HA, Bench CJ, Trimble MR, et al: Depression in Parkinson's disease: a positron emission study. Br J Psychiatry 1994; 165:333–339
79. Grasso MG, Pantano P, Ricci M, et al: Mesial temporal cortex hypoperfusion is associated with depression in subcortical stroke. Stroke 1994; 25:980–985
80. Laplane D, Levasseur M, Pillon B, et al: Obsessive-compulsive and other behavioural changes with bilateral basal ganglia lesions: a neuropsychological, magnetic resonance imaging and positron tomography study. Brain 1989; 12:699–725
81. Bromfield EB, Altschuler L, Leiderman DB, et al: Cerebral metabolism and depression in patients with complex partial seizures. Arch Neurol 1992; 49:617–623

82. Mayeux R, Stern Y, Sano M, et al: The relationship of serotonin to depression in Parkinson's disease. Mov Disord 1988; 3:237–244
83. Fibiger HC: The neurobiological substrates of depression in Parkinson's disease: a hypothesis. Can J Neurol Sci 1984; 11:105–107
84. Cummings JL: Depression and Parkinson's disease: a review. Am J Psychiatry 1992; 149:443–454
85. Zubenko GS, Sullivan P, Nelson JP, et al: Brain imaging abnormalities in mental disorders of late life. Arch Neurol 1990; 47:1107–1111
86. Dupont RM, Jernigan TL, Heindel W, et al: Magnetic resonance imaging and mood disorders: localization of white matter and other subcortical abnormalities. Arch Gen Psychiatry 1995; 52:747–755
87. Post RM, DeLisi LE, Holcomb HH, et al: Glucose utilization in the temporal cortex of affectively ill patients: positron emission tomography. Biol Psychiatry 1987; 22:545–553
88. Buchsbaum MS, Wu J, DeLisi LE, et al: Frontal cortex and basal ganglia metabolic rates assessed by positron emission tomography with 18F-2-deoxyglucose in affective illness. J Affect Disord 1986; 10:137–152
89. Bench CJ, Friston KJ, Brown RG, et al: The anatomy of melancholia: focal abnormalities of cerebral blood flow in major depression. Psychol Med 1992; 22:607–615
90. Drevets WC, Videen TO, Price JL, et al: A functional anatomical study of unipolar depression. J Neurosci 1992; 12:3628–3641
91. Lesser I, Mena I, Boone KB, et al: Reduction of cerebral blood flow in older depressed patients. Arch Gen Psychiatry 1994; 51:677–686
92. Bench CJ, Friston KJ, Brown RG, et al: Regional cerebral blood flow in depression measured by positron emission tomography: the relationship with clinical dimensions. Psychol Med 1993; 23:579–590
93. Dolan RJ, Bench CJ, Brown RG, et al: Regional cerebral blood flow abnormalities in depressed patients with cognitive impairment. J Neurol Neurosurg Psychiatry 1992; 55:768–773
94. Bauer M, Frazer A: Mood disorders and their treatment, in Biological Bases of Brain Function and Disease, 2nd edition, edited by Frazer A, Molinoff P, Winokur A. New York, Raven, 1994, pp 303–323
95. Hyman SE, Nestler EJ: Initiation and adaptation: a paradigm for understanding psychotropic drug action. Am J Psychiatry 1996; 153:151–162
96. Nibuya M, Nestler EJ, Duman RS: Chronic antidepressant administration increases the expression of cAMP response element binding protein (CREB) in rat hippocampus. J Neurosci 1996; 16:2365–2372
97. Duncan GE, Knapp DJ, Johnson KB, et al: Functional classification of antidepressants based on antagonism of swim stress-induced fos-like immunoreactivity. J Pharmacol Exp Ther 1996; 277:1076–1089
98. Ballenger JC: Biological aspects of depression: implications for clinical practice, in Review of Psychiatry, vol 7, edited by Frances AJ, Hales RE. Washington, DC, American Psychiatric Press, 1988, pp 169–187
99. Caldecott-Hazzard S, Morgan DG, Delison-Jones F, et al: Clinical and biochemical aspects of depressive disorders, II: transmitter/receptor theories. Synapse 1991; 9:251–301
100. Stancer HC, Cooke RG: Receptors in affective illness, in Receptors and Ligands in Psychiatry, edited by Sen AK, Lee T. New York, Cambridge University Press, 1988, pp 303–326
101. Drevetz WC, Raichle ME: Neuroanatomical circuits in depression: implications for treatment mechanisms. Psychopharmacol Bull 1992; 28:261–274
102. Swerdlow NR, Koob GF: Dopamine, schizophrenia, mania and depression: towards a unified hypothesis of cortico-striato-pallido-thalamic function. Behav Brain Sci 1987; 10:197–245
103. Delgado PL, Charney DS, Price LH, et al: Serotonin function and the mechanism of antidepressant action: reversal of antidepressant-induced remission by rapid depletion of plasma tryptophan. Arch Gen Psychiatry 1990; 47:411–418
104. Meltzer H (ed): Psychopharmacology: The Third Generation of Progress. New York, Raven, 1987
105. Arango V, Ernsberger P, Marzuk PM, et al: Autoradiographic demonstration of increased serotonin 5-HT$_2$ and β-adrenergic receptor binding sites in the brain of suicide victims. Arch Gen Psychiatry 1990; 47:1038–1047
106. Crino PB, Morrison JH, Hof PR: Monoamine innervation of cingulate cortex, in The Neurobiology of Cingulate Cortex and Limbic Thalamus: A Comprehensive Handbook, edited by Vogt BA, Gabriel M. Boston, Birkhauser, 1993, pp 285–310
107. Chaput Y, deMontigny C, Blier P: Presynaptic and postsynaptic modifications of the serotonin system by long-term administration of antidepressant treatments: an in vitro electrophysiologic study in the rat. Neuropsychopharmacology 1991; 5:219–229
108. Goldfarb J: Electrophysiologic studies of serotonin receptor activation. Neuropharmacology 1990; 3:435–466
109. Frazer A, Hensler JG: 5HT$_{1a}$ receptors and 5HT$_{1a}$-mediated responses: effect on treatments that modify serotonergic neurotransmission, in The Neuropharmacology of Serotonin, edited by Whitaker-Azmitia PM, Peroutka SJ. New York, NY Academy of Sciences, 1990, pp 460–475
110. Simon H, LeMoal M, Calas A: Efferents and afferents of the ventral tegmental-A$_{10}$ region studied after local injection of ^{3}H-leucine and horseradish peroxidase. Brain Res 1979; 178:17–40
111. Graybiel AM: Neurotransmitters and neuromodulators in the basal ganglia. Trends Neurosci 1990; 13:244–254
112. Wise RA: The dopamine synapse and the notion of "pleasure centers" in the brain. Trends Neurosci 1980; 4:91–95
113. Martin WR, Sloan JW, Sapira JD, et al: Physiologic subjective, and behavioural effects of amphetamine, methamphetamine, ephedrine, phenmetrazine, and methylphenidate in man. Clin Pharmacol Ther 1975; 32:632–637
114. Cantello R, Aguaggia M, Gilli M, et al: Major depression in Parkinson's disease and the mood response to intravenous methylphenidate: possible role of the "hedonic" dopamine synapse. J Neurol Neurosurg Psychiatry 1989; 52:724–731
115. Torack RM, Morris JC: The association of ventral tegmental area histopathology with adult dementia. Arch Neurol 1988; 45:211–218
116. Petty F, Kramer GL, Gullion CM, et al: Low plasma gamma-aminobutyric acid levels in male patients with depression. Biol Psychiatry 1992; 32:354–363
117. Gross-Iseseroff R, Dollon KA, Israeli M, et al: Regionally selective increases in mu opioid receptor density in the brains of suicide victims. Brain Res 1990; 530:312–316
118. Nemeroff CB, Widerlov E, Bissette G, et al: Elevated concentrations of CSF corticotropin-releasing factor-like immunoreactivity in depressed patients. Science 1984; 226:1342–1343
119. Young EA, Akil H, Haskett RF, et al: Evidence against changes in corticotroph CRF receptors in depressed patients. Biol Psychiatry 1995; 37:355–363
120. Pardo JV, Pardo PJ, Raichel ME: Neural correlates of self-induced dysphoria. Am J Psychiatry 1993; 150:713–719
121. George MS, Ketter TA, Parekh PI, et al: Brain activity during transient sadness and happiness in healthy women. Am J Psychiatry 1995; 152:341–351
122. Schneider F, Gur RE, Mozley LH, et al: Mood effects on limbic blood flow correlate with emotional self-rating: a PET study with oxygen-15 labeled water. Psychiatry Res 1995; 61:265–283
123. Martinot JL, Hardy P, Feline A, et al: Left prefrontal glucose hypometabolism in the depressed state: a confirmation. Am J Psychiatry 1990; 147:1313–1317
124. Bench CJ, Frackowiak RSJ, Dolan RJ: Changes in regional cerebral blood flow on recovery from depression. Psychol Med 1995; 25:247–251
125. Goodwin GM, Austin MP, Dougall N, et al: State changes in brain activity shown by the uptake of 99mTc-exametazime with single photon emission tomography in major depression before and after treatment. J Affect Disord 1993; 29:243–253

126. Wu JC, Gilin JC, Buchsbaum MS, et al: Effect of sleep deprivation on brain metabolism of depressed patients. Am J Psychiatry 1992; 149:538–543
127. Nobler MS, Sackeim HA, Prohovnik I, et al: Regional cerebral blood flow in mood disorders, III: treatment and clinical response. Arch Gen Psychiatry 1994; 51:884–897
128. Fulton JF: Frontal Lobotomy and Affective Behavior: A Neurophysiological Analysis. London, Chapman and Hall, 1951
129. Cosgrove GR, Rausch SL: Psychosurgery. Neurosurg Clin N Am 1995; 6:167–176
130. Malizia A: Frontal lobes and neurosurgery for psychiatric disorders. J Psychopharmacol 1997; 2:179–187
131. Livingston KE, Escobar A: Tentative limbic system models for certain patterns of psychiatric disorders, in Surgical Approaches in Psychiatry, edited by Laitinen V, Livingston KE. Baltimore, University Park Press, 1972, pp 245–252
132. Diorio D, Viau V, Meaney MJ: The role of the medial prefrontal cortex (cingulate gyrus) in the regulation of hypothalamic-pituitary-adrenal responses to stress. J Neurosci 1993; 13:3839–3947
133. Overstreet DH: The Flinders sensitive line rats: a genetic animal model of depression. Neurosci Biobehav Rev 1993; 17:51–68
134. Chalmers DT, Kwak SP, Mansour A, et al: Corticosteroids regulate brain hippocampal 5-HT$_{1A}$ receptor mRNA expression. J Neurosci 1993; 13:914–923
135. Meaney MJ, Bhatnagar S, Diorio J, et al: Molecular basis for the development of individual differences in the hypothalamic-pituitary-adrenal stress response. Cell Mol Neurobiol 1993; 13:321–347

13

The Neurobiology of Drug Addiction

George F. Koob, Ph.D., Eric J. Nestler, M.D., Ph.D.

Common to most descriptions of drug addiction or substance dependence is the idea of *a compulsion to take a drug, with a loss of control in limiting intake*.[1,2] The diagnostic criteria that are used to diagnose substance dependence incorporate changes in behavior that when represented in a person's daily repertoire are likely to reflect someone who is drug dependent or drug addicted.[2] These symptoms (three of which must be present) include tolerance; withdrawal; persistent desire or unsuccessful attempts to reduce substance use; use in larger amounts than intended; important social, occupational, or recreational activities reduced because of drug use; a great amount of time spent in obtaining the substance; and continued substance use despite recurrent problems resulting from substance use. Clearly, such criteria define a syndrome where behavioral repertoires are significantly narrowed toward substance use and what most would consider compulsive use. (We are using the word *compulsive* in a generic sense, to mean repetitive, driven behavior, rather than in the context of DSM diagnoses of obsessive-compulsive disorders.) For the purposes of this chapter, drug addiction will be equated with substance dependence as defined by the American Psychiatric Association.[2] However, it should be kept in mind that the term *dependence* has a different and more specific meaning in the basic pharmacology literature, as will be seen below. Drug abuse, in contrast to substance dependence, can be readily defined as a maladaptive pattern of drug use resulting in impairment or distress, and it is important to distinguish between the concepts of drug use, abuse, and addiction.[3] Although no animal model exists that incorporates all the signs and symptoms associated with substance dependence, it is becoming clear that many of the criteria used in DSM-IV can be reproduced in various animal models.[4]

DRUG ADDICTION AND REINFORCEMENT

Because drug addiction centers on compulsive, often excessive use of a substance, the concept of reinforcement or motivation is a crucial part of this syndrome. A reinforcer can be defined operationally as "any event that increases the probability of a response." This definition can also be used as a definition for reward, and the two words are often used interchangeably. However, *reward* often connotes some additional emotional value such as pleasure.[5] This contrasts with the concept of punishment, which would entail the ability of an event or drug to decrease the probability of a response.

Historically, most conceptualizations of drug addiction emphasized the development of *tolerance* and *withdrawal*, but recent discussions on this subject have reduced tolerance and withdrawal to optional criteria. However, some have emphasized selective aspects of tolerance and withdrawal, focusing on motivational measures, not physical signs.[6] The concepts of tolerance and withdrawal are key elements supporting the idea that neuroadaptive processes are initiated to counter the acute effects of a drug. Another neuroadaptive process that has been proposed as a key element in the development of drug addiction is *sensitization*. Sensitization can be defined as the opposite of tolerance: "the increased response to a drug that follows its repeated intermittent presentation."[7] These neuroadaptive processes can then persist long after the drug has cleared from the brain; such neuroadaptations have been explored at all levels of drug addiction research, from the behavioral to the molecular.[8] Motivational hypotheses involving both central nervous system counteradaptive changes[9] and sensitization[7] have particular relevance to drug addiction phenomena.[9]

Many sources of reinforcement contribute to compulsive drug use during the course of drug addiction. The primary pharmacological effect of a drug is thought to produce a direct effect through positive or negative reinforcement (Table 13–1). Examples of negative reinforcement would include self-medication of an existing aversive state or self-medication of a drug-generated aversive state (such as withdrawal).[9] The secondary pharmacological effects of a drug can also have powerful motivating properties (Table 13–1). Secondary positive reinforcing effects can be obtained through conditioned positive reinforcement (such as pairing of previously neutral stimuli with acute positive reinforcing effects of drugs). Secondary negative reinforcing effects can be obtained through removal of the conditioned negative reinforcing effects of conditioned abstinence. Using this framework, we can explore the neurobiological bases for the acute positive reinforcing effects of drugs, the negative reinforcing effects imparted by the dependent state, and the conditioned reinforcing effects associated with protracted abstinence and relapse.[10]

DRUG ADDICTION AND NEUROTRANSMISSION

All drugs of abuse interact initially with proteins located at the extracellular aspect of specific synapses[11] (summarized in Table 13–2). For example, opiates activate opioid receptors, and cocaine inhibits reuptake proteins for the monoamine neurotransmitters (dopamine, norepinephrine, and serotonin). Alcohol is thought to act at specific "ethanol-receptive elements," which include the ionotropic γ-aminobutyric acid, type A (GABA$_A$), and *N*-methyl-D-aspartate (NMDA) glutamate receptors as well as voltage-gated ion channels.[12] These initial effects lead, in the short term, to alterations in the functional levels of specific neurotransmitters or to different activation states of specific neurotransmitter receptors in the brain.

However, although the initial effects of drugs of abuse are extracellular, the many effects these drugs elicit are achieved ultimately via the intracellular messenger pathways that transduce these extracellular actions.[11] This mechanism is further discussed below.

POSITIVE REINFORCING EFFECTS OF DRUGS: NEURAL SUBSTRATES

The neural substrates of reward have long been hypothesized to involve the medial forebrain bundle, which contains both ascending and descending pathways that include most of the brain's monoamine systems.[13–15] Much of the early work focused on those structures that supported intracranial self-stimulation: the ventral tegmental area, the basal forebrain, and the medial forebrain bundle that connects these two areas.[13,14,16,17] Recent work on the neurobiology of addiction has provided significant insights into the neurochemical and neuroanatomical components of the medial forebrain bundle, which may provide the key not only to drug reward but also to natural rewards.

The origins and terminal areas of the mesocorticolimbic dopamine system have been the principal focus of research on the neurobiology of drug addiction, and there is now compelling evidence for the importance of this system in drug reward. The major components of this drug reward circuit are the ventral tegmental area (the site of dopaminergic cell bodies), the basal forebrain (the nucleus accumbens, olfactory tubercle, frontal cortex, and amygdala), and the dopaminergic connection between the ventral tegmental area and the basal forebrain. Other components are the opioid peptide, GABA, glutamate, serotonin, and presumably many other neural inputs that interact with the ventral tegmental area and the basal forebrain[18] (Figure 13–1). The functional significance of this circuitry for different types of drug reward will be discussed in the following sections, and a construct called the extended amygdala will be introduced that provides important insights into the relationship of drug reward to natural reward systems.

COCAINE AND AMPHETAMINE: THE MESOCORTICOLIMBIC DOPAMINE SYSTEM

Psychomotor stimulants of high abuse potential interact initially with monoamine transporter proteins. These transporter proteins, which have been cloned and characterized,[19-21] are located on monoaminergic nerve terminals and terminate a monoamine signal by transporting the monoamine from the synaptic cleft back into the terminals. Cocaine is a potent inhibitor of all three monoamine transporters, those for dopamine, serotonin, and norepinephrine, and thereby potentiates monoaminergic transmission. Amphetamine and its derivatives also potentiate monoaminergic transmission, but apparently via a distinct mechanism: by increasing monoamine release. It now appears that amphetamine itself serves as a substrate for all three monoamine transporters and is transported into monoaminergic nerve terminals, where it disrupts the storage of the monoamine transmitters. This disruption leads to an increase in extravesicular levels of the monoamines and to the reverse transport of the monoamine into the synaptic cleft via the monoamine transporters.[22] Certain amphetamine derivatives are toxic to monoaminergic nerve terminals via as-yet-unknown effects within the nerve terminal cytoplasm.[23]

Amphetamine and cocaine are psychomotor stimulants and have behavioral effects consistent with that class of drugs, including suppression of hunger and fatigue[24] and induction of euphoria[25] in humans. In animals, these drugs increase motor activity,[26] decrease food intake,[26] have psychomotor stimulant actions on operant behavior,[27] enhance conditioned responding,[27] decrease thresholds for reinforcing brain stimulation,[28,29] produce preferences for environments where the drugs have been previously experienced (place preferences),[30,31] and readily act as reinforcers for drug self-administration.[32]

Both the psychomotor stimulant effects of amphetamine and cocaine[33-35] and their reinforcing actions depend critically on the mesocorticolimbic dopamine system.[36,37] The most direct evidence implicating dopamine generally in the reinforcing actions of cocaine comes from studies of intravenous self-administration. Low doses of dopamine receptor antagonists, when injected systemically, reliably decrease the reinforcing effects of cocaine and amphetamine self-administration in rats.[36,37] Confirmation that dopamine antagonists block the reinforcing effects of cocaine comes from dose-effect studies where the antagonists shift the cocaine dose-effect function to the right (Figure 13–2).[38] In general, experiments investigating the effects of antagonists selective for dopamine receptor subtypes reveal that antagonists for the D_1,[39] D_2,[40,41] and D_3 receptors decrease the reinforcing properties of cocaine.[42]

The reinforcing actions of cocaine were linked specifically to the mesocorticolimbic dopamine system by a series of observations using neurotoxin-induced lesions of this subset of midbrain dopamine projections. Dopamine-selective lesions with 6-hydroxydopamine (6-OHDA) of the nucleus accumbens produced extinction-like responding and significant and long-lasting decreases in self-administration of cocaine and amphetamine over days.[43,44] These decreases in cocaine self-administration following dopamine-selective lesions of the nucleus accumbens have now been observed in a variety of different tests and conditions, including situations where animals show a decrease in the amount of work they would perform for cocaine[32] and situations where other reinforcers such as food were unaffected but cocaine self-administration was abolished.[45]

OPIATES AND OPIOID PEPTIDE SYSTEMS

Opiate drugs such as heroin have long been known to be readily self-administered intravenously by animals,[46] and if these drugs are provided in limited-access situations, rats and primates[47] will maintain stable levels of opiate intake on a daily basis without obvious signs of physical dependence.[48] In intravenous self-administration studies, systemic and central administration of competitive opiate antagonists decrease opiate reinforcement as measured by an increase in the number of infusions (decrease in the interval between injections) for opiate drugs.[37,48-51] This decrease in reinforcement appears to result from a competitive interaction between the antagonist and agonist at opioid receptors.

The opioid receptor subtype most important for the reinforcing actions of heroin and morphine appears to be the mu receptor. Mu opioid receptor agonists produce dose-dependent decreases in heroin self-administration, and irreversible mu-selective antagonists dose-dependently increase heroin self-administration.[52] Intracerebral injection of quaternary derivatives of opiate antagonists—charged, hydrophilic compounds that

TABLE 13–1. Relationship of addiction components, behavioral constructs, and treatment focus

Addiction Component	Behavioral Construct	Treatment Focus
Pleasure	Positive reinforcement	Motivational
Self-medication	Negative reinforcement	AA and motivational
Habit	Conditioned positive reinforcement	Cognitive/behavioral
Habit	Conditioned negative reinforcement	Cognitive/behavioral

Note: AA = Alcoholics Anonymous.

TABLE 13–2. Acute effects of abused drugs on neurotransmitters and receptors

Drug	Action
Opiates	Agonist at opioid receptors
Cocaine	Inhibits monoamine reuptake transporters
Amphetamine	Stimulates monoamine release
Alcohol	Facilitates GABA$_A$ receptor function and inhibits N-methyl-D-aspartate (NMDA) glutamate receptor function[a]
Nicotine	Agonist at nicotinic acetylcholine receptors
Cannabinoids	Agonist at cannabinoid receptors[b]
Hallucinogens	Partial agonist at 5-HT$_{2A}$[c] serotonin receptors
Phencyclidine (PCP)	Antagonist at NMDA glutamate receptors

[a]The mechanism by which alcohol produces these effects has not been established, but it does not appear to involve direct alcohol binding to the receptors as is the case for the other drugs listed in this table.
[b]Although a specific receptor for cannabinoids has been identified in the brain, the endogenous ligands for this receptor are under current active investigation.
[c]5-Hydroxytryptamine-2.

do not readily spread from the sites in the brain at which they are injected—dose-dependently block heroin self-administration in nondependent rats.[51,53–55] This antagonism was observed when the antagonists were injected into the ventral tegmental area[53] or the region of the nucleus accumbens.[54] However, rats also will self-administer opioid peptides directly in the region of the nucleus accumbens,[56] and heroin self-administration is not blocked by cocaine-blocking doses of dopamine antagonists,[37] nor by dopamine-selective lesions of the mesocorticolimbic dopamine system.[57] Chronic dopamine receptor blockade does not alter heroin self-administration, a finding that further demonstrates dopamine-independent mechanisms.[58] In addition, whereas opioid peptides injected into either the nucleus accumbens or the ventral tegmental area produce dose-related increases in locomotor activity,[59] the effects of nucleus accumbens injections appear to be independent of dopamine release.[60]

Nevertheless, evidence for a dopamine-dependent action for opiates in the ventral tegmental area is strong. Opiates can produce an increase in dopamine release in the nucleus accumbens similar to that of cocaine and ethanol[61] (but see Hemby et al.[62]). Opioid peptides are self-administered into the ventral tegmental area,[63] and microinjections of opioids into the ventral tegmental area will lower brain stimulation reward thresholds and produce robust place preferences.[64] Altogether, these results suggest that neural elements in the region of the ventral tegmental area and the nucleus accumbens are responsible for the reinforcing properties of opiates, and the findings implicate both dopamine-dependent and dopamine-independent mechanisms of opiate action.[57,65–67]

ALCOHOL: MULTIPLE NEUROCHEMICAL SUBSTRATES

Ethanol, barbiturates, and benzodiazepines all are considered sedative-hypnotics and produce a characteristic euphoria, disinhibition, anxiety reduction, sedation, and hypnosis. These drugs exert anti-anxiety effects that are reflected in a reduction of behavior that would be suppressed by punishment in conflict situations in laboratory animals. This anti-conflict effect correlates well with the capacity of sedative-hypnotics to act as anxiolytics in the clinic,[68] and the anti-conflict effect may be a major component of the reinforcing actions of these drugs.

The sedative and anti-punishment (anxiolytic) effects of sedative-hypnotics are mediated via facilitation of the GABA$_A$ receptor.[69] The actions of sedative-hypnotics on this receptor are complex: the drugs do not bind to the receptor at or near the GABA-binding site. Rather, they bind to other sites on the receptor complex and thereby facilitate, via allosteric effects, activation of the receptor by GABA. The net result is potentiation of GABA-induced Cl$^-$ flux through the receptor ionophore.[70] The GABA$_A$ receptor is a heteromeric complex, and the ability of sedative-hypnotics to facilitate receptor function depends on the actual subunit composition of the receptor, which differs markedly throughout the brain. Although benzodiazepines, barbiturates, and ethanol interact with the GABA$_A$ receptor at distinct sites, the fact that they converge on the functioning of the same protein complex no doubt explains the long-appreciated cross-tolerance and cross-dependence exhibited by these drugs.

Neuropharmacological studies of the anxiolytic properties of sedative-hypnotics provided some of the first clues to their reinforcing properties and abuse potential.[71] GABAergic antagonists were found to reverse many of the behavioral effects of ethanol, which led to the hypothesis that GABA has a role in the intoxicating effects of ethanol.[72,73] The partial inverse benzodiazepine agonist Ro 15-4513, which has been shown to reverse some of the behavioral effects of ethanol,[70] produces a dose-dependent reduction of oral ethanol (10%) self-administration in rats.[74,75] More recent studies have shown similar effects with potent GABA antagonists microinjected into the brain, with the most effective site to date being the central nucleus of the amygdala.[3]

FIGURE 13–1. Sagittal rat brain section illustrating a drug (cocaine, amphetamine, opiate, nicotine, and alcohol) neural reward circuit that includes a limbic-extrapyramidal motor interface. *Yellow dotted lines* indicate limbic afferents to the nucleus accumbens (N Acc.). *Orange* represents efferents from the nucleus accumbens thought to be involved in psychomotor stimulant reward. *Red* indicates projection of the mesocorticolimbic dopamine system thought to be a critical substrate for psychomotor stimulant reward. This system originates in the A10 cell group of the ventral tegmental area (VTA) and projects to the N Acc., olfactory tubercle, ventral striatal domains of the caudate-putamen (C-P), and amygdala (AMG). *Green* indicates opioid peptide-containing neurons, systems that may be involved in opiate, ethanol, and possibly nicotine reward. These opioid peptide systems include the local enkephalin circuits (short segments) and the hypothalamic midbrain beta-endorphin circuit (long segment). *Blue* indicates the approximate distribution of $GABA_A$ receptor complexes, some of which may mediate sedative/hypnotic (ethanol) reward, determined by both tritiated flumazenil binding and expression of the alpha, beta, and gamma subunits of the $GABA_A$ receptor. *Yellow solid structures* indicate nicotinic acetylcholine receptors hypothesized to be located on dopaminergic and opioid peptidergic systems. AC = anterior commissure; ARC = arcuate nucleus; Cer = cerebellum; DMT = dorsomedial thalamus; FC = frontal cortex; Hippo = hippocampus; IF = inferior colliculus; LC = locus coeruleus; LH = lateral hypothalamus; OT = olfactory tract; PAG = periaqueductal gray; RPn = raphe pontis nucleus; SC = superior colliculus; SNr = substantia nigra pars reticulata; VP = ventral pallidum. Modified, with permission, from Koob 1992.[18]

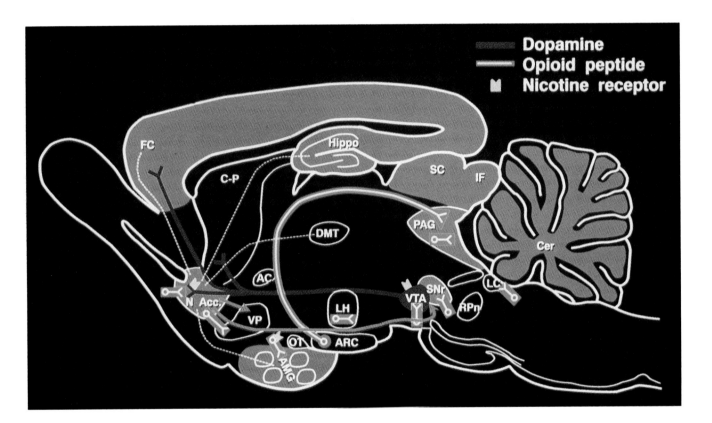

Ethanol, unlike other sedative-hypnotics, also exerts potent effects on the NMDA glutamate receptor. Ethanol inhibits the functioning of the receptor, again not by blocking the glutamate binding site but via a more complex allosteric effect on the receptor complex, which results in diminished glutamate-induced Na^+ and Ca^{2+} flux through the receptor ionophore.[76] Ethanol antagonism of the NMDA receptor appears to contribute to the intoxicating effects of ethanol,[77,78] and perhaps to the dissociative effects seen in people with high ethanol blood levels.[79] Whether ethanol antagonism of the NMDA receptor also contributes to its reinforcing effects remains to be established. At still higher doses, ethanol can exert more general inhibitory effects on voltage-gated ion channels, particularly Na^+ and Ca^{2+} channels.[76] These actions occur only with extreme concentrations seen clinically and would therefore not appear to be involved in the reinforcing actions of ethanol, although they may contribute to the severe nervous system depression, even coma, that are seen at these blood levels.

Via its initial effects on the $GABA_A$ and NMDA glutamate receptors, ethanol influences several additional neurotransmitter systems in the brain that are believed to mediate its reinforcing properties. Again, considerable focus has been placed on dopamine, which is implicated in the reinforcing actions of low, non–dependence-inducing doses of ethanol. Dopamine receptor antagonists have been shown to reduce lever-pressing for ethanol in nondeprived rats,[80] and extracel-

FIGURE 13–2. Shift of the cocaine dose-effect function to the right following pretreatment with dopamine D_1 antagonist SCH-23390. Left: Effects of pretreatment with the D_1 dopamine receptor antagonist SCH-23390 (0.01 mg/kg S.C.) on the cocaine (0.06–0.5 mg) self-administered dose-effect function measured using the within-session dose-effect paradigm ($n = 4$). Right: The same, but for an individual rat. Reprinted, with permission, from Caine and Koob 1995.[38]

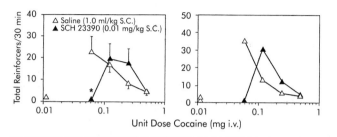

lular dopamine levels also have been shown to increase in nondependent rats orally self-administering low doses of ethanol.[81] However, virtually complete 6-OHDA denervation of the nucleus accumbens failed to alter voluntary responding for alcohol.[82] Thus, as with opiates, these results suggest that while mesocorticolimbic dopamine transmission may be associated with important aspects of ethanol reinforcement, it may not be critical in this regard, and that other, dopamine-independent neurochemical systems likely contribute to the mediation of ethanol's reinforcing actions. In fact, the view is emerging that multiple neurotransmitters combine to "orchestrate" the reward profile of alcohol.[83]

The brain's serotonin systems also have received attention. Modulation of various aspects of serotonergic transmission, including increases in the synaptic availability of serotonin with precursor loading, blockade of serotonin reuptake, or blockade of certain serotonin receptor subtypes, can decrease ethanol intake.[84] Consistent with a role for serotonergic transmission in ethanol abuse are several double-blind, placebo-controlled clinical studies in which serotonin reuptake inhibitors were reported to produce mild decreases in alcohol consumption in humans.[85] However, these findings remain controversial, and, in general, it is now believed that these compounds are of only limited utility in the treatment of nondepressed alcoholics.

Opioid peptide systems have been implicated in alcohol reinforcement by numerous reports that the opioid receptor antagonists naloxone and naltrexone reduce alcohol self-administration in several animal models.[86] Although opioid antagonists dose-dependently decrease consumption of sweet solutions of water as well as ethanol in operant, free-choice tests,[86] it is possible that antagonism of specific opioid receptor subtypes in specific brain regions might reveal more selective effects.[87] A role for opioid peptides in alcohol reinforcement has been further demonstrated by two double-blind, placebo-controlled clinical trials showing that naltrexone significantly reduces alcohol consumption, frequency of relapse, and craving for alcohol in humans.[88,89] Thus, alcohol interactions with opioid neurotransmission may contribute to certain aspects of alcohol reinforcement that may be of particular importance to the motivation associated with relapse.

NICOTINE: DOPAMINE AND OPIOID PEPTIDES

The initial molecular site of action for nicotine is likely to be a direct agonist action at the nicotinic acetylcholine receptors specifically in the brain mesolimbic dopamine system, although brain nicotinic acetylcholine receptors are widely distributed throughout the brain. Nicotine self-administration is blocked by dopamine antagonists and opioid peptide antagonists.[90,91] Nicotine is thus thought to activate both the mesolimbic dopamine system and opioid peptide systems in the same neural circuitry associated with other drugs of abuse[92] (see Figure 13–1).

EXTENDED AMYGDALA: A COMMON SUBSTRATE FOR DRUG REWARD

An interesting hypothesis that is gaining support from recent neuroanatomical data and new functional observations is that the neuroanatomical substrates for the reinforcing actions of drugs may involve a common neural circuitry that forms a separate entity within the basal forebrain, termed the *extended amygdala*.[93] The term represents a macrostructure, originally described by Johnston,[94] that is composed of several basal forebrain structures: the bed nucleus of the stria terminalis, the central medial amygdala, the medial part of the nucleus accumbens (e.g., the area labeled the shell),[95] and the area termed the sublenticular substantia innominata. These structures have similarities in morphology, immunohistochemistry, and connectivity.[93] They receive afferent connections from limbic cortices, hippocampus, basolateral amygdala, midbrain, and lateral hypothalamus. Efferent connections from this complex include the posterior medial (sublenticular) ventral pallidum, medial ventral tegmental area, various brainstem projections, and, perhaps most intriguing from a functional point of view, a considerable projection to the lateral hypothalamus.[96]

Recent studies have demonstrated selective effects of D_1 dopamine antagonists in blocking cocaine self-administration when the antagonist is administered directly into the shell of the nucleus accumbens.[97] Moreover, selective activation of dopaminergic transmission occurs in the shell of the nucleus accumbens in response to acute administration of virtually all major drugs of abuse.[98] In addition, the central nucleus of the amygdala has been implicated in the GABAergic and opioidergic influences on the acute reinforcing effects of ethanol,[99,100] as well as on the aversive stimulus effects of drug withdrawal.[101] This concept of the extended amygdala links the extensive recent developments in the neurobiology of drug reward with existing knowledge of the substrates for natural rewards, bridging what have been largely independent research pursuits.

DRUG DEPENDENCE: NEURAL SUBSTRATES

Common to most drugs of abuse is a withdrawal syndrome that is made up of two elements. There are the physical signs of withdrawal, which are characteristic for each drug, such as the well-known tremor and autonomic hyperactivity of alcohol withdrawal and the abdominal discomfort and pain associated with opiate withdrawal. There are also the "psychological" aspects of drug withdrawal, which may be considered more motivational; these signs consist of varying components of a negative emotional state including dysphoria, depression, anxiety, and malaise.[2,10]

The neural substrates for the physical signs of drug withdrawal are not well understood in general and probably involve many different brain sites and neurochemical systems for the various types of drugs of abuse. Such neural substrates are best established for opiates. Much evidence implicates the nucleus locus coeruleus (the brain's predominant noradrenergic nucleus, located in the pons) in the activational properties and stress-like effects of opiate withdrawal.[102–104] In addition, there is evidence that the changes in body temperature associated with opiate withdrawal may be due to interactions in the hypothalamus. Specific neural substrates for different aspects of ethanol withdrawal largely remain to be explored, but some neuropharmacological mechanisms have been identified, including a decrease in GABAergic function and an increase in glutamatergic function[12,105] as well as associated changes in cellular calcium levels.[76,106]

REWARD THRESHOLDS AND DRUG ABSTINENCE

Recent work has focused on the neural substrates and neuropharmacological mechanisms of the motivational effects of drug withdrawal, effects that may contribute to the negative reinforcement associated with drug dependence. For example, cocaine withdrawal in humans in the outpatient setting is characterized by severe depressive symptoms combined with irritability, anxiety, and anhedonia lasting several hours to several days (the "crash") and may be one of the motivating factors in the maintenance of the cocaine dependence cycle.[107] Inpatient studies have shown similar changes in mood and anxiety states, but the changes generally are much less severe.[108] Opiate withdrawal is characterized by severe dysphoria, and ethanol withdrawal produces dysphoria and anxiety. Recent work suggests the same neural systems implicated in the positive reinforcing effects of drugs of abuse may be involved in these aversive motivational effects of drug withdrawal. Using the technique of intracranial self-stimulation to measure reward thresholds throughout the course of drug dependence, recent studies have shown that reward thresholds are increased (reflecting a decrease in reward) following chronic administration of all major drugs of abuse, including opiates, psychostimulants, alcohol, and nicotine. These effects may reflect changes in the activity of the same mesocorticolimbic system (midbrain-forebrain system) implicated in the positive reinforcing effects of drugs and can last up to 72 hours, depending on the drug and dose administered (Table 13–3).[109–124]

DRUG DEPENDENCE: NEUROCHEMICAL SUBSTRATES

The neuropharmacological basis for the change in reward function associated with the development of drug

TABLE 13–3. Drug effects on thresholds for rewarding brain stimulation

Drug Class	Acute Administration	Withdrawal From Chronic Treatment	Reference
Psychostimulants (cocaine, amphetamines)	↓	↑	109–112, 122
Opiates (morphine, heroin)	↓	↑	113–115
Nicotine	↓	↑	116, 117, 124
Sedative-hypnotics	↓	↑	118, 119

dependence has largely followed two neuroadaptive models: sensitization and some form of homeostatic adaptive mechanism.[125] With drugs, sensitization is more likely to occur with intermittent exposure to a drug, in contrast to tolerance, which is more likely to occur with continuous exposure.

In a recent conceptualization of the role of sensitization in drug dependence, a shift in an incentive-salience state described as "wanting" was hypothesized to be progressively increased by repeated exposure to drugs of abuse,[126] and the transition to pathologically strong wanting or craving would define compulsive use. In contrast, a homeostatic adaptive mechanism would exist where the initial acute effect of the drug is opposed or counteracted by homeostatic changes in systems that mediate the primary drug effects.[127-129]

In one formulation, called *opponent process theory*, tolerance and dependence were inextricably linked.[127] Here it was proposed that affective states, pleasant or aversive, were automatically opposed by centrally mediated mechanisms that reduce the intensity of these affective states.

For neuroadaptive mechanisms described by both theoretical positions, several types of adaptation can be envisioned at the molecular, cellular, and system levels. *Within-system* adaptations have been hypothesized wherein neurochemical changes associated with the same neurotransmitters implicated in the acute reinforcing effects of drugs are altered during the development of dependence.[8] Examples of such homeostatic, within-systems adaptive neurochemical events include decreases in dopaminergic and serotonergic transmission in the nucleus accumbens during drug withdrawal as measured by in vivo microdialysis,[121,130] increased sensitivity of opioid receptors in the nucleus accumbens during opiate withdrawal,[131] decreased GABAergic and increased NMDA glutamatergic transmission during ethanol withdrawal,[76,132,133] and differential regional changes in nicotine receptor function.[134,135]

Other neurotransmitter systems may also be recruited in the adaptive responses to drugs of abuse. Such a neuroadaptation, wherein a neurotransmitter system not linked to the acute reinforcing effects of the drug is recruited or altered during chronic drug administration, has been termed a *between-system* adaptation.[8] Particular attention has focused on components of stress responses for between-system adaptations. Corticotropin-releasing factor function appears to be activated during acute withdrawal from alcohol or opiates and thus may mediate aspects of stress associated with abstinence.[136] A role for circulating glucocorticoids in adaptations to the reinforcing effects of a drug of abuse also has been hypothesized.[137]

DRUG DEPENDENCE: MOLECULAR AND CELLULAR MECHANISMS

Drug-induced adaptations in neurotransmitter systems would exert their functional effects on the brain ultimately through post-receptor intracellular messenger pathways that mediate neurotransmitter-receptor actions. In a similar way, the development of drug-induced adaptations in neurotransmitter systems occurs via perturbation of these intracellular messenger pathways.

The brain's intracellular messenger pathways are reviewed elsewhere[138] and summarized in Figure 13-3. Briefly, most neurotransmitter receptors in the brain belong to a family of G protein–coupled receptors, which produce all of their effects on brain function via activation of specific types of G proteins, guanine nucleotide-binding membrane proteins that couple the plasma membrane receptors to intracellular processes. Activated G proteins can then directly regulate the activity of certain ion channels as well as regulate the levels of specific second messengers in the brain, which include adenosine 3',5'-cyclic monophosphate (cAMP), Ca^{2+}, and metabolites of phosphatidylinositol. These second messengers then regulate the activity of enzymes called protein kinases and protein phosphatases, which add and remove, respectively, phosphate groups from other proteins.

Phosphate groups, because of their charge and size, alter a protein's conformation and therefore its function. Through the phosphorylation of virtually every type of neural protein, neurotransmitter receptor interactions can elicit myriad biological responses in their target neurons. For example, phosphorylation of ion channels alters their probability of opening, and phosphorylation of receptors alters their capability to be activated by ligand as well as to subsequently activate G proteins. Among the proteins regulated by phosphorylation are those that control gene expression and protein synthesis. An example can be seen in transcription factors, proteins that bind to specific DNA sequences in certain genes and thereby increase or decrease the rate at which those genes are transcribed. Phosphorylation of transcription factors is a critical control point in regulating the activity of these proteins.

This transduction of biological action is illustrated well by the acute actions of opiates on the brain (Figure 13-4).[11] Opiate activation of opioid receptors leads to recruitment of Gi and related G proteins. This, in turn, leads to activation of certain K^+ channels and inhibition of voltage-gated Ca^{2+} channels, although the two actions occur to varying extents in different neuronal cell types. Both are inhibitory actions (more K⁺ flows out of the cell and less Ca^{2+} flows into the cell) that mediate some of the relatively rapid inhibitory effects of opiates on the electrical properties of their target neurons. Recruitment of Gi also leads to the inhibition of adenylate cyclase and of the cAMP protein phosphorylation cascade. Similarly, reductions in cellular Ca^{2+} levels alter Ca^{2+}-dependent protein phosphorylation cascades. Altered activity of these protein phosphorylation cascades, which also can vary among different cell types, leads in turn to the regulation of still additional ion channels, which contribute further to the acute effects of the drug. Perhaps more important, these protein phosphorylation mechanisms can lead to changes in many other neural processes within target neurons, including those that trigger the long-term effects of the drugs that lead eventually to tolerance, dependence, withdrawal, sensitization, and, ultimately, addiction.

Repeated exposure to a drug of abuse results in repeated perturbation of intracellular messenger pathways, which eventually elicits long-term adaptations in the pathways that contribute to dependence and tolerance. Insights into the specific molecular and cellular adaptations involved in chronic drug action have been progressing at a rapid rate. Some of the most complete studies involve the locus coeruleus, the major noradrenergic nucleus in the brain, which plays an important role in physical dependence to opiates, as mentioned above.[102,103,139-141] Activation of the locus coeruleus has been shown to mediate many of the signs and symptoms of opiate withdrawal in rodents and nonhuman primates. This knowledge led to the introduction of clonidine, an alpha₂-adrenergic agonist, as the first nonopiate treatment of opiate withdrawal.[102,142]

Withdrawal-induced activation of the locus coeruleus has been hypothesized to be due to a combination of intrinsic and extrinsic factors. The intrinsic mechanisms involve regulation of the cAMP pathway in locus coeruleus neurons,[143,144] which has long been considered a molecular site of opiate neuroadaptation.[145-147] Acute administration of opiates inhibits the firing of the neurons via regulation of two ion channels: activation of K⁺ channels via direct Gi protein coupling, and inhibition of an Na⁺ current indirectly via inhibition of adenylate cyclase and of cAMP-dependent protein kinase[148] (Figure 13-5). In contrast, chronic exposure to opiates increases the amount of adenylate cyclase and cAMP-dependent protein kinase expressed in the neurons. This up-regulated cAMP pathway has been shown to contribute to the increase in the intrinsic electrical excitability of locus coeruleus neurons that underlies the cellular forms of tolerance and dependence exhibited by these neurons.[144,147] The up-regulated cAMP pathway would contribute to tolerance by opposing inhibition of the neurons by the continued presence of opiates and thereby help return firing rates toward control levels. It would also contribute to dependence: upon removal of opiate, the up-regulated cAMP pathway, now unopposed, would drive the firing of the neurons far above control levels. A major unanswered question is the precise mechanism (for instance, at the level of transcription, translation, or protein modification) by which chronic opiate exposure leads to the up-regulation of the cAMP pathway.[143] Recent work has implicated the transcription factor CREB (cAMP responseelement binding protein) as one mediator of these adaptations.[146,149,150]

Increased activation of the major glutamatergic input

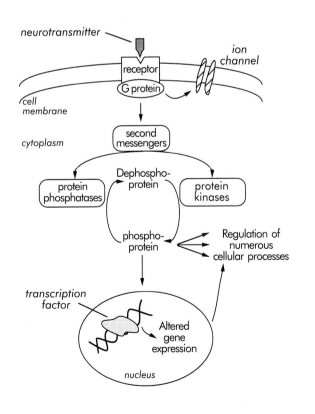

FIGURE 13-3. Schematic illustration of the brain's intracellular messenger pathways. Activation of neurotransmitter receptors leads to the activation of specific G proteins, second messengers, and protein phosphorylation systems, which produce multiple effects on neuronal function through the phosphorylation of numerous types of substrate proteins. Among the effects of these intracellular systems on neuronal function is the regulation of gene expression. This can be accomplished through the phosphorylation of transcription factors, which results in alterations in the expression of numerous target genes. G proteins can also exert effects independent of protein phosphorylation—for example, through the direct regulation of ion channels.

to the locus coeruleus that arises from a brainstem area called the paragigantocellularis appears to be one of the major extrinsic mechanisms of withdrawal-induced activation of the locus coeruleus.[151,152] The driving force for this increase in glutamatergic tone remains to be elucidated. Chronic opiate exposure could lead to intrinsic changes in the glutamatergic neurons of the paragigantocellularis themselves or in neurons that drive those neurons in some neural circuit.[143]

Much less is known about the molecular and cellular mechanisms of motivational dependence, although there is some evidence to suggest that similar mechanisms may be involved. Several drugs of abuse up-regulate the cAMP pathway in the nucleus accumbens after chronic administration.[144,153,154] This up-regulation could mediate some of the documented electrophysiological changes in the nucleus accumbens associated with chronic drug exposure, such as enhanced responsiveness of D_1 dopamine receptors after chronic cocaine treatment.[138] Moreover, studies involving direct admini-

FIGURE 13–4. Schematic illustration of opiate-regulated signal transduction pathways. Opiates produce their effects in target neurons via interactions with three major classes of receptors, termed mu, delta, and kappa opioid receptors. There are two major signal transduction pathways through which each of these receptors has been shown to influence target neuron functioning. First, opiates, via coupling to a pertussis toxin–sensitive G protein (presumably Gi), inhibit adenylate cyclase and thereby reduce cellular cAMP levels, the activity of cAMP-dependent protein kinase, and the phosphorylation state of numerous substrate proteins for the protein kinase. Such substrate proteins include ion channels and many other types of neuronal proteins known to be regulated by cAMP-dependent phosphorylation. Through this action, therefore, opiates induce many and diverse types of effects in target neurons. Second, opiates, via coupling to Gi and/or Go, increase the conductance of certain types of K^+ channels and decrease the conductance of voltage-dependent Ca^{2+} channels. In most cases, the opiate-regulated K^+ channel appears to be identical to the inward rectifying channel regulated by several other types of neurotransmitter receptors (such as D_2 dopaminergic, alpha$_2$ adrenergic, and 5-HT$_{1A}$ serotonergic receptors) also via coupling with a pertussis toxin–sensitive G protein. Regulation of the K^+ and Ca^{2+} channels inhibits the electrical properties of the target neurons. Such regulation also leads to decreases in intracellular Ca^{2+} levels and, consequently, to decreases in the activity of Ca^{2+}-dependent protein kinases (both Ca^{2+}/calmodulin-dependent protein kinases and protein kinase C) and the phosphorylation state of numerous types of substrate proteins for these protein kinases. As with regulation of the cAMP system, opiate-induced changes in Ca^{2+}-dependent protein phosphorylation lead to many changes in neuronal function.

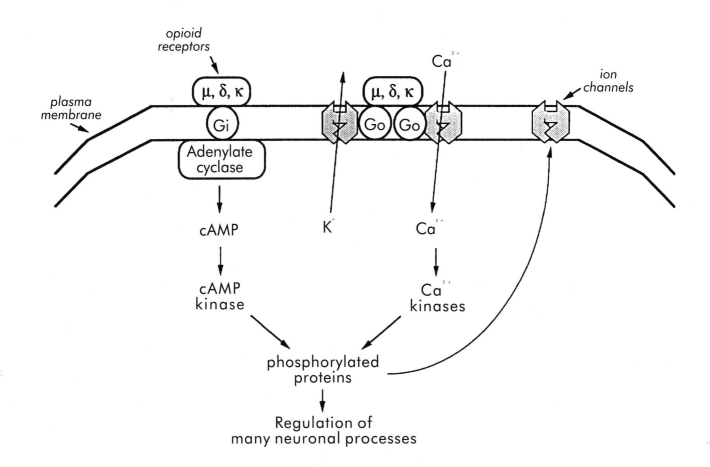

stration of activators or inhibitors of the cAMP pathway into the nucleus accumbens are consistent with the interpretation that up-regulation of the cAMP pathway in this brain region may contribute to an aversive state during drug withdrawal.[153] A major task now will be to explore these mechanisms directly in animal models of dependence that reflect both the positive and the negative reinforcing effects of dependence.

The persistence of changes in drug reinforcement mechanisms that characterize drug addiction suggests that the underlying molecular mechanisms are long-lasting, and indeed considerable attention has been given to drug regulation of gene expression. Current research focuses on two types of transcription factors, CREB and novel Fos-like proteins (termed chronic FRAs, or Fos-related antigens), as possible mediators of chronic drug action.[155–157] However, it has not yet been possible to relate regulation of a specific transcription factor to specific features of drug reinforcement.

FIGURE 13–5. Scheme illustrating opiate actions in the locus coeruleus. Opiates acutely inhibit locus coeruleus neurons by increasing the conductance of a K^+ channel (light cross-hatch) via coupling with subtypes of Gi and/or Go and by decreasing an Na^+-dependent inward current (dark cross-hatch) via coupling with Gi/o and the consequent inhibition of adenylate cyclase. Reduced levels of cAMP decrease protein kinase activity (PKA) and the phosphorylation of the responsible channel or pump. Inhibition of the cAMP pathway also decreases phosphorylation of numerous other proteins and thereby affects many additional processes in the neuron. For example, it reduces the phosphorylation state of cAMP response element binding protein (CREB), which may initiate some of the longer-term changes in locus coeruleus function. Upward bold arrows summarize effects of chronic morphine in the locus coeruleus. Chronic morphine increases levels of adenylate cyclase, protein kinase activity, and several phosphoproteins, including CREB. These changes contribute to the altered phenotype of the drug-addicted state. For example, the intrinsic excitability of locus coeruleus neurons is increased via enhanced activity of the cAMP pathway and Na^+-dependent inward current, which contributes to the tolerance, dependence, and withdrawal exhibited by these neurons. This altered phenotypic state may be maintained in part by up-regulation of CREB expression. Reprinted, with permission, from Nestler 1996.[146]

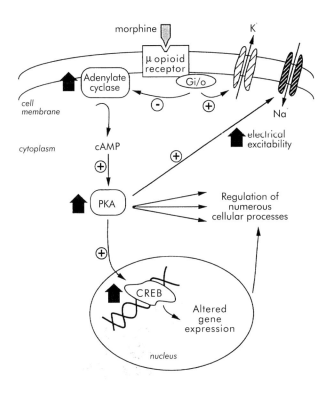

DRUG TOLERANCE: NEURAL SUBSTRATES

Tolerance to the reinforcing actions of drugs of abuse may also be an important mechanism for drug addiction. However, the measurement in animal models with such phenomena as intravenous self-administration has been limited.[158] One would hypothesize that the neural substrates for drug tolerance would overlap significantly with those associated with acute withdrawal, since tolerance and withdrawal appear to be components of the same neuroadaptive process. However, tolerance also depends on learning processes, and this has been most explored in the context of opiate drugs and sedative-hypnotics such as alcohol.[159] Mechanisms for these associative processes may involve several neurotransmitters independently of their role in acute withdrawal. Norepinephrine and serotonin have long been shown to be involved in the development of tolerance to ethanol and barbiturates.[12] More recently, co-administration of glutamate receptor antagonists and opiates has been shown to block the development of tolerance to opiates.[160] This is again consistent with an associative component of tolerance.

Mechanisms at the molecular level for tolerance also probably overlap with those of dependence.[144] For example, up-regulation of the cAMP pathway could be a mechanism of tolerance, as mentioned earlier: the changes would be expected to oppose the acute actions of opiates of inhibiting adenylate cyclase. In addition, tolerance appears to involve the functional uncoupling of opioid receptors from their G proteins. The mechanisms underlying this uncoupling remain unknown, but they could involve drug-induced changes in the phosphorylation state of the receptors or G proteins that reduce their affinity for each other. Such phosphorylation of the receptor could occur via cAMP- or Ca^{2+}-dependent protein kinases known to be regulated by opiate exposure, or by other types of protein kinases (termed G protein receptor kinases or GRKs) that phosphorylate and desensitize receptors only when they are in their ligand-bound conformation.[146] Another possible mechanism of tolerance is drug-induced changes in the ion channels that mediate the acute effects of the drugs. For example, alterations in the phosphorylation state, amount, or even type of channel could conceivably contribute to drug tolerance.[143,161]

RELAPSE: NEURAL SUBSTRATES

Limited animal models exist for the study of relapse,[162] and this has significantly hampered the study of neurobiological mechanisms associated with this process. Neuropharmacological probes have been employed to reinstate self-administration in animals trained and then extinguished on intravenous drug self-administration; these probes have shown that drugs that activate the mesolimbic dopamine system rapidly reinstate intravenous self-administration.[163,164]

There are a limited number of observations using other models. Acamprosate, a drug being marketed in Europe to prevent relapse in alcoholics, has potential glutamate modulatory action.[165] It blocks the increase in drinking observed in rodents after a forced abstinence, again in nondependent rats,[166] and has efficacy in preventing relapse in detoxified human alcoholics.[167] Similarly, opioid antagonists were shown to prevent the increase in drinking of ethanol in animals after stress,[168] and subsequently naltrexone was shown to have efficacy in preventing relapse in detoxified human alcoholics.[88,89] Finally, a recent study reports that agonists selective for D_1 dopamine receptors, but not for D_2-like receptors, can block reinstatement of lever-pressing inferred to represent cocaine-seeking behavior.[169]

One thing that is clear from current studies of relapse is the need for better animal models. As improved animal models are developed, it will become possible to obtain a more complete understanding of the neurobiological mechanisms underlying this critical feature of drug addiction at the molecular, cellular, and system levels.

NEUROBIOLOGICAL SUBSTRATES AND CLINICAL IMPLICATIONS

To summarize, drug addiction centers on alteration of the neurobiological substrates of reinforcement. Much is known about the substrates for the acute positive reinforcing effects of drugs of abuse. The mesolimbic dopamine system and its connections form a focal point for our understanding of both dopamine-dependent and dopamine-independent effects. Indirect sympathomimetics such as cocaine and amphetamine are critically dependent on increased dopaminergic activity in the terminal areas of the mesolimbic dopamine system. Nicotine, opiates, and ethanol all activate the mesolimbic dopamine system but also recruit the actions of other neurotransmitters such as opioid peptides, serotonin, GABA, and glutamate. The neural substrates associated with the motivational aspects of dependence—tolerance, acute drug withdrawal, protracted abstinence, and vulnerability—remain largely to be determined but may involve molecular, cellular, and system-level adaptations to the same neurochemical elements implicated in the acute reinforcing actions of drugs of abuse. A subsystem of the basal forebrain termed the extended amygdala may play a particularly important role in the motivational aspects of drug reinforcement, both positive and negative.

The clinical implications of this work are numerous. Understanding the biological basis of a disease provides important insight for therapeutic intervention. Disturbances in specific signal transduction pathways in the brain that underlie addiction provide a conceptual anchor for psychotherapeutic as well as pharmacotherapeutic intervention. There are already new treatments for opiate and ethanol dependence that are based on a neurobiological understanding of these disorders, and considerable activity is now centered on developing treatments for cocaine dependence as well. Basic research in the neurobiology of addiction also provides the key elements for identifying the biological basis of vulnerability to drug abuse and relapse that will guide the development of sound prevention interventions in vulnerable individuals.

REFERENCES

1. World Health Organization: International Statistical Classification of Diseases and Related Health Problems, 10th revision. Geneva, World Health Organization, 1990
2. American Psychiatric Association: Diagnostic and Statistical Manual of Mental Disorders, 4th edition. Washington, DC, American Psychiatric Association, 1994
3. Institute of Medicine: Pathways of Addiction. Washington, DC, National Academy Press, 1996
4. Markou A, Weiss F, Gold LH, et al: Animal models of drug craving. Psychopharmacology (Berl) 1993; 112:163–182
5. White FJ, Wolf ME: Psychomotor stimulants, in The Biological Bases of Drug Tolerance and Dependence, edited by Pratt JA. London, Academic Press, 1991, pp 153–197
6. Koob GF: Dopamine, addiction and reward. Seminars in the

Neurosciences 1992; 4:139–148
7. Stewart J, Badiani A: Tolerance and sensitization to the behavioral effects of drugs. Behav Pharmacol 1993; 4:289–312
8. Koob GF, Bloom FE: Cellular and molecular mechanisms of drug dependence. Science 1988; 242:715–723
9. Wikler A: Dynamics of drug dependence: implications of a conditioning theory for research and treatment. Arch Gen Psychiatry 1973; 28:611–616
10. Koob GF, Markou A, Weiss F, et al: Opponent process and drug dependence: neurobiological mechanisms. Seminars in the Neurosciences 1993; 5:351–358
11. Nestler EJ, Fitzgerald LW, Self DW: Substance abuse: neurobiology, in The American Psychiatric Press Review of Psychiatry, vol 14, edited by Oldham JM, Riba MB. Washington, DC, American Psychiatric Press, 1995, pp 51–81
12. Tabakoff B, Hoffman PL: Alcohol: neurobiology, in Substance Abuse: A Comprehensive Textbook, edited by Lowinson JH, Ruiz P, Millman RB. Baltimore, Williams and Wilkins, 1992, pp 152–185
13. Olds J, Milner P: Positive reinforcement produced by electrical stimulation of septal area and other regions of rat brain. Journal of Comparative and Physiological Psychology 1954; 47:419–427
14. Stein L: Chemistry of reward and punishment, in Psychopharmacology: A Review of Progress (1957–1967) (US Public Health Service publ no 1836), edited by Efron DH. Washington, DC, US Government Printing Office, 1968, pp 105–123
15. Nauta WJH, Haymaker W: Hypothalamic nuclei and fiber connections, in The Hypothalamus, edited by Haymaker W, Anderson E, Nauta WJH. Springfield, IL, Charles C Thomas, 1969, pp 136–209
16. Liebman JL, Cooper SJ: The Neuropharmacological Basis of Reward. Oxford, Clarendon, 1989
17. Valenstein ES, Campbell JF: Medial forebrain bundle-lateral hypothalamic area and reinforcing brain stimulation. Am J Physiol 1966; 210:270–274
18. Koob GF: Drugs of abuse: anatomy, pharmacology, and function of reward pathways. Trends Pharmacol Sci 1992; 13:177–184
19. Kilty JE, Lorang D, Amara SG: Cloning and expression of a cocaine-sensitive rat dopamine transporter. Science 1991; 254:578–579
20. Blakely RD, Berson HE, Fremeau RT, et al: Cloning and expression of a functional serotonin transporter from rat brain. Nature 1991; 354:66–70
21. Giros B, El Mestikawy S, Bertrand L, et al: Cloning and functional characterization of a cocaine-sensitive dopamine transporter. FEBS Lett 1991; 295:149–154
22. Rudnick G, Clark J: From synapse to vesicle: the reuptake and storage of biogenic amine neurotransmitters. Biochim Biophys Acta 1993; 1144:249–263
23. O'Callaghan JP, Miller DB: Neurotoxicity profiles of substituted amphetamines in the C57BL/6J mouse. J Pharmacol Exp Ther 1994; 270:741–751
24. Angrist B, Sudilovsky A: Central nervous system stimulants: historical aspects and clinical effects, in Handbook of Psychopharmacology, vol 11, edited by Iversen LL, Iversen SD, Snyder SH. New York, Plenum, 1976, pp 99–165
25. Fischman MW, Schuster CR, Hatano Y: A comparison of the subjective and cardiovascular effects of cocaine and lidocaine in humans. Pharmacol Biochem Behav 1983; 18:123–127
26. Groppetti A, Zambotti F, Biazzi A, et al: Amphetamine and cocaine on amine turnover, in Frontiers in Catecholamine Research, edited by Usdin E, Snyder SH. New York, Pergamon, 1973, pp 917–925
27. Spealman RD, Goldberg SR, Kelleher RT, et al: Some effects of cocaine and two cocaine analogs on schedule-controlled behavior of squirrel monkeys. J Pharmacol Exp Ther 1977; 202:500–509
28. Kornetsky C, Esposito RU: Reward and detection thresholds for brain stimulation: dissociative effects of cocaine. Brain Res 1981; 209:496–500
29. Kornetsky C, Esposito RU: Euphorigenic drugs: effects on the reward pathways of the brain. Federation Proceedings 1979; 38:2473–2476
30. Mucha RF, van der Kooy D, O'Shaughnessy M, et al: Drug reinforcement studied by the use of place conditioning in rat. Brain Res 1982; 243:91–105
31. Carr GD, Fibiger HC, Phillips AG: Conditioned place preference as a measure of drug reward, in The Neuropharmacological Basis of Reward, edited by Liebman JM, Cooper SJ. New York, Oxford University Press, 1989, pp 264–319
32. Koob GF, Vaccarino FJ, Amalric M, et al: Positive reinforcement properties of drugs: search for neural substrates, in Brain Reward Systems and Abuse, edited by Engel J, Oreland L. New York, Raven, 1987, pp 35–50
33. Kelly PH, Seviour PW, Iversen SD: Amphetamine and apomorphine responses in the rat following 6-OHDA lesions of the nucleus accumbens septi and corpus striatum. Brain Res 1975; 94:507–522
34. Kelly PH, Iversen SD: Selective 6OHDA-induced destruction of mesolimbic dopamine neurons: Abolition of psychostimulant-induced locomotor activity in rats. Eur J Pharmacol 1976; 40:45–55
35. Pijnenburg AJJ, Honig WMM, Van Rossum JM: Inhibition of d-amphetamine-induced locomotor activity by injection of haloperidol into the nucleus accumbens of the rat. Psychopharmacologia 1975; 41:87–95
36. Yokel RA, Wise RA: Increased lever pressing for amphetamine after pimozide in rats: implications for a dopamine theory of reward. Science 1975; 187:547–549
37. Ettenberg A, Pettit HO, Bloom FE, et al: Heroin and cocaine intravenous self-administration in rats: mediation by separate neural systems. Psychopharmacology (Berl) 1982; 78:204–209
38. Caine SB, Koob GF: Pretreatment with the dopamine agonist 7-OH-DPAT shifts the cocaine self-administration dose-effect function to the left under different schedules in the rat. Behav Pharmacol 1995; 6:333–347
39. Koob GF, Le HT, Creese I: The D_1 dopamine receptor antagonist SCH 23390 increases cocaine self-administration in the rat. Neurosci Lett 1987; 79:315–320
40. Woolverton WL, Virus RM: The effects of a D_1 and a D_2 dopamine antagonist on behavior maintained by cocaine or food. Pharmacol Biochem Behav 1989; 32:691–697
41. Bergman J, Kamien JB, Spealman RD: Antagonism of cocaine self-administration by selective dopamine D_1 and D_2 antagonists. Behav Pharmacol 1990; 1:355–363
42. Caine SB, Koob GF: Modulation of cocaine self-administration in the rat through D-3 dopamine receptors. Science 1993; 260:1814–1816
43. Roberts DCS, Koob GF, Klonoff P, et al: Extinction and recovery of cocaine self-administration following 6-hydroxydopamine lesions of the nucleus accumbens. Pharmacol Biochem Behav 1980; 12:781–787
44. Lyness WH, Friedle NM, Moore KE: Destruction of dopaminergic nerve terminals in nucleus accumbens: effect of d-amphetamine self-administration. Pharmacol Biochem Behav 1979; 11:553–556
45. Caine SB, Koob GF: Effects of mesolimbic dopamine depletion on responding maintained by cocaine and food. J Exp Anal Behav 1994; 61:213–221
46. Schuster CR, Thompson T: Self administration of and behavioral dependence on drugs. Annu Rev Pharmacol 1969; 9:483–502
47. Deneau G, Yanagita T, Seevers MH: Self-administration of psychoactive substances by the monkey: a measure of psychological-dependence. Psychopharmacologia 1969; 16:30–48
48. Koob GF, Pettit HO, Ettenberg A, et al: Effects of opiate antagonists and their quaternary derivatives on heroin self-administration in the rat. J Pharmacol Exp Ther 1984; 229:481–486
49. Goldberg SR, Woods JH, Schuster CR: Nalorphine-induced changes in morphine self-administration in rhesus monkeys. J Pharmacol Exp Ther 1971; 176:464–471
50. Weeks JR, Collins RJ: Changes in morphine self-administration in

rats induced by prostaglandin E₁ and naloxone. Prostaglandins 1976; 12:11–19

51. Vaccarino FJ, Pettit HO, Bloom FE, et al: Effects of intracerebroventricular administration of methyl naloxonium chloride on heroin self-administration in the rat. Pharmacol Biochem Behav 1985; 23:495–498

52. Negus SS, Henriksen SJ, Mattox SR, et al: Effects of antagonists selective for mu, delta and kappa opioid receptors on the reinforcing effects of heroin in rats. J Pharmacol Exp Ther 1993; 265:1245–1252

53. Britt MD, Wise RA: Ventral tegmental site of opiate reward: antagonism by a hydrophilic opiate receptor blocker. Brain Res 1983; 258:105–108

54. Vaccarino FJ, Bloom FE, Koob GF: Blockade of nucleus accumbens opiate receptors attenuates intravenous heroin reward in the rat. Psychopharmacology (Berl) 1985; 86:37–42

55. Schroeder RL, Weinger MB, Vakassian L, et al: Methylnaloxonium diffuses out of the rat brain more slowly than naloxone after direct intracerebral injection. Neurosci Lett 1991; 121:173–177

56. Goeders NE, Lane JD, Smith JE: Self-administration of methionine enkephalin into the nucleus accumbens. Pharmacol Biochem Behav 1984; 20:451–455

57. Pettit HO, Ettenberg A, Bloom FE, et al: Destruction of dopamine in the nucleus accumbens selectively attenuates cocaine but not heroin self-administration in rats. Psychopharmacology (Berl) 1984; 84:167–173

58. Stinus L, Cador M, Le Moal M: Interaction between endogenous opioids and dopamine within the nucleus accumbens. Ann NY Acad Sci 1992; 654:254–273

59. West TE: The effects of nucleus accumbens injections of receptor-selective opiate agonists on brain stimulation reward. Doctoral dissertation, Concordia University, Montreal, Quebec, Canada, 1991

60. Pert A, Sivit C: Neuroanatomical focus for morphine and enkephalin-induced hypermotility. Nature 1977; 265:645–647

61. Di Chiara G, Imperato A: Opposite effects of mu and kappa opiate agonists on dopamine release in the nucleus accumbens and in the dorsal caudate of freely moving rats. J Pharmacol Exp Ther 1988; 244:1067–1080

62. Hemby SE, Martin TJ, Co C, et al: The effects of intravenous heroin administration on extracellular nucleus accumbens dopamine concentrations as determined by in vivo microdialysis. J Pharmacol Exp Ther 1995; 273:591–598

63. Bozarth MA, Wise RA: Intracranial self-administration of morphine into the ventral tegmental area in rats. Life Sci 1981; 28:551–555

64. Di Chiara G, North RA: Neurobiology of opiate abuse. Trends Pharmacol Sci 1992; 13:185–193

65. Stinus L, Nadaud D, Deminiere JM, et al: Chronic flupentixol treatment potentiates the reinforcing properties of systemic heroin administration. Biol Psychiatry 1989; 26:363–371

66. Spyraki C, Fibiger HC, Phillips AG: Attenuation of heroin reward in rats by disruption of the mesolimbic dopamine system. Psychopharmacology (Berl) 1983; 79:278–283

67. Shippenberg TS, Herz A, Spanagel R, et al: Conditioning of opioid reinforcement: neuroanatomical and neurochemical substrates. Ann NY Acad Sci 1992; 654:347–356

68. Sepinwall J, Cook L: Behavioral pharmacology of anti-anxiety drugs, in Handbook of Psychopharmacology, vol 13, edited by Iversen LL, Iversen SD, Snyder SH. London, Plenum, 1978, pp 345–393

69. Richards G, Schoch P, Haefely W: Benzodiazepine receptors: new vistas. Seminars in the Neurosciences 1991; 3:191–203

70. Suzdak PD, Glowa JR, Crawley JN, et al: A selective imidazobenzodiazepine antagonist of ethanol in the rat. Science 1986; 234:1243–1247

71. Koob GF, Britton KT: Neurobiological substrates for the anti-anxiety effects of ethanol, in The Pharmacology of Alcohol and Alcohol Dependence, edited by Begleiter H, Kissin B. New York, Oxford University Press, 1996, pp 477–506

72. Frye GD, Breese GR: GABAergic modulation of ethanol-induced motor impairment. J Pharmacol Exp Ther 1982; 223:750–756

73. Liljequist S, Engel J: Effects of GABAergic agonists and antagonists on various ethanol-induced behavioral changes. Psychopharmacology (Berl) 1982; 78:71–75

74. Samson HH, Tolliver GA, Pfeffer AO, et al: Oral ethanol reinforcement in the rat: effect of the partial inverse benzodiazepine agonist Ro 15-4513. Pharmacol Biochem Behav 1987; 27:517–519

75. Rassnick S, D'Amico E, Riley E, et al: GABA antagonist and benzodiazepine partial inverse agonist reduce motivated responding for ethanol. Alcohol Clin Exp Res 1993; 17:124–130

76. Fitzgerald LW, Nestler EJ: Molecular and cellular adaptations in signal transduction pathways following ethanol exposure. Clin Neurosci 1995; 3:165–173

77. Hoffman PL, Rabe C, Moses F, et al: N-methyl-D-aspartate receptors and ethanol: inhibition of calcium flux and cyclic GMP production. J Neurochem 1989; 52:1937–1940

78. Lovinger DM, White G, Weight FF: Ethanol inhibits NMDA-activated ion current in hippocampal neurons. Science 1989; 243:1721–1724

79. Tsai G, Gastfriend DR, Coyle JT: The glutamatergic basis of human alcoholism. Am J Psychiatry 1995; 152:332–340

80. Pfeffer AO, Samson HH: Haloperidol and apomorphine effects on ethanol reinforcement in free feeding rats. Pharmacol Biochem Behav 1988; 29:343–350

81. Weiss F, Lorang MT, Bloom FE, et al: Ethanol self-administration stimulates dopamine release in the rat nucleus accumbens: genetic and motivational determinants. J Pharmacol Exp Ther 1993; 267:250–258

82. Rassnick S, Stinus L, Koob GF: The effects of 6-hydroxydopamine lesions of the nucleus accumbens and the mesolimbic dopamine system on oral self-administration of ethanol in the rat. Brain Res 1993; 623:16–24

83. Engel JA, Enerback C, Fahlke C, et al: Serotonergic and dopaminergic involvement in ethanol intake, in Novel Pharmacological Interventions for Alcoholism, edited by Naranjo CA, Sellers EM. New York, Springer, 1992, pp 68–82

84. Sellars EM, Higgins GA, Sobell MB: 5-HT and alcohol abuse trends. Pharmacol Sci 1992; 13:69–75

85. Naranjo C, Kadlec K, Sanhueza P, et al: Fluoxetine differentially alters alcohol intake and other consummatory behaviors in problem drinkers. Clin Pharmacol Ther 1990; 47:490–498

86. Hubbell CL, Marglin SH, Spitalnic SJ, et al: Opioidergic, serotonergic, and dopaminergic manipulations and rats' intake of a sweetened alcoholic beverage. Alcohol 1991; 8:355–367

87. Hyytia P: Involvement of μ-opioid receptors in alcohol drinking by alcohol-preferring AA rats. Pharmacol Biochem Behav 1993; 45:697–701

88. O'Malley SS, Jaffe AJ, Chang G, et al: Naltrexone and coping skills therapy for alcohol dependence: a controlled study. Arch Gen Psychiatry 1992; 49:881–887

89. Volpicelli JR, Alterman AI, Hayashida M, et al: Naltrexone in the treatment of alcohol dependence. Arch Gen Psychiatry 1992; 49:876–880

90. Malin DH, Lake JR, Carter VA, et al: Naloxone precipitates nicotine abstinence syndrome in the rat. Psychopharmacology (Berl) 1993; 112:339–342

91. Malin DH, Lake JR, Carter VA, et al: The nicotine antagonist mecamylamine precipitates nicotine abstinence syndrome in the rat. Psychopharmacology (Berl) 1994; 115:180–184

92. Corrigall WA, Franklin KBJ, Coen KM, et al: The mesolimbic dopaminergic system is implicated in the reinforcing effects of nicotine. Psychopharmacology (Berl) 1992; 107:285–289

93. Alheid GF, Heimer L: New perspectives in basal forebrain organization of special relevance for neuropsychiatric disorders: the striatopallidal, amygdaloid, and corticopetal components of substantia

innominata. Neuroscience 1988; 27:1–39
94. Johnston JB: Further contributions to the study of the evolution of the forebrain. J Comp Neurol 1923; 35:337–481
95. Heimer L, Alheid G: Piecing together the puzzle of basal forebrain anatomy, in The Basal Forebrain: Anatomy to Function, edited by Napier TC, Kalivas PW, Hanin I. New York, Plenum, 1991, pp 1–42
96. Heimer L, Zahm DS, Churchill L, et al: Specificity in the projection patterns of accumbal core and shell in the rat. Neuroscience 1991; 41:89–125
97. Caine SB, Heinrichs SC, Coffin VL, et al: Effects of the dopamine D-1 antagonist SCH 23390 microinjected into the accumbens, amygdala or striatum on cocaine self-administration in the rat. Brain Res 1995; 692:47–56
98. Pontieri FE, Tanda G, Di Chiara G: Intravenous cocaine, morphine, and amphetamine preferentially increase extracellular dopamine in the "shell" as compared with the "core" of the rat nucleus accumbens. Proc Natl Acad Sci USA 1995; 92:12304–12308
99. Hyytia P, Koob GF: $GABA_A$ receptor antagonism in the extended amygdala decreases ethanol self-administration in rats. Eur J Pharmacol 1995; 283:151–159
100. Heyser CJ, Roberts AJ, Schulteis G, et al: Central administration of an opiate antagonist decreases oral ethanol self-administration in rats (abstract). Society for Neuroscience Abstracts 1995; 21:1698
101. Pich EM, Lorang M, Yeganeh M, et al: Increase of extracellular corticotropin-releasing factor-like immunoreactivity levels in the amygdala of awake rats during restraint stress and ethanol withdrawal as measured by microdialysis. J Neurosci 1995; 15:5439–5447
102. Aghajanian GK: Tolerance of locus coeruleus neurons to morphine and suppression of withdrawal response by clonidine. Nature 1978; 276:186–188
103. Taylor JR, Elsworth JD, Garcia EJ, et al: Clonidine infusions into the locus coeruleus attenuate behavioral and neurochemical changes associated with naloxone-precipitated withdrawal. Psychopharmacology (Berl) 1988; 96:121–134
104. Maldonado R, Stinus L, Gold LH, et al: Role of different brain structures in the expression of the physical morphine withdrawal syndrome. J Pharmacol Exp Ther 1992; 261:669–677
105. Grant KA, Valverius P, Hudspith M, et al: Ethanol withdrawal seizures and the NMDA receptor complex. Eur J Pharmacol 1990; 176:289–296
106. Littleton J, Little H, Laverty R: Role of neuronal calcium channels in ethanol dependence: from cell cultures to the intact animal. Ann NY Acad Sci 1992; 654:324–334
107. Gawin FH, Kleber HD: Abstinence symptomatology and psychiatric diagnosis in cocaine abusers: clinical observations. Arch Gen Psychiatry 1986; 43:107–113
108. Weddington WW, Brown BS, Haertzen CA, et al: Changes in mood, craving, and sleep during short-term abstinence reported by male cocaine addicts: a controlled, residential study. Arch Gen Psychiatry 1990; 47:861–868
109. Esposito RU, Motola AHD, Kornetsky C: Cocaine: acute effects on reinforcement thresholds for self-stimulation behavior to the medial forebrain bundle. Pharmacol Biochem Behav 1978; 8:437–439
110. Kokkinidis L, Zacharko RM, Predy PA: Post-amphetamine depression of self-stimulation responding from the substantia nigra: reversal by tricyclic antidepressants. Pharmacol Biochem Behav 1980; 13:379–383
111. Kokkinidis L, Zacharko RM, Anisman H: Amphetamine withdrawal: a behavioral evaluation. Life Sci 1986; 38:1617–1623
112. Frank RA, Martz S, Pommering T: The effect of chronic cocaine on self-stimulation train-duration thresholds. Pharmacol Biochem Behav 1988; 29:755–758
113. Schaefer GJ, Michael RP: Changes in response rates and reinforcement thresholds for intracranial self-stimulation during morphine withdrawal. Pharmacol Biochem Behav 1986; 25:1263–1269
114. Schulteis G, Markou A, Cole M, et al: Decreased brain reward produced by ethanol withdrawal. Proc Natl Acad Sci USA 1995; 92:5880–5884
115. Hubner CB, Kornetsky C: Heroin, 6-acetylmorphine, and morphine effects on threshold for rewarding and aversive brain stimulation. J Pharmacol Exp Ther 1992; 260:562–567
116. Huston-Lyons D, Kornetsky C: Effects of nicotine on the threshold for rewarding brain stimulation in rats. Pharmacol Biochem Behav 1992; 41:755–759
117. Bauco P, Wise RA: Potentiation of lateral hypothalamic and midline mesencephalic brain stimulation reinforcement by nicotine: examination of repeated treatment. J Pharmacol Exp Ther 1994; 271:294–301
118. Kornetsky C, Moolten M, Bain G: Ethanol and rewarding brain stimulation, in Neuropharmacology of Ethanol, edited by Meyer RE, Koob GF, Lewis MJ, et al. Boston, Birkhauser, 1991, pp 179–199
119. Schulteis G, Markou A, Cole M, et al: Decreased brain reward produced by ethanol withdrawal. Proc Natl Acad Sci USA 1995; 92:5880–5884
120. Leith NJ, Barrett RJ: Amphetamine and the reward system: evidence for tolerance and post-drug depression. Psychopharmacologia 1976; 46:19–25
121. Parsons LH, Koob GF, Weiss F: Serotonin dysfunction in the nucleus accumbens of rats during withdrawal after unlimited access to intravenous cocaine. J Pharmacol Exp Ther 1995; 274:1182–1191
122. Markou A, Koob GF: Post-cocaine anhedonia: an animal model of cocaine withdrawal. Neuropsychopharmacology 1991; 4:17–26
123. Markou A, Koob GF: Construct validity of a self-stimulation threshold paradigm: effects of reward and performance manipulations. Physiol Behav 1992; 51:111–119
124. Legault M, Wise RA: Effects of withdrawal from nicotine on intracranial self-stimulation (abstract). Society for Neuroscience Abstracts 1994; 20:1032
125. Koob GF: Drug addiction: the yin and yang of hedonic homeostasis. Neuron 1996; 16:893–896
126. Robinson TE, Berridge KC: The neural basis of drug craving: an incentive-sensitization theory of addiction. Brain Res Rev 1993; 18:247–291
127. Solomon RL, Corbit JD: An opponent-process theory of motivation, I: temporal dynamics of affect. Psychol Rev 1974; 81:119–145
128. Siegel S: Evidence from rats that morphine tolerance is a learned response. Journal of Comparative and Physiological Psychology 1975; 89:498–506
129. Poulos CX, Cappell H: Homeostatic theory of drug tolerance: a general model of physiological adaptation. Psychol Rev 1991; 98:390–408
130. Weiss F, Markou A, Lorang MT, et al: Basal extracellular dopamine levels in the nucleus accumbens are decreased during cocaine withdrawal after unlimited-access self-administration. Brain Res 1992; 593:314–318
131. Stinus L, Le Moal M, Koob GF: Nucleus accumbens and amygdala as possible substrates for the aversive stimulus effects of opiate withdrawal. Neuroscience 1990; 37:767–773
132. Roberts AJ, Cole M, Koob GF: Intra-amygdala muscimol decreases operant ethanol self-administration in dependent rats. Alcohol Clin Exp Res 1996; 20:1289–1298
133. Weiss F, Parsons LH, Schulteis G, et al: Ethanol self-administration restores withdrawal-associated deficiencies in accumbal dopamine and 5-hydroxytryptamine release in dependent rats. J Neurosci 1996; 16:3474–3485
134. Collins AC, Bhat RV, Pauly JR, et al: Modulation of nicotine receptors by chronic exposure to nicotinic agonists and antagonists, in The Biology of Nicotine Dependence, edited by Bock G, Marsh J. New York, Wiley, 1990, pp 87–105
135. Dani JA, Heinemann S: Molecular and cellular aspects of nicotine abuse. Neuron 1996; 16:905–908
136. Koob GF, Heinrichs SC, Menzaghi F, et al: Corticotropin-releasing

factor, stress and behavior. Seminars in the Neurosciences 1994; 7:221–229
137. Piazza PV, Le Moal M: Pathophysiological basis of vulnerability to drug abuse: role of an interaction between stress, glucocorticoids, and dopaminergic neurons. Annu Rev Pharmacol Toxicol 1996; 36:359–378
138. Henry DJ, White FJ: Repeated cocaine administration causes persistent enhancement of D_1 dopamine receptor sensitivity within the rat nucleus accumbens. J Pharmacol Exp Ther 1991; 258:882–890
139. Maldonado R, Koob GF: Destruction of the locus coeruleus decreases physical signs of opiate withdrawal. Brain Res 1993; 605:128–138
140. Rasmussen K, Beitner-Johnson DB, Krystal JH, et al: Opiate withdrawal and rat locus coeruleus: behavioral, electrophysiological, and biochemical correlates. J Neurosci 1990; 10:2308–2317
141. Koob GF, Maldonado R, Stinus L: Neural substrates of opiate withdrawal. Trends Neurosci 1992; 15:186–191
142. Gold MS, Redmond DE, Kleber HD: Clonidine blocks acute opiate-withdrawal symptoms. Lancet 1978; 2:599–602
143. Nestler EJ: Molecular mechanisms of drug addiction. J Neurosci 1992; 12:2439–2450
144. Nestler EJ, Hope BT, Widnell KL: Drug addiction: a model for the molecular basis of neural plasticity. Neuron 1993; 11:995–1006
145. Collier HOJ: Cellular site of opiate dependence. Nature 1980; 283:625–629
146. Nestler EJ: Under siege: The brain on opiates. Neuron 1996; 16:897–900
147. Sharma SK, Klee WA, Nirenberg M: Dual regulation of adenylate cyclase accounts for narcotic dependence and tolerance. Proc Natl Acad Sci USA 1975; 72:3092–3096
148. Alreja M, Aghajanian GK: Opiates suppress a resting sodium-dependent inward current and activate an outward potassium current in locus coeruleus neurons. J Neurosci 1993; 13:3525–3532
149. Widnell KL, Russell D, Nestler EJ: Regulation of expression of cAMP response element-binding protein in the locus coeruleus in vivo and in a locus coeruleus-like cell line in vitro. Proc Natl Acad Sci USA 1994; 91:10947–10951
150. Maldonado R, Blendy JA, Tzavara E, et al: Reduction of morphine abstinence in mice with a mutation in the gene encoding CREB. Science 1996; 273:657–659
151. Rasmussen K, Aghajanian GK: Withdrawal-induced activation of locus coeruleus neurons in opiate-dependent rats: attenuation by lesion of the nucleus paragigantocellularis. Brain Res 1989; 505:346–350
152. Akaoka H, Aston-Jones G: Opiate withdrawal–induced hyperactivity of locus coeruleus neurons is substantially mediated by augmented excitatory acid input. J Neurosci 1991; 11:3830–3839
153. Self DW, Nestler EJ: Molecular mechanisms of drug reinforcement and addiction. Annu Rev Neurosci 1995; 18:463–495
154. Nestler EJ: Molecular neurobiology of drug addiction. Neuropsychopharmacology 1994; 11:77–87
155. Hope BT, Nye HE, Kelz MB, et al: Induction of a long-lasting AP-1 complex composed of altered Fos-like proteins in brain by chronic cocaine and other chronic treatments. Neuron 1994; 13:1235–1244
156. Hyman SE: Addiction to cocaine and amphetamine. Neuron 1996; 16:901–904
157. Widnell K, Self DW, Lane SB, et al: Regulation of CREB expression: in vivo evidence for a functional role in morphine action in the nucleus accumbens. J Pharmacol Exp Ther 1996; 276:306–315
158. Li DH, Depoortere RY, Emmett-Oglesby MW: Tolerance to the reinforcing effects of cocaine in a progressive ratio paradigm. Psychopharmacology (Berl) 1994; 116:326–332
159. Young AM, Goudie AJ: Adaptive processes regulating tolerance to the behavioral effects of drugs, in Psychopharmacology: The Fourth Generation of Progress, edited by Bloom FE, Kupfer DJ. New York, Raven, 1995, pp 733–742
160. Trujillo KA, Akil H: Inhibition of morphine tolerance and dependence by the NMDA receptor antagonist MK-801. Science 1991; 251:85–87
161. Kovoor A, Henry DJ, Chavkin C: Agonist-induced desensitization of the mu opioid receptor-coupled potassium channel (GIRK1). J Biol Chem 1995; 270:589–595
162. Koob GF: Animal models of drug addiction, in Psychopharmacology: The Fourth Generation of Progress, edited by Bloom FE, Kupfer DJ. New York, Raven, 1995, pp 759–772
163. deWit H, Stewart J: Reinstatement of cocaine-reinforced responding in the rat. Psychopharmacology (Berl) 1981; 75:134–143
164. Stewart J, deWit H: Reinstatement of drug-taking behavior as a method of assessing incentive motivational properties of drugs, in Methods of Assessing the Reinforcing Properties of Abused Drugs, edited by Bozarth MA. New York, Springer-Verlag, 1987, pp 211–227
165. O'Brien CP, Eckardt MJ, Linnoila VMI: Pharmacotherapy of alcoholism, in Psychopharmacology: The Fourth Generation of Progress, edited by Bloom FE, Kupfer DJ. New York, Raven, 1995, pp 1745–1755
166. Heyser CJ, Schulteis G, Durbin P, et al: Chronic acamprosate decreases deprivation-induced ethanol self-administration in rats. Neuropsychopharmacology (in press)
167. Sass H, Soyka M, Mann K, et al: Relapse prevention by acamprosate: results from a placebo-controlled study on alcohol dependence. Arch Gen Psychiatry 1996; 53:673–680
168. Volpicelli JR, Davis MA, Olgin JE: Naltrexone blocks the post-shock increase of ethanol consumption. Life Sci 1986; 38:841–847
169. Self DW, Barnhart WJ, Lehman DA, et al: Opposite modulation of cocaine-seeking behavior by D_1- and D_2-like dopamine receptor agonists. Science 1996; 271:1586–1589

14

The Neural Substrates of Religious Experience

Jeffrey L. Saver, M.D., John Rabin, M.D.

Religious experience is brain-based. This should be taken as an unexceptional claim. All human experience is brain-based, including scientific reasoning, mathematical deduction, moral judgment, and artistic creation, as well as religious states of mind. Determining the neural substrates of any of these states does not automatically lessen or demean their spiritual significance.[1] The external reality of religious percepts is neither confirmed nor disconfirmed by establishing brain correlates of religious experience. Indeed, it has been argued that demonstrating the existence of a neural apparatus sustaining religious experience can reinforce belief because it provides evidence that a higher power has so constructed humans as to possess the capacity to experience the divine.[2] For the behavioral neurologist and neuropsychiatrist, the challenge is to delineate the distinctive neural substrates of religious experience and their alteration in brain disorders.

Investigation of the neural ground of religious experience is hampered by the absence of a widely accepted animal correlate that would allow laboratory experimentation. Evidence with neuroanatomic import is largely derived from clinical observations in patients with focal brain lesions, especially epileptic disorders, and much of this is anecdotal. We will review data that have been collected on religious experience in normal individuals and in different neurologic and neuropsychiatric syndromes. From this scattered literature, a preliminary unifying model of the brain basis of religious experience may be constructed.

CIRCUMSCRIPTION OF THE TOPIC

Most religious experience parallels ordinary experience. The religious sentiments include religious joy, religious love, religious fear, and religious awe. These religious emotions are analogues of ordinary emotions of joy, love, fear, and awe, differing not in their emotional tone, but only in being directed to a religious object. Their neural substrate is likely to contain nothing of a specifically religious nature, but instead to rest upon the same limbic and subcortical networks that support nonre-

---

The authors thank Jeffrey Cummings, M.D., and David Bear, M.D., for thoughtful comments and suggestions. This work was supported in part by a Frederick Sheldon Memorial Traveling Fellowship from Harvard University (J.L.S.).

ligious joy, love, fear, and awe, directed by dorsolateral and orbital frontal cortices to religious rather than nonreligious targets. The neural substrates of human emotionality have been extensively delineated.[3,4]

Similarly, religious language depends upon the customary dominant-hemisphere perisylvian language cortices for its production, differing only in taking sacred rather than nonsacred topics as linguistic themes. Focal left-hemisphere lesions produce aphasia for religious discourse that parallels aphasia for nonreligious discourse, as common clinical experience attests. Prosody and other emotional contributions to discourse of the right hemisphere apply to both religious and nonreligious themes, and nondominant-hemisphere lesions have been reported to produce parallel impairments of religious and nonreligious emotional processing.[5] Similarly, scholastic/talmudic reasoning is ordinary reasoning applied to religious problems and is undoubtedly mediated by the same neural networks in frontal and parietal multimodal association areas.

A first general observation, then, is that the neural substrate for the preponderance of religious affect and cognition is the whole human brain, employing processing that is parallel, distributed, affective, and symbolic, with contributions of large-scale neurocognitive networks subspecialized for linguistic, prosodic, logical, and affective processing.

What might be peculiarly distinctive to religious experience would appear, on first inspection, to reside not in the domains of affect, language, or cognition, but in perception. It is the direct sensory awareness of God or the divine that is a quintessential mark of specifically religious experience. There is, however, no identifiable separate organ of religious perception.[6-8] Accordingly, sensory apprehension of the divine is likely to be mediated, at least in part, by the neural systems for ordinary tactile, visual, auditory, and olfactory perception. William James and others have suggested, and we concur, that perception of the divine occurs not through the operation of a distinctive sensory faculty, but through the superimposition upon ordinary sensations of a unique numinous-mystical feeling, a feeling of direct awareness of a sacred or divine presence.[8,9] Studies of healthy individuals and neuropsychiatric populations support this hypothesis and suggest a distinctive neurolimbic substrate.

RELIGIOUS EXPERIENCE IN NORMAL INDIVIDUALS

Surveys suggest that religious-numinous experiences are common in both children and adults, across different historical eras, and across all cultures.[10-15] In national surveys in the United States, Britain, and Australia, 20% to 49% of individuals report having personally had numinous experiences, and this figure rises to more than 60% when in-depth interviews of randomly selected individuals are conducted.[8] Hardy and his colleagues[8,16] identified eight major types of numinous experience in British individuals, encompassing, in descending order of frequency, 1) a patterning of events in a person's life that convinces him or her that in some strange way they were meant to happen, 2) an awareness of the presence of God, 3) an awareness of receiving help in answer to a prayer, 4) an awareness of being looked after or guided by a presence not called God, 5) an awareness of being in the presence of someone who has died, 6) an awareness of a sacred presence in nature, 7) an awareness of an evil presence, and 8) experiencing in an extraordinary way that all things are "One." Studies in identical and fraternal twin pairs raised apart suggest that genetic factors account for 50% of interindividual variance in religious interests and attitudes.[17] The cross-cultural ubiquity of numinous experiences and the heritability of religious dispositions argue strongly for a biologic basis, but fail to indicate the specific neural mechanisms involved. Clues to neural substrate must be gleaned from the sites of brain disorders that provoke qualitatively similar experiences.

EPILEPSY AND RELIGIOUS EXPERIENCE

Humanity has long recognized a direct link between epilepsy and the divine. The early Greeks viewed epilepsy as a "sacred disease," a visitation from the gods, until the notion of divine genesis of seizures was dispelled by Hippocrates.[18] In the Medieval and early Renaissance periods, wide currency was given the biblical view that epileptic seizures are manifestations of demonic possession (Mark 9:14–29).[19] Esquirol[20] in 1838 and Morel[21] in 1860 recognized a heightened "religiosity" of epileptics, which they attributed to disability, social isolation, and greater need for religious consolation. The first reported conversion experience directly related to a seizure was noted by Howden,[22] whose patient experienced being transported to Heaven during a fit. Spratling[23] in 1904 reported a religious aura in 52 of 1,325 patients with epilepsy (4%). In the early 20th century, Turner[24] and others suggested that epileptics develop a distinctive interictal character, among the features of which is religious fervor. A substantial number of founders of major religions, prophets, and leading religious figures have been documented as having or suggested to have epilepsy (Table 14–1).

Modern investigations of epilepsy-related religious experience have been marked by contradiction and conflict. This confusion in part reflects failure to distinguish among epilepsy-related religious experiences of ictal, subacute postictal, and chronic interictal occurrence. The phenomenology and neurobiology of each differ in important aspects.

Ictal Events

Ictal events of any type may be the subject of religious or cosmological explanation. Seizures are paroxysmal, riveting, and unexpected—sudden intrusions of unanticipated and often extraordinary experience into the ordinary daily flow of consciousness. Patients who have culturally acquired explanatory systems of a religious character naturally tend to interpret any ictal experience as possessing religious significance. Studies have demonstrated that experiences that are personal, important, negative, and medical, like most seizures, are particularly likely to be interpreted in a religious framework.[25]

Some psychologists of religion have attempted to explain religious experience by employing classical attribution theory. Experiments demonstrate that individuals not only interpret, but also inwardly experience, the same physiologic stimuli in strikingly different ways according to the cognitive expectations they carry. In laboratory studies, identical, pharmacologically induced sympathetic arousals are variably labeled and are experienced across the spectrum of valence, from rewarding to distressful, according to the individual's cognitive set at onset.[26] Attribution theory predicts that religious individuals with epilepsy will often experience intrinsically neutral ictal physiologic events as having a religious-numinous character. This mechanism may indeed mediate some ictal religious experiences.

Human experience, however, is clearly not simply the product of an interaction of nonspecific physiologic arousal with set cognitive schemata. Epileptic auras themselves are the most convincing evidence of the insufficiency of attribution theory, instead demonstrating the existence of distinct physiologic neural substrates for several specific emotional states. Several "psychic" auras, including depersonalization, derealization, dreamy states, autoscopy, and ecstasy, are particularly likely to engender religious interpretation and experience, and merit detailed review.

Among individuals with partial complex seizures, the frequency of auras ranges from 23% to 83%, and up to one-quarter of auras are psychic in content.[27–29] The most common psychic or experiential ictal manifestations of temporal lobe epilepsy are fear, *déjà vu*, *jamais vu*, memory recall, and visual and auditory hallucinations.

Hughlings Jackson[30] was among the first to identify and characterize less common ictal "intellectual auras" ("dreamy states," "cognitive auras") in which the experience of the immediacy and liveliness of one's own or external reality is altered.[31] The intellectual auras include depersonalization, derealization, and double consciousness. Depersonalization auras produce an alteration or loss of the sense of one's own reality, often accompanied by a sense of detachment from others and the environment, and of acting like an automaton. Derealization auras generate an alteration or loss of the sense of the reality of the external environment—for example, the feeling that the external surround is just a dream—and also are often associated with a sense of detachment. Double consciousness ("mental diplopia") auras create a simultaneous experience of persisting remnants of one's normal consciousness and of a new quasi-parasitical consciousness with a different perception of reality. Auras of depersonalization, derealization, and dreamy state account for approximately one-quarter of psychic auras.[32]

Well-documented localizations of spontaneous discharges or electrical stimulations producing intellectual auras are extremely rare. Available evidence, however, suggests that mesolimbic structures, the hippocampus and especially the amygdala, are likely the critical generators of a feeling of unreality about the self or external reality.[33,34] The experience of unreality occurred in 9% of patients with temporal lobe epilepsy in one series, often accompanied by a sensation of fear.[35] One patient in Penfield's series[36] had a sense of "not being in this world" each time the first temporal gyrus was stimulated, reflecting either direct lateral temporal cortical excitation or rapid spread of afterdischarge to mesolimbic structures. One patient in Gloor and colleagues' more recent series[37] repeatedly experienced a faraway feeling during stimulation of temporal mesolimbic structures.

Deep similarities are readily apparent between these intellectual auras and alterations in the experience of reality that are a common feature of intense, nonepileptic religious experience. Individuals undergoing sudden religious awakening or conversion often report abruptly perceiving their ordinary, unenlightened selves as hollow, empty, and unreal (depersonalization) as a prelude to finding a truer, more authentic, religiously grounded self. Similarly, a sense of suddenly seeing through a veil of appearances previously taken for real (derealization) to a deeper, supernatural, genuine reality is a frequent aspect of mystical-numinous experience. Also, doubling of consciousness—a simulta-

TABLE 14–1. Historical-religious figures suggested in the medical literature to have had epilepsy

Person	Description of Spells	Frequency	Likelihood of Epilepsy	Differential Diagnosis[a]	Religious Aspects
Saint Paul (?–65 C.E.)	Conversion on road to Damascus: sudden bright light, falling to the ground, hearing the voice of Jesus, blindness for 3 days with inability to eat or drink. Ecstatic visions	Unknown	+	CPS with generalization[52,133,134] Psychogenic blindness Burns of cornea/retina Vertebrobasilar ischemia Occipital contusion Lightning-stroke Digitalis poisoning Vitreous hemorrhage Migraine equivalent	Father of Catholic Church Possible ecstatic aura, interictal hypermoralism, hyperreligiosity, hypergraphia
Muhammad (570–632)	Pallor, appearance of intoxication, falling, profuse sweating, visual and auditory hallucinations	At least several	+	Complex partial seizures[135,136]	Islamic prophet
Margery Kempe (ca. 1373–1438)	A cry, falling with convulsive movements, turning blue, nausea, psychotic behavior	Recurrent	+	Epilepsy[137] Hysteria Postpartum psychosis Migraine	14th-century Christian mystic and autobiographer
Joan of Arc (1412–1431)	"I heard this Voice [of an angel]... accompanied also by a...great light.... there is never a day when I do not hear this Voice; and I have much need of it."	At least daily by the time of her execution in 1431	+	Ecstatic partial seizure and musciogenic epilepsy[138] Intracranial tuberculoma	Extraordinary, deeply held, idiosyncratic religious beliefs motivating martial prowess in the defense of Orléans
St. Catherine of Genoa (1447–1510)	Extreme sense of heat or cold, whole-body tremor, transient aphasia, automatisms, sense of passivity, hyperesthesia, regression to childhood, dissociation, sleepwalking, transient weakness, transient suggestibility, inability to open eyes	Unknown	+	Complex partial seizure[52] Hysteria	Christian mystic
St. Teresa of Ávila (1515–1582)	Visions, chronic headaches, transient LOC, tongue-biting	1 major LOC spell; frequent headaches	++	Complex partial seizure[52] Hysteria	Catholic saint
St. Catherine dei Ricci (1522–1590)	LOC, visual hallucinations, mystical states	Every Thursday at noon with recovery by Friday at 4:00 P.M.	+	Complex partial seizure[52]	Catholic saint
Emanuel Swedenborg (1688–1772)	Acute psychosis, foaming at the mouth, olfactory, gustatory, and somatic hallucinations; ecstatic aura, falling, LOC, convulsions, hallucinations, postictal trance states	Recurrent	++	Complex partial seizure[139,140] Mania Schizophrenia	Founder of the New Jerusalem Church
Ann Lee (1736–1784)	Visual, auditory hallucinations	From childhood until at least 1774	+	Epilepsy[22]	Founder of the Shaker movement

Name	Symptoms	Frequency	Rating	Diagnosis[a]	Notes
Joseph Smith (1805–1844)	Speech arrest, fear, "pillar of light," hearing voices, "When I came to... I found myself lying on my back looking up at heaven."	One clear conversion event (1820)	+	Complex partial seizure[53]	Founder of Mormonism
Fyodor Mikhailovitch Dostoievsky (1821–1881)	Sense of bliss, then a cry, a fall, generalized tonic-clonic seizure with frothing at the mouth and injuries. Postictal intense depression and guilt, lasting several days	Every few days to every few months	+++	Complex partial seizure with generalization[41,43] Primary generalized seizures[41] Hysteria	Influential Russian novelist Ecstatic auras: "I have really touched God. He came into me myself; yes, God exists, I cried, and I don't remember anything else." Interictal religiosity, increasing with age
Hieronymous Jaegen (1841–1919)	Mystical experiences, visual hallucinations, headaches	Unknown	+	Complex partial seizure[53] Migraine	German mystic
Dr. Z (Arthur Thomas Myers; 1851–1894)	Pallor, vacant look, perseveration of "yes" to any remark, tongue smacking, déjà vu, right-sided motor signs, postictal passivity	Multiple episodes 1871–1894	+++	Complex partial seizures[141] Left temporal lobe lesion on autopsy	Late-life interest in afterlife, reincarnation; prominent in the Society for Psychical Research
Vincent Van Gogh (1853–1890)	Sense of vertigo, tinnitus, hyperacusis, xanthopsia, restlessness, delirium	About 1 dozen spells between the ages of 35 and 37	+	Complex partial seizure with postictal psychosis[42] Ménière's disease[142] Digitalis intoxication Meningoencephalitis luetica Schizophrenia	Renowned painter Hyperreligiosity
St. Thérèse of Lisieux (1873–1897)	Violent trembling, visual hallucinations, wounding by a "shaft of fire," mystical conversions	Several spells after age 9	++	Complex partial seizure[53]	Catholic saint

Note: CPS=complex partial seizures; LOC = loss of consciousness.
[a] All diagnoses listed in this column have been advanced in the medical literature.

neous perception of a higher, purer, religiously oriented self and a baser, irreligious self contesting for control of one's actions and spirit—is a recurrent leitmotif of religious experience. Although many individuals do not associate their intellectual auras with religious experience, it seems likely that repeated, intense, visceral experiences of the self or the external world as unreal would tend to foster a belief in a supranatural ground of reality and a religious outlook.

Another experiential ictal phenomenon that may become affiliated with religious experience is autoscopy, the experience of seeing oneself. Autoscopic experiences may be divided into two categories: 1) seeing one's double, a hallucinatory percept of one's own body visualized in external space, and 2) out-of-body experience, a feeling of leaving one's body and viewing it from another perspective, often from above. Autoscopic phenomena occur in healthy individuals, especially in settings of extreme stress and anxiety, and in a variety of neurologic and psychiatric disorders.[38,39] In the series of Devinsky et al.,[40] 6% of consecutive patients with epilepsy reported autoscopy as an ictal or postictal experience. Among patients with identifiable seizure foci, 88% had temporal lobe involvement. No preponderant trend in laterality of seizure focus was evident. Illustrating the potential for this type of epileptic experience to evolve a religious character, one patient with out-of-body transport experienced passage from a department store floor to a celestial realm (Hécaen,[38] case 7).

Ecstatic seizure experiences, ictal and peri-ictal, have been especially closely linked to religious experience. Dostoyevsky provided riveting and influential descriptions of his own ecstatic-religious epileptic experiences in letters, in other autobiographical writings, and in his fiction:[41–43]

> The air was filled with a big noise, and I thought it had engulfed me. I have really touched God. He came into me myself; yes, God exists, I cried, and I don't remember anything else. You all, healthy people, he said, can't imagine the happiness which we epileptics feel during the second before our attack. I don't know if this felicity lasts for seconds, hours, or months, but believe me, for all the joys that life may bring, I would not exchange this one. Such instants were characterized by a fulguration of the consciousness and by a supreme exaltation of emotional subjectivity. (Dostoyevsky, *The Idiot*)

The term *ecstatic seizure* has been applied to two types of experience, often coexisting. The first is primarily an emotional experience of deep pleasure, the cognitive content of which may vary. The second is primarily a cognitive experience of insight into the unity, harmony, joy, and/or divinity of all reality, usually with pleasurable accompanying affect. Ecstatic seizures of either type are rare. Pleasure as an ictal emotion is reported by less than 0.5% of epileptics.[44,45]

The available evidence suggests a temporolimbic origin for ecstatic seizures. Pleasurable emotions may be induced in humans by depth electrode stimulation of the amygdala, albeit rarely,[46,47] and has been associated with hippocampal-septal hypersynchrony.[48] The few well-studied modern clinical cases of ecstatic seizures all appear to have had a temporolimbic substrate. All 7 of Williams's patients with ictal pleasure had temporal or peritemporal foci.[49] Morgan[50] described a patient whose seizures consisted of feelings of detachment, ineffable contentment, and fulfillment; visualizing a bright light recognized as the source of knowledge; and sometimes visualizing a bearded young man resembling Jesus Christ. Computed tomography disclosed a right anterior temporal astrocytoma. Following anterior temporal lobectomy, ecstatic seizures resolved. Cirignotta et al.[51] described a patient whose seizures were characterized by estrangement from the immediate environment, ineffable joy, and total bliss (without a religious content). Continuous EEG monitoring capturing such a seizure demonstrated a right temporolimbic discharge. Naito and Matsui[52] described an older woman whose seizures were characterized by joyous visions of God and the sun: "My mind, my whole being was pervaded by a feeling of delight." Interictal EEG demonstrated spike discharges in the left anterior and middle temporal region during sleep.

Postictal Experiences

In an influential paper, Dewhurst and Beard[53] described 6 patients who experienced dramatic and often lasting religious conversions in the postictal period. The conversions occurred during the first hours or days following episodes of increased seizure activity, usually bouts of complex partial seizures of temporal origin. More recently, several groups have demonstrated that postictal psychoses, in contrast to acute or chronic interictal psychoses, are particularly likely to have a religious character.[54–56] Postictal psychoses generally emerge after an exacerbation in seizure activity. Contrary to classical teaching, they usually appear in patients with a clear, not clouded, consciousness, after a lucid interval of hours or days since the last seizure (average 1.2 days). Grandiosity and elevated mood are common accompaniments. In a study of 91 patients with epilepsy-related psychosis, Kanemoto et al.[56] found that 23% of postictal psychoses had a religious content, versus only 3% of acute interictal psychoses and 0% of chronic interictal psychoses. Illusion of familiarity, mental diplopia, and feeling of impending death occurred almost exclusively in the postictal psychosis group. Complex partial seizures of temporal origin predominated in all three groups and accounted for 80% of the patients with postictal psychoses.

Interictal Experiences

Since Esquirol in the early 19th century, religious preoccupation has been suggested to be a common feature of the interictal character of individuals with epilepsy. Waxman and Geschwind[57] reawakened interest in this topic when they suggested that hyperreligiosity and intense philosophical and cosmological concerns were leading features of a distinctive interictal personality syndrome of temporal lobe epilepsy. Additional features of this putative syndrome are hypermoralism, deepened affects, circumstantiality, humorlessness, interpersonal viscosity, aggressive irritability, and hypergraphia.

Quantitative support for this syndrome was provided by Bear and Fedio,[58] among whose findings was that religiosity trait scores were significantly higher in patients with temporal lobe epilepsy than in normal or neuromuscular patient control subjects. In a subsequent study, Bear and colleagues[59] found that religiosity trait scores distinguished temporal lobe epileptics from aggressive character disorder patients, affective disorder patients, and generalized or extratemporal focal epilepsy patients. Religiosity scores were similar between patients with temporal lobe epilepsy and patients with schizophrenia, but these groups differed in viscosity, hypergraphia, and other features.

Bear,[60] enlarging upon Gastaut and colleagues' original hypothesis,[61] suggested that interictal spiking and kindling in temporal lobe epileptics leads to intensified sensory-limbic connections. Neutral stimuli are endowed with exaggerated affective tone and significance. Religious and cosmological beliefs are natural responses of patients continually encountering objects and events of heightened meaningfulness and relevance. Animal models of interictal behavioral change have been created, and several additional plausible neurobiological mechanisms for seizure-induced alterations in interictal behavior have been described.[62–64]

Several groups, however, have failed to confirm findings of heightened interictal religiosity among temporal lobe epileptics.[65–67] Differences in religiosity measures and in control group selection account for some of the conflict among studies. An attractive unifying hypothesis is that religiosity is not a universal interictal personality trait among individuals with temporal lobe epilepsy, but emerges in a subgroup, especially those with more active seizure disorders. Several mechanisms may singly or in combination drive the development of religiosity. These may include a desire for religious solace; a need to explain abrupt, sometimes bizarre seizure experiences (attribution theory); a response to ictal numinous experiences; lesional disruptions of the temporal lobe, giving rise to seizures and hyperreligiosity as independent outcomes; abnormal religious interests arising as products of interictal psychopathology;[68] and seizure-induced alterations and intensification of sensory-limbic integration.

NEAR-DEATH EXPERIENCES

Individuals in severe, life-threatening circumstances often experience a variety of unusual mental phenomena. Anecdotal autobiographical and literary descriptions of near-death experiences date back centuries. The first systematic study was performed by Heim[69] in 1892, collecting the accounts of over 30 survivors of falls in the Alps, and subsequent investigators have surveyed patients surviving cardiac arrest, near drowning, and a variety of other conditions.[70–72] Although some near-death experiences are distressing or hellish,[73] most are serene and joyful and may produce profound and long-lasting changes in beliefs and values. The most common features of near-death experience reports are a sensation of peace; a lack of emotion and a feeling of detachment; a sense of being out of the body viewing the self; a sensation of traveling in a darkness or void at the end of which is an encounter with light; a life review;[74] altered passage of time, usually slowed; and a sense of harmony or unity. Thus, depersonalization, derealization, autoscopy, and time distortion are frequent features.

The brain basis of near-death experiences is conjectural. Sociocultural factors and expectations clearly influence the content of near-death experiences,[75] but an underlying driving biologic mechanism appears likely.

One speculation supported by scattered available evidence is endorphin-induced limbic system activity.[76,77] The limbic system is richly endowed with opiate receptors.[78] In an animal model, beta-endorphin administration into the cerebrospinal fluid produces marked increases in limbic neuronal activity at the same time that the animals become outwardly immobile.[79,80] A sudden increase in beta-endorphin in brain tissue, cerebrospinal fluid, and serum occurs in dogs conscious at the moment of cardiac arrest, but not in anesthetized animals.[77]

A second intriguing hypothesis relates near-death experiences to N-methyl-D-aspartate (NMDA) receptor blockade by putative endogenous neuroprotective molecules that dampen glutamate excitotoxicity in hypoxic-ischemic settings.[81] Exogenously administered ketamine, phencyclidine, and other NMDA blockers have been reported to induce several aspects of the near-death experience in normal individuals, including depersonalization, hallucination, sensory deprivation, and elated mood.

Several authors have suggested that depersonalization phenomena may be hard-wired, predetermined responses of the nervous system to extreme stress, and that they may have adaptive benefit.[40,70] For prey trapped by a predator, passive immobilization, feigning death, may promote survival. More generally, the clarity of perception and insight associated with dissociation might allow individuals to identify and carry out previously unrecognized strategies to escape desperate, life-threatening circumstances.

HALLUCINOGENS

The classic hallucinogenic agents include lysergic acid diethylamide (LSD), psilocybin, and mescaline.[82] Prototypical experiences after hallucinogen ingestion include visual illusions and hallucinations, often vividly colored; depersonalization and autoscopy; euphoria; and awareness of a larger intelligence or presence.[83] These experiences closely parallel religious numinous and mystical experience. Mescaline-containing peyote has been employed to foster religious reveries in the Native American Church, and psilocybin has been demonstrated to prompt mystical religious experiences during Protestant church services.[84-86] In one series of 206 observed hallucinogen ingestion sessions, chiefly of LSD and peyote, 96% of subjects experienced religious imagery of some kind, 91% saw religious buildings, and 58% encountered religious figures.[87]

Although the hallucinogenic drugs share dopaminergic and noradrenergic properties, current views affiliate their psychedelic properties with serotonergic agonist activity, particularly at $5-HT_2$, $5-HT_{1A}$, and $5-HT_{1C}$ receptors, widely distributed in basal ganglia, cortex, and temporolimbic structures.[82,88] The precise anatomic substrates for various features of hallucinogen-induced experience have not been elucidated, but the temporal lobe plays an important role. Patients given LSD before and after temporal lobectomy showed reduced richness of induced perceptual experience after temporal lobe resection.[89] The cognitive expectation set of the user strongly influences the quality and interpretation of the experience induced by hallucinogens.

DELUSIONAL DISORDERS

Religious delusions are a feature of schizophrenia and other psychotic illnesses. Distinguishing culturally accepted religious-mystical beliefs from bizarre psychotic delusions poses both a practical challenge to clinicians and an epistemologic-theoretical challenge to the foundations of psychiatric nosology.[90-94]

Mystical states and psychotic states are both characterized by apparent delusions, hallucinations, strange behavior, and social withdrawal. Ultimately, no diagnostic system can irrefutably characterize certain beliefs as delusional, as having no correspondence to objective reality. Several features, however, do tend to differentiate culturally idiosyncratic psychotic states from culturally validated mystical states (Table 14–2). In general, religious hallucinations and delusions of psychotic origin are more likely to be auditory, to focus on illness and deviance, to involve personal grandiosity, to produce affectual distress or indifference, to lead to progressive social isolation, to involve thought blocking, and to be recognized as bizarre and invalid by members of the patient's own subculture.

Delusions occur at some stage of the course of illness in 90% of schizophrenic patients.[95] Delusions of reference, of persecution, of self-depreciation, and of being controlled are the most typical aberrant beliefs in schizophrenia; they are present respectively in 12%, 10%, 8%, and 4% of unselected patients. But religious delusions are common, being present in 3.2% of unselected patients.[96] Religious delusions are one of the 13 categories of delusional belief in the Present State Examination.[97]

The specific content of religious delusions in schizophrenia is shaped by an individual's culture.[98] These effects vary not only across contemporary cultures, but also across historical time. In Britain, the frequency of preoccupations and delusions of a religious character among individuals with schizophrenia declined from 65% in the mid–19th century to 23% in the mid–20th century, while those of a sexual character increased.[99]

Theories of the neurobiological genesis of delusions in schizophrenia have emphasized several distinct mechanisms. The process of belief formation can be divided into several component operations: initial perception of the surround, logical and probabilistic reasoning about initial perceptions, generation of beliefs, and information search to corroborate or refute such beliefs. Defects in each of these operations have been

TABLE 14–2. Features tending to distinguish mystical and psychotic states

Feature	Mystical State	Psychotic State
Hallucinations	Often visual; typically elderly, wise counselors	Predominantly auditory; often accusatory
Vocabulary	Religiously imbued word choice, generally harmonious connotations: God, Christ, soul, peace, spirit	Frequent themes of illness and deviance
Personal role	Individual as self-negating vessel for higher power	Personal grandiosity and omnipotence
Affect	Ecstatic, joyful	Indifferent or terrified
Duration of state	Transient, usually hours, resolves completely	May persist for months or years and leave residual delusions, reduced social function
Withdrawal	Facultative: eventual return to share experience with others	Obligatory: progressive isolation
Disordered speech output	Glossolalia (speaking in tongues): output language is unknown and incomprehensible to speaker; fluency retained	Thought disorder: output may contain neologisms and bizarre associations, but is predominantly in known language; thought blocking may occur
Cultural compatibility	Beliefs are recognized as valid by others in the patient's culture or subculture	Beliefs are rejected by others in the patient's culture or subculture

Note: Based, in part, on references 91, 143–145.

identified in schizophrenic populations,[100,101] including the occurrence of anomalous perceptual experiences,[102] defects in formal logical reasoning[103] and Bayesian inference,[104] and bias in search for confirmatory or disconfirmatory evidence. Some authors have emphasized the role of "delusional mood"—an affectual change that precedes the appearance of overt cognitive delusions.[105] The patient is anxious, restless, and suspicious and feels that some important event is about to happen. The world feels strange, transformed, unreal. Attention and concentration may be impaired. This altered mood state drives the development of delusional thinking.

Neuroanatomically, these psychologic mechanisms implicate frontal systems (deductive and Bayesian reasoning),[106] parietal systems (perceptual distortion),[107] and temporolimbic systems (altered mood and affective charging of stimuli)[107] in the genesis of schizophrenic delusions.

Mood congruence is the distinctive feature of hallucinations and delusions with a religious content in mood disorders. As many as 75% of patients with mania exhibit delusions, and these are typically grandiose.[108,109] Grandiose delusions have been classified into delusions of 1) special abilities (inventor, athlete, spiritual healer), 2) grandiose identity (descendant of a famous historical figure, incarnation of God), 3) wealth, and 4) special mission (fated to solve the problem of war, selected by God to carry a message to world).[110] Conversely, delusions in depressed patients are often guilt-ridden or accusatory and may include reproachful voices or visions perceived as divine in origin.

Both manic and depressive delusions illustrate the vulnerability of perceptual and cognitive systems to affective influence. Overwhelming positive or negative moods generated in ascending aminergic and serotonergic neurotransmitter systems lead to 1) misinterpretation of chance anomalous perceptual experiences and 2) cognitive bias—acceptance only of mood-congruent outcomes of logical and probabilistic reasoning.

CORTICAL DEGENERATIVE DEMENTIAS

The religious experience of patients with dementing disorders has been little studied, and extensive empirical data are lacking. Relevant preliminary observations from our experience and recent literature are of interest for two disorders: Alzheimer's disease and frontotemporal dementia.

Patients with Alzheimer's disease typically demonstrate a progressive loss of religious interest and behaviors, illustrating that the neurobiologic substrates of religious experience can be revealed by conditions that produce hyporeligiosity as well as hyperreligiosity. Passivity and decreased spontaneity are common in moderate to advanced Alzheimer's disease.[111] In one study of 80 Alzheimer's patients, the most frequent personality changes were diminished initiative with growing apathy in more than 60% and relinquishment of hobbies in more than 50%.[112] In our experience, this decreased pursuit of lifelong interests commonly extends to and compromises religious concerns. The likely mechanisms are diminution of temporolimbic-cortical connections, leading to decreased affective charging of cognitions and stimuli, and disrupted cortical association area connections, producing impaired intellectual ability to construct religious cognitions. In addition, pathologic experiences in Alzheimer's disease are not usually of a religious character. Delusions in Alzheimer's disease are typically of the simple persecutory type, without religious content.[113] Memory impairments foster suspiciousness, and cortical association area loss precludes the elaboration of more complex delusional themes.

In contrast, a subset of patients with frontotemporal dementias exhibit hyperreligiosity. In the index family for chromosome 17–linked frontal dementias, 3 of 12 affected individuals showed hyperreligious behavior.[114] Among patients with asymmetric atrophy, increased religiosity appears to occur more frequently with right rather than left frontotemporal atrophy (Bruce Miller, M.D., personal communication, January 1997). We speculate that two neurobehavioral processes heighten religious experience and interests in these patients. First, daily experiences receive increased positive valuation as a result of predominance of left hemisphere (positive) over right hemisphere (negative) valence systems.[115] Second, orbitofrontal atrophy promotes impulsivity, leading to greater willingness to accept and proclaim religious explanations for frequently encountered positive, harmonious phenomena.

A UNIFYING HYPOTHESIS: LIMBIC MARKERS AND RELIGIOUS EXPERIENCE

The core qualities of religious and mystical experience, assented to by a wide variety of psychologists of religion, are the noetic and the ineffable—the sense of having touched the ultimate ground of reality and the sense of the unutterability or incommunicability of the experience.[13,14,116–121] Frequent additional features are an experience of unity, an experience of timelessness and spacelessness, and a feeling of positive affect, of peace and joy. We suggest that the primary substrate for this experience is the limbic system. Temporolimbic epileptic discharges can produce each of these components in fragmentary or complete form: distancing from apparent reality (depersonalization, derealization), timelessness and spacelessness (autoscopy, time distortion), or positive affect (ecstatic auras).[122]

The limbic system integrates external stimuli with internal drives and is part of a distributed neural network that marks stimuli and events with positive or negative value.[123–125] This role of the limbic system is of great evolutionary value, marking the valence and the importance of a novel stimulus or experience for accurate memory encoding and automatic future retrieval.

Moreover, in addition to simple positive or negative valence, limbic discharges can produce experiences that are intermediate between customary divisions between affects and cognitions. For example, a sense of "familiarity" arises in the limbic system as a quasi-emotional marker of experience. Usually the limbic familiarity jibes with explicit recall, but it can appear discordantly, producing *déjà vu* or *jamais vu* experiences. We suggest that, similarly, limbic charges may mark experiences as 1) depersonalized or derealized, 2) crucially important and self-referent, 3) harmonious—indicative of a connection or unity between disparate elements, and 4) ecstatic—profoundly joyous. This limbic activity underlies certain psychic seizure auras, near-death experiences, and the religious and mystical experiences of normal individuals.

An attractive feature of this hypothesis is that it offers a new account of the brain basis of the ineffability of religious experience. Several theorists in the past have advanced an explanation of ineffability that was based on hemispheric specialization.[126–131] In a variety of guises, the hypothesis was advanced that numinous religious experience is a right hemisphere event and consequently holistic and nonverbal. When transferred to the left hemisphere, the experience is translated into an analytic and verbal version that is inherently incomplete, and consequently the experience is reported as ineffable. These theories have several fundamental defects, chiefly that they rest upon metaphorical rather than process-specific accounts of dominant and nondominant hemisphere functions and that they do not clearly apply to individuals with an intact corpus callosum and unitary conscious experience.

The limbic-marker hypothesis provides an entirely different explanation for the ineffability of religious experience. The perceptual and cognitive contents of numinous experience are seen as similar to those of ordinary experience, except that they are tagged by the limbic system as of profound importance, as detached, as united into a whole, and/or as joyous. Consequently, descriptions of the contents of the numinous experience resemble descriptions of the contents of ordinary experience, and the distinctive feelings appended to them cannot be captured fully in words. Like strong emotions, these limbic markers can be named but cannot be communicated in their full visceral intensity, resulting in a report of ineffability.

The limbic marker theory is testable in several respects. It predicts that functional neuroimaging during numinous experiences in individuals who have repeated religious transports would reveal alterations in limbic system activity. It predicts that loss of the ability to have numinous experiences will occur in individuals with bilateral limbic pathology (Klüver-Bucy syndrome) but not other focal lesions. It predicts that more detailed depth electrode studies will map numinous experience to stimulations at limbic and not other sites.

We note in conclusion that although the psychology of religion has a long and honorable history,[132] investigations of the neural substrates of religious experience are in their infancy. Humanity has been called *homo religio*—the religious animal. Behavioral neuroscience must encompass a fully realized account of the neural substrates of religious experience if it is to achieve a systematic understanding of the brain basis of all human behavior. The task before neuropsychiatrists and behavioral neurologists is to fully understand brain disorders that promote, intensify, or alter religious experience—unique clues to the neural basis of the spiritual nature of humanity.

REFERENCES

1. James W: Religion and neurology, in The Varieties of Religious Experience. New York, Longmans, Green and Co, 1902
2. Mackay DM: International biologists' attitudes toward science and religion project (oral interview archive). Los Angeles, CA, UCLA-Reed Neurologic Research Center, October 1981
3. Ross ED, Homan RW, Buck R. Differential hemispheric specialization of primary and social emotions. Neuropsychiatry, Neuropsychology, and Behavioral Neurology 1994; 7:1–19
4. Heilman KM, Bowers D, Valenstein E: Emotional disorders associated with neurological diseases, in Clinical Neuropsychology, 3rd edition, edited by Heilman KM, Valenstein E. New York, Oxford University Press, 1993, pp 461–498
5. Miller DA: Neuropsychology and the emotional component of religion. Pastoral Psychology 1995; 33:267–272
6. Feuerbach L: Lectures on the Essence of Religion (1851), translated by Mannheim R. New York, Harper and Row, 1967
7. James W: The reality of the unseen, in The Varieties of Religious Experience. New York, Longmans, Green and Co, 1902
8. Hay D: "The biology of god": What is the current status of Hardy's hypothesis? International Journal for the Psychology of Religion 1994; 4:1–23
9. Gendlin E: Experiencing and the Creation of Meaning. Glencoe, IL, Free Press, 1962
10. Hardy A: The Biology of God. New York, Taplinger, 1975
11. Hay D: Religious Experience Today: Studying the Facts. London, Mowbray, 1990
12. Tamminen K: Religious experiences in childhood and adolescence: a viewpoint of religious development between the ages of 7 and 20. International Journal for the Psychology of Religion 1994; 4:61–85
13. Wach J: Types of Religious Experience. Chicago, University of Chicago Press, 1951, pp 30–47
14. Stace WT: Mysticism and Philosophy. New York, Macmillan, 1960
15. Carmody DL, Carmody JY: Mysticism. New York, Oxford University Press, 1996
16. Hardy A: The Spiritual Nature of Man. Oxford, Clarendon Press, 1979
17. Waller NG, Kojetin BA, Bouchard TJ, et al: Genetic and environmental influences on religious interests, attitudes, and values: a study of twins reared apart and together. Psychol Sci 1990; 1:138–142
18. Hippocrates: The Genuine Works of Hippocrates. London, Sydenham Society, 1846, pp 843–858
19. Taxil J: Traicté de l'épilepsie [Treatise on epilepsy]. Tournon, 1602
20. Esquirol JED: Des maladies mentales [On mental illness]. Paris, Balliere, 1838
21. Morel BA: Traité des maladies mentales [Treatise on mental illness]. Paris, Masson, 1860
22. Howden JC: The religious sentiments in epileptics. J Ment Sci 1872/1873; 18:491–497
23. Spratling WP: Epilepsy and Its Treatment. Philadelphia, WB Saunders, 1904
24. Turner WA: Epilepsy. London, Macmillan, 1907
25. Spilka B, Schmidt G: General attribution theory for the psychology of religion: the influence of event-character on attributions to God. Journal for the Scientific Study of Religion 1983; 22:326–339
26. Schachter S, Singer JE: Cognitive, social, and physiological determinants of emotional states. Psychol Rev 1962; 69:379–399
27. Palmini A, Gloor P: The localizing value of auras in partial seizures: a prospective and retrospective study. Neurology 1992; 42:801–806
28. So NK: Epileptic auras, in The Treatment of Epilepsy: Principles and Practice, edited by Wyllie E. Philadelphia, Lea and Febiger, 1993, pp 369–377
29. Mendez MF, Cherrier MM, Perryman KM: Epileptic forced thinking from left frontal lesions. Neurology 1996; 47:79–83
30. Jackson JH, Stewart JP: Epileptic attacks with a warning of a crude sensation of smell and with the intellectual aura (dreamy state) in a patient who had symptoms pointing to a gross organic disease of the right temporo-sphenoidal lobe. Brain 1899; 22:534–549
31. Devinsky O, Luciano D: Psychic phenomena in partial seizures. Semin Neurol 1991; 11:100–109
32. Mendez MF, Engebrit B, Doss R, et al: The relationship of epileptic auras and psychological attributes. J Neuropsychiatry Clin Neurosci 1996; 8:287–292
33. Gloor P: Experiential phenomena of temporal lobe epilepsy. Brain 1990; 113:1673–1694
34. Fish DR, Gloor P, Quesney FL, et al: Clinical responses to electrical brain stimulation of the temporal and frontal lobes in patients with epilepsy. Brain 1993; 116:397–414
35. Weil AA: Ictal emotions occurring in temporal lobe dysfunction. Arch Neurol 1959; 1:87–97
36. Penfield W, Jasper H: Epilepsy and the Functional Anatomy of the Human Brain. Boston, Little, Brown, 1954
37. Gloor P, Olivier A, Quesney LF, et al: The role of the limbic system in experiential phenomena of temporal lobe epilepsy. Ann Neurol 1982; 23:129–144
38. Hécaen H, de Ajuriaguerra J: Méconnaissances et hallucinations corporelles [Somatic delusions and hallucinations]. Paris, Masson, 1952
39. Dening TR, Berrios GE: Autoscopic phenomena. Br J Psychiatry 1994; 165:808–817
40. Devinsky O, Feldmann E, Burrowes K, et al: Autoscopic phenomena with seizures. Arch Neurol 1989; 46:1080–1088
41. Voskuil PHA: The epilepsy of Fyodor Mikhailovitch Dostoevsky (1821–1881). Epilepsia 1983; 24:658–667
42. Gastaut H: New comments on the epilepsy of Fyodor Dostoevsky. Epilepsia 1984; 25:408–411
43. Geschwind N: Dostoievsky's epilepsy, in Psychiatric Aspects of Epilepsy, edited by Blumer D. Washington, DC, American Psychiatric Press, 1984, pp 325–334
44. Williams D: The structure of emotions reflecting in epileptic experiences. Brain 1956; 79:29–67
45. Devinsky O, Feldmann E, Bromfield E, et al: Structured interview for simple partial seizures: clinical phenomenology and diagnosis. Journal of Epilepsy 1991; 4:107–116
46. Stevens JR, Mark VH, Ervin F, et al: Deep temporal stimulation in man. Arch Neurol 1969; 21:157–169

47. Obrador S, Delgado JMR, Martin-Rodriguez JG: Emotion areas of the human brain and their therapeutic stimulation, in Cerebral Localization, edited by Zulch KJ. Berlin, Springer, 1973, pp 171–183
48. Mandell AJ: Toward a psychobiology of transcendence: God in the brain, in The Psychobiology of Consciousness, edited by Davidson JM, Davidson RJ. New York, Plenum, 1980, pp 379–464
49. Williams D: The structure of emotions reflected in epileptic experience. Brain 1956; 79:29–67
50. Morgan H: Dostoyevsky's epilepsy: a case report and comparison. Surg Neurol 1990; 33:413–416
51. Cirignotta F, Todesco CV, Lugaresi E: Temporal epilepsy with ecstatic seizures (so-called Dostoyevsky epilepsy). Epilepsia 1980; 21:705–710
52. Naito H, Matsui N: Temporal lobe epilepsy with ictal ecstatic state and interictal behavior of hypergraphia. J Nerv Ment Dis 1988; 176:123–124
53. Dewhurst K, Beard AW: Sudden religious conversions in temporal lobe epilepsy. Br J Psychiatry 1970; 117:497–507
54. Logsdail SJ, Toone BK: Post-ictal psychoses. Br J Psychiatry 1988; 152:246–252
55. Savard G, Andermann F, Olivier A, et al: Postictal psychosis after complex partial seizures: a multiple case study. Epilepsia 1991; 32:225–231
56. Kanemoto K, Kawasaki J, Kawai I: Postictal psychosis: a comparison with acute interictal and chronic psychoses. Epilepsia 1996; 37:551–556
57. Waxman SG, Geschwind N: The interictal behavior syndrome of temporal lobe epilepsy. Arch Gen Psychiatry 1975; 32:1580–1586
58. Bear DM, Fedio P: Quantitative analysis of interictal behavior in temporal lobe epilepsy. Arch Neurol 1977; 34:454–467
59. Bear D, Levin K, Blumer D, et al: Interictal behaviour in hospitalised temporal lobe epileptics: relationship to idiopathic psychiatric syndromes. J Neurol Neurosurg Psychiatry 1982; 45:481–488
60. Bear DM: Temporal lobe epilepsy: a syndrome of sensory-limbic hyperconnection. Cortex 1979; 15:357–384
61. Gastaut H, Morin G, Leserve N: Étude du comportement des épileptiques psychomoteurs dans l'intervalle de leurs crises [Study of the interictal behavior of psychomotor epileptics]. Annales Medico-Psychologiques 1955; 113:1–27
62. Adamec RE: Partial kindling of the ventral hippocampus: identification of changes in limbic physiology which accompany changes in feline aggression and defense. Physiol Behav 1991; 49:443–453
63. Griffith N, Engel J, Bandler R: Ictal and enduring interictal disturbances in emotional behaviour in an animal model of temporal lobe epilepsy. Brain Res 1987; 400:360–364
64. Engel J, Bandler R, Griffith NC, et al: Neurobiological evidence for epilepsy-induced interictal disturbances. Adv Neurol 1991; 55:97–111
65. Willmore LJ, Heilman KM, Fennell E: Effect of chronic seizures on religiosity. Transactions of the American Neurological Association 1980; 105:85–87
66. Sensky T, Wilson A, Petty R, et al: The interictal personality traits of temporal lobe epileptics: religious belief and its association with reported mystical experiences, in Advances in Epileptology, edited by Porter R. New York, Raven, 1984, pp 545–549
67. Tucker DM, Novelly RA, Walker PJ: Hyperreligiosity in temporal lobe epilepsy: redefining the relationship. J Nerv Ment Dis 1987; 175:181–184
68. Roberts JKA, Guberman A: Religion and epilepsy. Psychiatric Journal of the University of Ottawa 1989; 14:282–286
69. Heim A: Remarks on fatal falls. Yearbook of the Swiss Alpine Club 1892; 27:327–337
70. Noyes R, Kletti R: Depersonalization in the face of life-threatening danger: a description. Psychiatry 1976; 39:19–27
71. Rodin E: The reality of death experience: a personal perspective. J Nerv Ment Dis 1980; 168:259–263
72. Owens JE, Cook EW, Stevenson I: Features of "near-death experience" in relation to whether or not patients were near death. Lancet 1990; 336:1175–1177
73. Greyson B, Bush NE: Distressing near-death experiences. Psychiatry 1992; 55:95–110
74. Stevenson I, Cook EW: Involuntary memories during severe physical illness or injury. J Nerv Ment Dis 1995; 183:452–458
75. Kellehear A: Culture, biology, and the near-death experience. J Nerv Ment Dis 1993; 181:148–156
76. Carr DB: Endorphins at the approach of death (letter). Lancet 1981; 1:390
77. Sotelo J, Perez R, Guevara P, et al: Changes in brain, plasma, and cerebrospinal fluid contents of beta-endorphin in dogs at the moment of death. Neurol Res 1995; 17:223–225
78. Kuhar MJ, Pert CB, Snyder SH: Regional distribution of opiate receptor binding in monkey and human brain. Nature 1973; 245:447–450
79. Henriksen S, Bloom F, McCoy F, et al: Beta-endorphin induces nonconvulsive limbic seizures. Proc Natl Acad Sci USA 1978; 75:5221–5225
80. Ableitner A, Schulz R: Neuroanatomical sites mediating the central actions of beta-endorphin as mapped by changes in glucose utilization: involvement of mu opioid receptors. J Pharmacol Exp Ther 1992; 262:415–423
81. Jansen K: Neuroscience, ketamine, and the near-death experience, in The Near-Death Experience Reader, edited by Bailey LW, Yates J. New York, Routledge, 1996, pp 265–282
82. Strassman RJ: Hallucinogenic drugs in psychiatric research and treatment. J Nerv Ment Dis 1995; 183:127–138
83. Strassman RJ, Qualls CR, Uhlenhuth EH, et al: Dose-response study of N,N-dimethyltryptamine in humans. Arch Gen Psychiatry 1994; 41:98–108
84. LaBarre W: The Peyote Cult, 5th edition. Norman, OK, University of Oklahoma Press, 1989
85. Pahnke WN: An analysis of the relationship between psychedelic drugs and the mystical consciousness. Doctoral dissertation, Harvard University, 1963
86. Doblin R: Pahnke's "Good Friday Experiment": a long-term follow-up and methodological critique. J Transpers Psychol 1991; 23:1–28
87. Masters REL, Houston J: The Varieties of Psychedelic Experience. New York, Holt, Rinehart, and Winston, 1966
88. Joyce JN, Share A, Lexow N, et al: Serotonin uptake sites and serotonin receptors are altered in the limbic system of schizophrenics. Neuropsychopharmacology 1993; 8:315–336
89. Serafetinides EA: The significance of the temporal lobes and of hemisphere dominance in the production of the LSD-25 symptomatology in man. Neuropsychologia 1965; 95:53–63
90. Heise DR: Delusions and the construction of reality, in Delusional Beliefs, edited by Oltmanns TF, Maher BA. New York, Wiley, 1988, pp 259–272
91. Greenberg D, Witzum E, Buchbinder JT: Mysticism and psychosis: the fate of Ben Zoma. Br J Med Psychol 1992; 65:223–235
92. Barnhouse RT: How to evaluate patients' religious ideation, in Psychiatry and Religion: Overlapping Concerns, edited by Robinson LH. Washington, DC, American Psychiatric Press, 1986, pp 89–106
93. Gaines A: Delusions: culture, psychosis, and the problem of meaning, in Delusional Beliefs, edited by Oltmanns TF, Maher BA. New York, Wiley, 1988, pp 230–258
94. Winters KC, Neale JM: Delusions and delusional thinking in psychotics: a review of the literature. Clin Psychol Rev 1983; 3:227–253
95. Cutting J: Descriptive psychopathology, in Schizophrenia, edited by Hirsch SR, Weinberger DR. Cambridge, MA, Blackwell Science, 1995, pp 15–27
96. World Health Organization: Schizophrenia: An International Follow-up Study. New York, Wiley, 1979
97. Wing JK, Cooper JE, Sartorius N: The Description and Classification

of Psychiatric Symptoms: An Instruction Manual for the PSE and CATEGO system. Cambridge, England, Cambridge University Press, 1974
98. Westermeyer J: Some cross-cultural aspects of delusions, in Delusional Beliefs, edited by Oltmanns TF, Maher BA. New York, Wiley, 1988, pp 212–229
99. Klaf FS, Hamilton JG: Schizophrenia: a hundred years ago and today. Journal of Mental Science 1961; 107:819–827
100. Garety PA, Hemsley DR: Delusions: Investigations Into the Psychology of Delusional Reasoning. New York, Oxford University Press, 1994
101. Bentall RP: The syndromes and symptoms of psychosis, in Reconstructing Schizophrenia, edited by Bentall RP. London, Routledge, 1990, pp 23–60
102. Maher BA, Spitzer M: Delusions, in Comprehensive Handbook of Psychopathology, 2nd edition, edited by Sutker P, Adams H. New York, Plenum, 1993, pp 263–293
103. Von Damurus E: The specific laws of logic in schizophrenia, in Language and Thought in Schizophrenia, edited by Kasanin JS. Berkeley, CA, University of California Press, 1944, pp 30–44
104. Hemsley DR, Garety PA: The formation and maintenance of delusions: a Bayesian analysis. Br J Psychiatry 1986; 149:51–56
105. Wiggins OP, Schwartz MA, Northoff G: Toward a Husserlian phenomenology of the initial stages of schizophrenia, in Philosophy and Psychopathology, edited by Spitzer M, Maher BA. New York, Springer-Verlag, 1990, pp 21–34
106. Benson DF, Stuss DT: Frontal lobe influences on delusions: a clinical perspective. Schizophr Bull 1990; 16:403–411
107. Cutting J: The phenomenology of acute organic psychosis: comparison with acute schizophrenia. Br J Psychiatry 1987; 151:324–332
108. Carlson GA, Goodwin FK: The stages of mania: a longitudinal analysis of the manic episode. Arch Gen Psychiatry 1973; 28:221–228
109. World Health Organization: The International Pilot Study of Schizophrenia. Geneva, WHO, 1973
110. Leff JP, Fischer M, Bertelsen AC: A cross-national epidemiological study of mania. Br J Psychiatry 1976; 129:428–442
111. Petry S, Cummings JL, Hill MA, et al: Personality alterations in dementia of the Alzheimer type. Arch Neurol 1988; 45:1187–1190
112. Chatterjee A, Strauss ME, Smyth KA, et al: Personality changes in Alzheimer disease. Arch Neurol 1992; 49:486–491
113. Cummings JL: Organic delusions: phenomenology, anatomical correlations, and review. Br J Psychiatry 1985; 146:184–197
114. Lynch T, Sano M, Marder KS, et al: Clinical characteristics of a family with chromosome 17–linked disinhibition-dementia-parkinsonism-amyotrophy complex. Neurology 1994; 44:1878–1884
115. Davidson RJ: Anterior cerebral asymmetry and the nature of emotion. Brain Cogn 1992; 20:125–151
116. James W: The Varieties of Religious Experience. New York, Longmans, Green and Co, 1902
117. Otto R: The Idea of the Holy. New York, Oxford University Press, 1923
118. Byrnes JF: The Psychology of Religion. New York, The Free Press, 1984
119. Brown LB: The Psychology of Religious Belief. London, Academic Press, 1987
120. Clark WH: The phenomena of religious experience, in Religious Experience: Its Nature and Function in the Human Psyche, edited by Clark WH, Malony HN, Daane J, et al. Springfield, IL, Charles C Thomas, 1973, pp 21–40
121. Spika B, Hood RW, Gorsuch RL: The Psychology of Religion: An Empirical Approach. Engelwood Cliffs, NJ, Prentice-Hall, 1985
122. Joseph R: Neuropsychiatry, Neuropsychology, and Clinical Neuroscience, 2nd edition. Baltimore, Williams and Wilkins, 1996
123. Aggleton JP: The contribution of the amygdala to normal and abnormal emotional states. Trends Neurosci 1993; 16:328–333
124. Damasio AR, Tranel D, Damasio H: Somatic markers and the guidance of behavior, in Frontal Lobe Function and Dysfunction, edited by Levin H, Eisenberg H, Benton A. New York, Oxford University Press, 1991, pp 217–228
125. Saver JL, Damasio AR: Preserved access and processing of social knowledge in a patient with acquired sociopathy due to ventromedial frontal damage. Neuropsychologia 1991; 29:1241–1249
126. Jaynes J: The Origin of Consciousness in the Breakdown of the Bicameral Mind. Boston, Houghton Mifflin, 1976
127. Lex BW: The neurobiology of ritual trance, in The Spectrum of Ritual, edited by D'Aquili EG, Laughlin CD, Mcmanus J. New York, Columbia University Press, 1979, pp 117–151
128. Ashbrook JB: The Brain and Belief: Faith in the Light of Brain Research. Bristol, IN, Wyndham Hall Press, 1988
129. Persinger MA: Vectorial cerebral hemisphericity as differential sources for the sensed presence, mystical experiences, and religious conversions. Percept Mot Skills 1993; 76:915–930
130. D'Aquili E: The myth-ritual complex, a biogenetic structural analysis, in Brain, Culture, and the Human Spirit, edited by Asbrook JB. Lanham, MD, University Press of America, 1993, pp 45–75
131. Turner V: Body, brain, and culture, in Brain, Culture, and the Human Spirit, edited by Asbrook JB. Lanham, MD, University Press of America, 1993, pp 77–108
132. Wulff DM: Psychology of Religion, 2nd edition. New York, Wiley, 1996
133. Landsborough D: St. Paul and temporal lobe epilepsy. J Neurol Neurosurg Psychiatry 1987; 50:659–664
134. Vercelletto P: La maladie de Saint Paul. Rev Neurol 1994; 150:835–839
135. Trimble MR: The Psychoses of Epilepsy. New York, Raven, l991
136. Temkin O: The Falling Sickness: A History of Epilepsy From the Greeks to the Beginning of Modern Neurology. Baltimore, Johns Hopkins University Press, 1971
137. Drucker T: Malaise of Margery Kempe. New York State Journal of Medicine 1972; 72:2911–2917
138. Foote-Smith E, Bayne L: Joan of Arc. Epilepsia 1991; 32:810–815
139. Johnson J: Henry Maudslay on Swedenborg's messianic psychosis. Br J Psychiatry 1994; 165:690–691
140. Foote-Smith E, Smith TJ: Emanuel Swedenborg. Epilepsia 1996; 37:211–218
141. Taylor DC, Marsh SM: Hughlings Jackson's Dr Z: the paradigm of temporal lobe epilepsy revealed. J Neurol Neurosurg Psychiatry 1980; 43:758–767
142. Arenberg IK, Countryman LLF, Bernstein LH, et al: Vincent's violent vertigo. Acta Otolaryngol 1991; 485:84–103
143. Buckley P: Mystical experience and schizophrenia. Schizophr Bull 1981; 7:516–521
144. Oxman TE, Rosenberg SD, Schnurr PP, et al: The language of altered states. J Nerv Ment Dis 1988; 176; 401–408
145. Group for the Advancement of Psychiatry: Mysticism: Spiritual Quest or Psychic Disorder? (Mental Health Materials Center, vol 9, publ no 97). New York, Mental Health Materials Center, 1976

Index

Page numbers printed in **boldface** type refer to tables or figures.

A

Abstinence, from drugs, reward thresholds and, 185, **185**
Abuse, recovered memories of. *See* Recovered memory
Acamprosate, 190
Accumbens, 43–66, **45**
　compartmental infrastructure of, 44–45
　core of, **45**, 45–46
　　shell differentiated from, 54
　development of, 46–50, **47–49**
　drug abuse studies of, 45
　extended amygdala and, 46, 61–65, **62**
　　interstitial nucleus of posterior limb of anterior commissure and, **64**, 64–65
　　transitions between, 62, **63**, **64**, 64
　mesencephalic dopamine neurons and, 59, **59**
　projections of, 59–60, **60**
　projections to, **60**, 60–61
　neurons of, 50–54
　　convergence and segregation of afferents and, **52**, 53, **53**
　　functional aspects of, 50
　　projections to multiple targets and, **52**, 53–54
　　relationships of thalamic and amygdaloid afferents with prefrontal cortex-ventral striatal system and, 50–51, **51**, 53
　shell of, **45**, 45–46, 54, **55**, 56
　　core differentiated from, 54
　　response to antipsychotic and psychoactive drugs, 56–57, **57**, **58**
　　ventral striatum in primate and, **58**, 58–59
Addiction. *See* Drug addiction
Adrenocorticotropic hormone (ACTH), amygdala lesions and unconditioned fear and, 79
Aggression, in temporolimbic syndromes, 128
Akinetic episodes, psychogenic seizure-induced, 101–102, **102**
Alcohol addiction, neurochemical substrates and, 183–184
Alzheimer's disease
　religious experiences in, 203–204
　ventromedial temporal areas in, 24–26
　　amygdala, 24–25
　　hippocampal formation, 24
　　parahippocampal gyrus, **25**, 25–26, **26**
Amnesia
　global, transient
　　distinguishing features of, **109**
　　paroxysmal limbic disorders with, 106, **107**
　hypolimbic syndromes and, 15
　infantile, 147
　psychiatric disorder-induced, 100–101
　source, 147
　in temporolimbic syndromes, 126–127
Amphetamine
　addiction to, mesocorticolimbic dopamine system and, 180–181, **184**
　striatal dopamine receptors and, 57, **58**
Amygdala, 8–9, **10**, **12**
　afferents of, relationship with prefrontal cortex-ventral striatal system, 50–51, **51**, 53
　in Alzheimer's disease, **25**, 25–26, **26**
　centromedial, 61
　extended. *See* Extended amygdala
　fear responses and. *See* Fear
　feedback projections to cortex, 23
Angular bundle, 22
Antipsychotic drugs, accumbens shell reactions to, 56–57, **57**, **58**
Anxiety. *See also* Fear
　in dysfunctional limbic syndromes, 16
　with migraines, 104–106
　psychiatric disorder-induced, 99–100
　in temporolimbic syndromes, 128
Anxiety attacks, due to combination of epilepsy and psychiatric illness, 103
Aphasia, with migraines, 104–106
Approach-avoidance, modular theory of emotional experience and, 139–140
Archicortical cortex, **6**, 6–8, **8**

Arousal
    attention linked to, 138
    excessive, autonomic, psychiatric disorder-induced, 99–100
    modular theory of emotional experience and, 137–139
Attention
    amygdala lesions and conditioned fear and, 77–78
    anatomical connections between amygdala and brain areas involved in fear and anxiety and, 74
    arousal linked to, 138
    electrical or chemical stimulation of amygdala and, 75
Auditory auras, with temporal lobe seizures, 115
Auditory hallucinations, epilepsy-induced, 95–96, **97**
Auras, epilepsy and, 98–99, **99**, 197
    temporal lobe seizures and, 115–117
        classification of, 115–116
        electrical stimulation and, 117–118
        localization of, 116–117
Automatic behavior, epilepsy-induced, 95–96, **97**
    complex partial seizures and, 114
Autonomic nervous system
    amygdala and
        anatomical connections between amygdala and brain areas involved in fear and anxiety and, 73
        conditioned fear and, 77
        corticotropin-releasing hormone infusion in, 80
        electrical or chemical stimulation of, 75
        unconditioned fear and, 78–79
    hyperarousal of, psychiatric disorder-induced, 99–100
    peripheral, level of activity of, 138–139
Autoscopy, 200

B

Basal nucleus of Meynert, 13
Basolateral complex, 61
Behavioral arrest, complex partial seizures and, 114
Benzodiazepines, infusion in amygdala, fear and, 80, 81
Bicuculline, infusion in amygdala, 75
    fear and, 80–81
Body image, ventrobasal complex and, 34
Brain. *See also specific regions of brain*
    depression model of, 168, **169**
    imaging of, in schizophrenia, 156–158
        functional, 157–158
        structural, 156–157
    memory organization in, 145–148
        dynamic view of, 145–147
        infantile amnesia and, 147
        multiple memory systems and, 147–148
        source amnesia and, 147
        static view of, 145
    phylogenetic development of, 5–8, **5–8**
    postmortem studies of, in schizophrenia, 158–161
        cytoarchitectural changes in medial temporal lobe and, 159
        macroscopic and microscopic morphometric changes in medial temporal lobe and, 158–159
        neurochemical and receptor changes in medial temporal lobe and, 160
        neurodevelopment, neurodegeneration, and neural injury studies and, 161
        synaptic and connectional neuroanatomy and, 159–160
    three-layered model of organization, structure, and function of, **5**, 5–6, **6**
    triune, 6, **6**

Brain injury
    frontotemporal, 98–99, **99**
    temporolimbic syndromes and, 129
    tentorium cerebelli, **26**, 26–27

C

Calcium channels, voltage-gated, drug dependence and, 186–187
Catastrophic-depressive reaction, with left hemisphere lesions, 136
Central theories, of emotional experience, 135
Cerebral tumors, temporolimbic syndromes and, 129
c-fos, reaction following administration of antipsychotic drugs, 57
Chemical stimulation, of amygdala, fear elicitation by, 75–76
Chlordiazepoxide, infusion in amygdala, fear and, 81
Cholinergic system, limbic, 13–14
Cingulate
    anatomy of, 8, **9**
    anterior, 11, **12**
        hypofunctioning of, 14–15
        subcortical unit of, 12
    posterior, 11, **11, 12**
    rostral
        in depression, 171–172, **172**
        in depression model, 168, **169**
Cocaine addiction, mesocorticolimbic dopamine system and, 180–181, **184**
Cognitive auras, epilepsy and, 197
Coincident-current selection technique, 33
Confusion
    with migraines, 104–106
    with nonconvulsive status epilepticus, 106–107, **108**
Conscious memory, 147
Consciousness, impairment of, epilepsy-induced, 95–96, **97**, 98–99, **99**
Cortex, connections between amygdala and, in fear and anxiety, 82–83

Cortical-association division, of amygdala, 9
Cortical degenerative dementias, religious experiences in, 203–204
Corticoamygdaloid projections, 21–22, **22**
Corticosubcortial reentrant circuits, 44
Corticotropin-releasing hormone (CRH), in amygdala, fear and, 80, 81

D

Declarative memory, 147
Déjà entendu, with temporal lobe seizures, 116
Déjà vu, epilepsy and, 98–99, **99,** 197
   with temporal lobe seizures, 116, 117–118
Delusions
   religious, 202–203, **203**
   in schizophrenia, neurobiological genesis of, 202–203
   in temporolimbic syndromes, 127–128
Dementias, religious experiences in, 203–204
Dependence, on drugs
   molecular and cellular mechanisms of, 186–189, **187–189**
   neural substrates of, 185
   neurochemical substrates of, 185–186
Depersonalization
   distinguishing features of, **109**
   psychiatric disorder-induced, 99–100, **101**
Depression
   catastrophic-depressive reaction with left hemisphere lesions and, 136
   limbic-cortical dysregulation in, 167–174, **169**
      depression in neurological disease and, 169–170
      neurochemical markers and, 170–171
      primary depression and, 170
      psychosurgical parallels with, 173
      resting state patterns in depressed patients and, 171–172, **172**
      similarity between depression recovery and induced sadness and, 173
      transient sadness in healthy subjects and, 171, **171**
      treatment effects and, 172–173
   recovery from, similarity with induced sadness, 173
   with stroke, 137
   in temporolimbic syndromes, 128
Diencephalic-hypothalamic model, of emotional experience, 135
Diencephalic-limbic model, of emotional experience, 135
Dissociative disorders, in temporolimbic syndromes, 128
Distributed processing
   memory organization and, 146
   parallel, temporolimbic, 124
Dizziness, with temporal lobe seizures, 115
Dopamine
   accumbens shell versus core and, **55,** 56
   anatomical connections between amygdala and brain areas involved in fear and anxiety and, 74
   dorsal thalamic regulation of, 38
   nicotine addiction and, 184
Dopamine cells
   mesotelencephalic, projections to striatum, 59–60, **60**
   midbrain, striatal projections to, **60,** 60–61
   of substantia nigra pars compacta, 59, **59**
Dopaminergic system
   limbic, 14
   stress-induced activation of, 45–46
Dorsal compartment, in depression model, 168, **169**
Dorsal thalamic projection, 37–39, **38**

Dreamy states
   epilepsy and, 197
   with temporal lobe seizures, 116
Drug addiction, 179–190
   accumbens and, 45
   to alcohol, neurochemical substrates and, 183–184
   clinical implications of neurobiological substrates and, 190
   to cocaine and amphetamine, mesocorticolimbic dopamine system and, 180–181, **184**
   drug dependence and
      molecular and cellular mechanisms of, 186–189, **187–189**
      neural substrates of, 185
      neurochemical substrates of, 185–186
   drug tolerance and, neural substrates of, 189–190
   extended amygdala as common substrate for drug reward and, 184–185
   neurotransmission and, 180, **182**
   to nicotine, dopamine and opioid peptides and, 184
   to opiates, opioid peptide systems and, 181–182
   reinforcement and, 179–180, **181**
   neural substrates of, 180, **183**
   neurobiological substrates of, 190
   relapse and, neural substrates of, 190
   reward thresholds and drug abstinence and, 185, **185**
   to sedative-hypnotics, neurochemical substrates and, 182
Dysthymia, psychiatric disorder-induced, 100–101

E

Ecstatic seizures, 200
Electrical stimulation
   of amygdala, fear elicitation by, 75–76
   temporal lobe seizures and, 117–118

Electroencephalogram (EEG), in schizophrenia, 157
Electrophysiology, in schizophrenia, 157–158
Emotional experience, 133–141
  central theories of, 135
  feedback theories of, 133–135
    facial feedback hypothesis, 133–134
    visceral feedback hypothesis, 134–135
  modular theory of, 135–140
    arousal and, 137–139
    motor activation and approach-avoidance and, 139–140
    valence and, 136–137
Emptiness, psychiatric disorder-induced, 100–101
Encephalitis, temporolimbic syndromes and, 130
Entorhinal cortex, 22, **23**
  in Alzheimer's disease, 25, **25**, **26**
  in schizophrenia
    cytoarchitectural studies of, 159
    neurochemical changes in, 160
Epigastric auras, with temporal lobe seizures, 115
  localization of, 116
Epilepsy. *See also* Seizures; Status epilepticus
  paroxysmal limbic disorders due to, 95–99
  partial complex, 98–99, **99**
  psychiatric illness combined with, paroxysmal limbic disorders due to, 103–104
  religious experience and, 196–201, **198–199**
    ictal events and, 197, 200
    interictal experiences and, 201
    postictal experiences and, 200–201
Epithalamus, 37
Ethanol addiction, neurochemical substrates and, 183–184
Event-related potentials (ERPs), in schizophrenia, 157–158
Excitatory amino acid receptor dysfunction hypothesis, of schizophrenia, 160

Experiential auras, temporal lobe seizures and, 116
  electrical stimulation and, 117–118
  localization of, 116
Extended amygdala, 9, **10**
  accumbens and, 46, **62**
  interstitial nucleus of posterior limb of anterior commissure and, **64**, 64–65
  transitions between, 62, **63**, **64**, 64
  as common substrate for drug reward, 184–185
  temporolimbic syndromes and, 124
Eye-tracking, thalamus and, 37

F
Facial feedback hypothesis, of emotional experience, 133–134
Facial recognition, classical fear conditioning and, 77
False memories, creation of, 149–150
False memory syndrome. *See* Recovered memory
Fasciculus retroflexus of Meynert, 37–38
Fear. *See also* Anxiety
  amygdala and, 71–83, **72**
    anatomical connections between central nucleus and other brain areas involved in fear and anxiety and, **73**, 73–75, **74**
    conditioned fear and, 76–78
    connections between amygdala and cortex and, 82–83
    electrical stimulation of, 75–76
    in Klüver-Bucy syndrome, 76–77
    local drug infusion and, 79–82
    unconditioned fear and, 78–79
  conditioned, 72, **72**
    amygdala lesions and, 76–78
  epilepsy-induced, 95–96, **97**

  psychogenic seizure-induced, 101–102, **102**
  unconditioned, 72
    amygdala lesions and, 78–79
Feedback theories, of emotional experience, 133–135
  facial feedback hypothesis, 133–134
  visceral feedback hypothesis, 134–135
Feedforward systems, 21
Flashbacks, psychogenic seizure-induced, 102–103, **103**
Flashbulb memories, 149
Fluoxetine, in depression, 173
Franklin, Eileen, 144–145, 151
Frontal lobe, dorsolateral, motor activation and, 139
Frontotemporal dementias, religious experiences in, 204
Frontotemporal head injury, 98–99, **99**
Functional magnetic resonance imaging (fMRI), in schizophrenia, 157
Fundus striati, 64

G
Gamma-aminobutyric acid (GABA) agonists and antagonists, infusion in amygdala, fear and, 80–81
Gamma-aminobutyric acid (GABA)$_A$ receptor
  alcohol addiction and, 183
  sedative-hypnotic addiction and, 182
Ganglionic eminence, medial, derivatives of, 46–47
Gastaut-Geschwind syndrome, 15, **125**, 125–126
"Gatelet" concept, of reticular function, 34
Gaze mechanisms, thalamus and, 37
G protein-coupled receptors, drug dependence and, 186
Grand lobe limbique, 3, **4**
Gustatory auras, with temporal lobe seizures, 115
  localization of, 116
Gyrus ambiens, **20**, 20–21

H

Habenula, 37–39, **38**
Hallucinations
    auditory, epilepsy-induced, 95–96, **97**
    in temporolimbic syndromes, 127
Hallucinogens, religious experiences and, 202
Haloperidol, neurotensin immunoreactivity and, 57
Head injury
    frontotemporal, 98–99, **99**
    temporolimbic syndromes and, 129
    tentorium cerebelli, **26**, 26–27
Herpes encephalitis, 96–98, **98**
Hippocampal formation
    in Alzheimer's disease, 24
    cortical association input to, 22, **22**
    feedback projections to cortex, 23, **24**
Hippocampal paralimbic division, of limbic system, 10–12, **11**
Hippocampus, **12**
Historical truth, 144
Hormones
    amygdala lesions and conditioned fear and, 77
    amygdala lesions and unconditioned fear and, 78–79
    anatomical connections between amygdala and brain areas involved in fear and anxiety and, 73
    corticotropin-releasing hormone infusion in amygdala and, 80
    electrical or chemical stimulation of amygdala and, 75
Hyperarousal, autonomic, psychiatric disorder-induced, 99–100
Hypergraphia, in Gastaut-Geschwind syndrome, 126
Hyperlimbic syndromes, **14**, 15
Hyperoral behaviors, in temporolimbic syndromes, 128–129
Hypersexuality, in temporolimbic syndromes, 129
Hypoalgesia, amygdala and
    amygdala lesions and conditioned fear and, 78
    amygdala lesions and unconditioned fear and, 79
    drug infusion in, 82
Hypolimbic syndromes, **14**, 14–15
Hyposexuality, in temporolimbic syndromes, 129

I

Imitation behavior, in dysfunctional limbic syndromes, 15–16
Infantile amnesia, 147
Intellectual aurae, epilepsy and, 197
    with temporal lobe seizures, 116
Interictal syndromes, Gastaut-Geschwind, 15
Interstitial nucleus of posterior limb of anterior commissure (IPAC), **64**, 64–65
Intracellular messenger pathways, drug dependence and, 186–189, **187–189**
Intracranial pressure, increased in tentorium cerebelli injury, **26**, 26–27
    with uncal herniation, 27, **27**

J

Jamais entendu, with temporal lobe seizures, 116
Jamais vu, epilepsy and, 197
    temporal lobe, 116

K

Klüver-Bucy syndrome, 15, 76–77, 124–125, **125**

L

Left-right social emotional theories, 135
Limbic-cortical dysregulation, in depression. *See* Depression
Limbic-marker hypothesis, of religious experience, 204–205
Limbic system, 3–16, **4**. *See also* Temporolimbic syndromes
    anatomy and neurotransmitters of, 8–14, **9**
    chemoarchitecture and, 12–14
    of hippocampal paralimbic division, 10–12, **11**
    of orbitofrontal paralimbic division, 8–10, **10**
    of subcortical structures, 12, **12**
    clinical syndromes involving, **14**, 14–16
    dysfunctional, **14**, 15–16
    hyperlimbic, **14**, 15
    hypolimbic, **14**, 14–15
    hippocampal paralimbic division of, 10–12, **11**
    historical background of, 3–5, **4**
    motor activation and, 139
    paroxysmal disorders involving. *See* Paroxysmal limbic disorders
    phylogenetic development of, 5–8, **5–8**
Locus coeruleus
    and fear response, 74, 100
    withdrawal-induced activation of, 187–188, **189**
Lysergic acid diethylamide (LSD), religious experiences and, 202

M

Magnocellular portion, of mediodorsal nucleus of thalamus, 36, **36**
Mania
    in hyperlimbic syndromes, 15
    in temporolimbic syndromes, 128
Medial circuit of Papez, **4**, 4–5, 123–124
Medial temporal lobe
    memory organization and, 146
    in schizophrenia. *See* Schizophrenia
Mediodorsal nucleus, of thalamus, **36**, 36–37
Mediothalamic frontocortical system, 35

Memory. *See also* Amnesia
  declarative (conscious), 147
  flashbulb, 149
  nondeclarative (nonconscious), 147–148
  organization in brain, recovered memory and. *See* Recovered memory
  recovered. *See* Recovered memory
Mesencephalic reticular formation (MRF), arousal and, 138
Mesial temporal sclerosis, FLAIR in, 110, **110**
Mesocorticolimbic dopamine system, cocaine and amphetamine addiction and, 180–181, **184**
Methylenedioxymethylamphetamine, accumbens and, 45
Meynert, basal nucleus of, 13
Midazolam, infusion in amygdala, fear and, 80
Migraines
  complicated, distinguishing features of, **109**
  paroxysmal limbic disorders with, 104–106
Minnesota Multiphasic Personality Inventory (MMPI), emotional experience defects and, 136–137
Modular theory, of emotional experience, 135–140
  arousal and, 137–139
  motor activation and approach-avoidance and, 139–140
  valence and, 136–137
Morphine, infusion in amygdala, fear and, 82
Motor activation, modular theory of emotional experience and, 139–140
Motor behavior
  amygdala lesions and conditioned fear and, 78
  amygdala lesions and unconditioned fear and, 79
  anatomical connections between amygdala and brain areas involved in fear and anxiety and, 74–75
  drug infusion in amygdala and, 80–82
  electrical or chemical stimulation of amygdala and, 75–76
Muscimol, infusion in amygdala, fear and, 80, 81
Mystical states, psychotic states compared with, 202, **203**

N
Naltrexone, in drug addiction, 190
Narrative truth, 144
Near-death experiences, 201–202
Neural activation, memory organization and, 146
Neural injury model, of schizophrenia, 161
Neuritic plaques, in hippocampal formation, 24
Neurochemical markers, in depression, 170–171
Neurodegenerative model, of schizophrenia, 161
Neurodevelopmental model, of schizophrenia, 161
Neuroepithelium, regions of, 46–47
Neurofibrillary tangles
  in hippocampal formation, 24
  in parahippocampal gyrus, 25–26
Neuroimaging, in schizophrenia, 156–158
  functional, 157–158
  structural, 156–157
Neurological disease, depression in, 169–170
Neurotensin, haloperidol and, 57
Neurotransmission, drug addiction and, 180, **182**
Neurovegetative disorders, in temporolimbic syndromes, 128–129
Nicotine addiction, dopamine and opioid peptides and, 184
N-methyl-D-aspartate (NMDA) receptors
  blockade of, near-death experiences and, 202
  glutamate, alcohol addiction and, 183
Nonconscious memory, 147
Nondeclarative memory, 147
Noradrenergic mechanisms, in depression, 170–171
Norepinephrine, anatomical connections between amygdala and brain areas involved in fear and anxiety and, 74
Norepinephrine system, limbic, 14
Normalization, forced (paradoxical), 127
Nucleus reticularis thalami, 34–36, **35**
Numbness, left-sided, epilepsy-induced, 96–98, **98**

O
Obsessive-compulsive symptoms, hyperlimbic syndromes and, 15
Olfactory auras, with temporal lobe seizures, 115
  localization of, 116
Opiate addiction, opioid peptide systems and, 181–182
Opioid antagonists, in drug addiction, 190
Opioid peptide systems
  alcohol addiction and, 184
  nicotine addiction and, 184
  opiate addiction and, 181–182
Opponent process theory, of drug dependence, 186
Oral behaviors, in Klüver-Bucy syndrome, 125
Orbitofrontal cortex, 6, **6–8**, 7, **12**
  anatomy of, 8, 9
  hypofunctioning of, 14
  paralimbic division of, 8–10, **10**
Orbitofrontal paralimbic division, of limbic system
  lateral portion of, **01**, 9–10
  medial portion of, 9, **10**
Orbitofrontal subcortical circuit, 12, **13**
Oscillatory mode, of thalamocortical activity, 35

P

Pain, thalamic pain syndrome and, 35–36
Pallidum, **12**
  ventral, 44
Palpitations, epilepsy-induced, 96–98, **98**
Panic disorder, 100
  distinguishing features of, **109**
  in dysfunctional limbic syndromes, 16
Papez's circuit, **4**, 4–5, 123–124
Parahippocampal gyrus, in Alzheimer's disease, **25**, 25–26, **26**
Paralaminar cells, of mediodorsal nucleus of thalamus, 36, **36**
Parallel distributed processing (PDP), temporolimbic, 124
Paroxysmal limbic disorders, 95–110, **109**, **110**
  due to combination of epilepsy and psychiatric illness, 103–104
  epilepsy-induced, 95–99
  with migraines, 104–106
  with nonconvulsive status epilepticus, 106–107, **108**
  psychiatric disorder-induced, 99–101
  psychogenic seizure-induced, 101–103
  with REM sleep disorder, 107–109
  with transient global amnesia, 106, **107**
Parvicellular basal nucleus, projection to accumbens, **52**, 53
Parvocellular division, of mediodorsal nucleus of thalamus, 36, **36**
Perceptual distortion, epilepsy-induced, 98–99, **99**
Perforant pathway, 21, 22
Perirhinal cortex, in Alzheimer's disease, 25
Picrotoxin, infusion in amygdala, fear and, 80–81
Positron-emission tomography (PET), in schizophrenia, 157

Proton magnetic resonance imaging, in schizophrenia, 157
Psychiatric disorders. *See also specific disorders*
  epilepsy combined with, paroxysmal limbic disorders due to, 103–104
  paroxysmal limbic disorders due to, 99–101
Psychoactive drugs, accumbens shell reactions to, 56–57, **57**, **58**
Psychogenic seizures, paroxysmal limbic disorders due to, 101–103
Psychosis. *See also* Schizophrenia
  alternating, 127
  interictal
    distinguishing features of, **109**
    in temporal lobe epilepsy, 127
  mystical states compared with, 202, **203**
  postictal, 104, **105**
    distinguishing features of, **109**
Psychotic symptoms, in dysfunctional limbic syndromes, 16

R

Rage
  in hyperlimbic syndromes, 15
  in temporolimbic syndromes, 128
Recovered memory, 143–152
  behavioral evidence concerning traumatic memory and, 148–150
    change of memories with time and, 148–149
    creation of false memories and, 149–150
  Eileen Franklin case and, 144–145, 151
  memory organization in brain and. *See* Brain
  narrative versus historical truth and, 144
Regional cerebral blood flow (rCBF), in schizophrenia, 157
Reinforcement
  drug addiction and, 179–180, **181**
  neural substrates of, 180, **183**

neurobiological substrates of, 190
Relapse, in drug addiction, neural substrates of, 190
Religious experience, 195–205
  in cortical degenerative dementias, 203–204
  in delusional disorders, 202–203, **203**
  epilepsy and, 196–201, **198–199**
    ictal events and, 197, 200
    interictal experiences and, 201
    postictal experiences and, 200–201
  hallucinogens and, 202
  limbic markers and, 204–205
  near-death experiences, 201–202
  in normal individuals, 196
REM sleep disorder
  distinguishing features of, **109**
  paroxysmal limbic disorders with, 107–109
Rhinal sulcus, inferior, 27
Right hemisphere, motor activation/inhibition and, 139–140

S

Sadness. *See also* Depression
  induced, similarity of depression with, recovery from, 173
  transient, 171, **171**
Schizophrenia
  delusions in, neurobiological genesis of, 202–203
  left temporolimbic lobe in, 127
  medial temporal lobe in, 155–162
    animal studies relevant to, 161–162
    neuroimaging of, 156–158
    neuropsychological studies of, 156
    postmortem studies of, 158–161
  thalamus in, 31–32, 35, 37, 38
Seizures. *See also* Epilepsy; Status epilepticus
  ecstatic, 200

Seizures *(continued)*
  psychogenic
    distinguishing features of, **109**
    Seizures
    paroxysmal limbic disorders
      due to, 101–103
  following psychosis, 104, **105**
  temporal lobe. *See* Temporal
    lobe seizures
Self, sense of, ventrobasal complex
  and, 34
Sensitization, drug addiction and,
  180, 186
Sentiment de l'inconnu, with
  temporal lobe seizures, 116
Serotonergic system, limbic, 13
Serotonin
  alcohol addiction and, 184
  in depression, 170–171
  dorsal thalamic regulation of, 38
  infusion in amygdala, fear and,
    81–82
Sexuality changes, in
  temporolimbic syndromes, 129
Shaking
  due to combination of epilepsy
    and psychiatric illness, 103
  psychogenic seizure-induced,
    102–103, **103**
Single photon emission computed
  tomography (SPECT), in
  schizophrenia, 157
Sleep disorder, REM
  distinguishing features of, **109**
  paroxysmal limbic disorders
    with, 107–109
Source amnesia, 147
Status epilepticus, nonconvulsive
  distinguishing features of, **109**
  paroxysmal limbic disorders
    with, 106–107, **108**
Stria medullaris thalami, 37, **38**
Striatum, **12**. *See also* Accumbens
  central, 44
  matrix compartment of, 46,
    47–48
  formation of, 47–48, **48**
  patch compartment of, 46, 47, **47**
Strokes
  emotional changes associated
    with, 137
  temporolimbic syndromes and,
    129

Subcallosal streak, 46
Subcommissural striatal pocket, 64
Substantia nigra pars compacta,
  dopamine cells of, 59, **59**
Sulcus, collateral, 20, **20**
Superficial nuclei, 61

T

Temporal lobe
  medial
    memory organization and,
      146
    in schizophrenia. *See*
      Schizophrenia
    ventromedial. *See* Ventromedial
      temporal lobe
Temporal lobe experiential
  responses, 118–120
  electrical stimulation and,
    117–118
  fragments of, 119–120
  pathophysiology of, 118–119
Temporal lobe seizures, 113–118
  auras and, 115–117
    classification of, 115–116
    localization of, 116–117
  complex partial, 113–114
  distinguishing features of, **109**
  electrical stimulation and,
    117–118
  interictal psychosis in, 127
  interictal syndrome of, **125**,
    125–126
  localization of, 114–115
  temporolimbic syndromes and,
    129
Temporo-ammonic pathway, 21
Temporolimbic syndromes,
  123–130
  amnestic, 126–127
  anxiety and dissociative
    disorders in, 128
  delusions and hallucinations in,
    127–128
  emotional and mood disorders
    in, 128
  extended amygdala and related
    concepts and, 124
  Gastaut-Geschwind syndrome,
    15, **125**, 125–126
  Klüver-Bucy syndrome, 15,
    76–77, 124–125, **125**

  medial and lateral limbic
    circuits and, 124
  neuroanatomical networks and
    circuits and, 123–124
  neurovegetative disorders in,
    128–129
  pathology of, 129–130
Tentorium cerebelli, injury to, **26**,
  26–27
Thalamic pain syndrome, 35–36
Thalamus, **12**, 31–40
  afferents of, relationship with
    prefrontal cortex-ventral
    striatal system, 50–51, **51**,
    53
  habenula and dorsal thalamic
    projection of, 37–39, **38**
  mediodorsal nucleus of, **36**,
    36–37
  nucleus reticularis thalami of,
    34–36, **35**
  in schizophrenia, 31–32, 35, 37,
    38
  ventrobasal complex of, 32–34,
    33
Tolerance, to drugs, 179–180
Transfer mode, of thalamocortical
  activity, 35
Transient global amnesia
  distinguishing features of, **109**
  paroxysmal limbic disorders
    with, 106, **107**
Triune brain, 6, **6**

U

Uncal herniation, 27, **27**
Uncus, 20
Unresponsiveness, complex
  partial seizures and, 113
Utilization behavior, in
  dysfunctional limbic
  syndromes, 15–16

V

Valence, modular theory of
  emotional experience and,
  136–137
Ventral compartment, in
  depression model, 168, **169**
Ventrobasal complex, 32–34, **33**
Ventromedial temporal lobe, 19–28
  in Alzheimer's disease, 24–26

amygdala and, 24–25
hippocampal formation and, 24
parahippocampal gyrus and, **25,** 25–26, **26**
cortical connections of, 21–24
cortical input and, 21–22, **22, 23**
cortical output and, 22–24, **24**
temporal injury and, 26–27
of tentorium cerebelli, **26,** 26–27
uncal herniation and raised intracranial pressure and, 27, **27**
topography in humans and nonhuman primates, 19–21
boundaries and, 19–20, **20**
component structures and, 19
sulcal landmarks and, 20, **20**
surface features and, 20–21
Verrucae, 21
Vigilance
amygdala lesions and conditioned fear and, 77–78
anatomical connections between amygdala and brain areas involved in fear and anxiety and, 74
electrical or chemical stimulation of amygdala and, 75

Visceral feedback hypothesis, of emotional experience, 134–135
Visceral-subcortical division, of amygdala, 9
Viscerosensory auras, with temporal lobe seizures, 115
Viscosity, in Gastaut-Geschwind syndrome, 126
Visual auras, with temporal lobe seizures, 115

W
Wada test, 136
Withdrawal, from drugs, 179–180

The Journal of Neuropsychiatry and Clinical Neurosciences
The Official Journal of the American Neuropsychiatric Association

The Journal of Neuropsychiatry and Clinical Neurosciences is the one journal in the field which speaks clearly and promptly on the most urgent emerging issues in neuropsychiatry. Features include:

Special and Regular Articles which showcase original research important to the assessment and treatment of neuropsychiatric disorders. A brief abstract introduces each.

Clinical and Research Reports gather together new research findings with data from pilot studies, worthwhile replications studies, and clinical studies involving more than one patient.

A Practice and Opinion Column features advice from colleagues on specific diagnostic and clinical challenges in neuropsychiatry.

Letters to the Editor capture brief, single case reports from your colleagues, and invites their commentary on articles and issues in neuropsychiatry.

Book Reviews spark lively discussion of current literature in the field of neuropsychiatry.

Occasional Special Issues target one subject of absorbing interest and give it a full, rounded editorial treatment. Recent examples: background papers for DSM-IV, new discoveries in frontal lobe function.

Subscribe Now! Or see the following information about receiving the Journal as a benefit of membership in the American Neuropsychiatric Association.

As a member of the American Neuropsychiatric Association you'll enjoy the benefits of reading the authoritative source of information on Neuropsychiatry because as a perquisite of membership you'll receive *The Journal of Neuropsychiatry and Clinical Neurosciences*. Psychiatrists interested in joining the association should contact: Patti Arnold, The American Neuropsychiatric Association, 4510 West 89th St., Suite 110, Prairie Village, KS 66207 Phone: (913) 341-0765   Fax: (913) 341-6912.

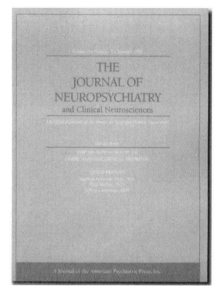

Editor: Stuart Yudofsky, M.D.
Deputy Editor: Robert E. Hales, M.D.
ISSN: 0895-0172
Frequency: Quarterly

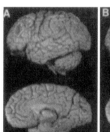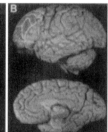

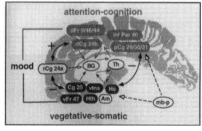

Subscription Category:
One year   [ ] $99 U.S. Individual   [ ] $125 Foreign Individual
           [ ] $189 U.S. Institution   [ ] $209 Foreign Institution
[ ] Check enclosed, payable to: **American Psychiatric Press, Inc.**
[ ] Charge my: [ ] VISA [ ] MasterCard [ ] American Express
Account Number _____ Expiration Date_____ Signature_____
Daytime Telephone _____ E-mail_____
[ ] Bill me

Five Easy Ways To Order:
Phone: 1-800-368-5777 (Mon.-Fri., 9am-5pm EST) **Fax:** (202) 789-2648 (24 hours a day)
Mail: American Psychiatric Press, Inc.
             1400 K St NW   Suite 1100 Washington, DC 20005
E-Mail: circulat@psych.org   **World Wide Web:** www.appi.org/orderform.html